Elementary Statistics Using SAS®

Sandra D. Schlotzhauer

The correct bibliographic citation for this manual is as follows: Schlotzhauer, Sandra D. 2009. *Elementary Statistics Using SAS®*. Cary, NC: SAS Institute Inc.

Elementary Statistics Using SAS®

ISBN 978-1-60764-379-1

1st printing, November 2009

Contents

Part 3   Comparing Groups   187

Acknowledgments

Most authors would tell you that writing a book involves a large support team, and this book is no exception. Many people contributed in many ways to this book. Thank you to the individuals and publishing companies who gave permission to use their data in the examples. Special thanks go to the reviewers who took time away from their jobs and families to provide detailed comments. I am grateful for your revisions to text, corrections to statistical and SAS discussions, and suggestions for topics to include or omit. Thanks to Dr. Julie Jones, who ran all the SAS programs, and provided her perspective as a statistician and as a programmer. Thanks to the anonymous industry, academic, and SAS reviewers. Your feedback improved the book.

This book started as the third edition of the *SAS System for Elementary Statistical Analysis,* which Dr. Ramon Littell coauthored. Initially, he planned to coauthor this book as well, but found that his consulting projects did not allow time to focus on the book. I first worked with him as a graduate student, and am, as ever, grateful for his insight and input on this book. He generously agreed to give permission to use the data from our previous book.

After all of the text is written, and all of the software screen captures are complete, a SAS Press book still involves the efforts of many people before it is published. Thanks to George McDaniel, Acquisitions Editor, for shepherding the book through every stage of the publishing process. Thanks to Amy Wolfe, Technical Editor, who I specifically requested after our last successful effort together, for correcting mistakes (any remaining are mine, Amy!) and learning some statistics along the way. Thanks to Jennifer Dilley, Technical Publishing Specialist, who created several of the figures. Thanks to Patrice Cherry, Designer, for the cover design, and to Galen Koch for producing the photo. Thanks to Candy Farrell, Technical Publishing Specialist; Shelly Goodin and Stacey Hamilton, Marketing Specialists; and Mary Beth Steinbach, Managing Editor.

Finally, a special thank you goes to my husband, Galen Koch, for his patience and support.

x

Part 1

The Basics

Chapter 1

Getting Started

Elementary Statistics Using SAS helps you analyze your data in SAS. This chapter discusses the following topics:

- gives an overview of the book that explains the purpose, audience, organization, and how to use the book

- introduces basic concepts for using SAS

- describes how to start SAS, get help, and exit

- gives a simple example of using SAS

- introduces several SAS statements

Introducing This Book

This section summarizes how to use this book.

Purpose

This book shows how to use SAS for basic statistical analyses and explains the statistical methods used—when to use them, what the assumptions are, and how to interpret the results.

Audience

This book is for people who have minimal knowledge of SAS or statistics or both. You might be a manager, researcher, or analyst. Or, you might be a student or professor. The range of example data used in the chapters and sample data in the exercises is relevant to a broad audience. The book assumes only that you want to learn to use SAS to perform analyses.

What This Book Is and Isn't

Consider this book a task-oriented tool for basic data analysis. The book integrates analysis methods with SAS steps, and then explains the results. Through the use of examples, the book describes how the analysis leads to decisions about the data.

This book is intended to be a companion to SAS. As you use the chapters, follow along with the examples. Looking at the SAS results as you read through the book will help you learn.

If you are familiar with statistical terms, the book discusses descriptive summaries and graphs, testing for normality, hypothesis testing, two-sample *t*-tests, paired-difference *t*-tests, Wilcoxon Rank Sum tests, Wilcoxon Signed Rank tests, analysis of variance (ANOVA), basic regression, and basic contingency table analysis. The book does not discuss analysis of variance for designs other than completely randomized designs. It does not discuss multivariate analysis, time series methods, quality control, operations research analysis, or design of experiments.

The book concentrates on data analysis. Although you can perform analyses using only the information in this book, you might want to refer to statistical texts for more theory or details. The book includes very few formulas. Formulas for the statistical methods are available in many texts and in SAS documentation. See "Further Reading" at the end of the book for a list of references.

How This Book Is Organized

This book is divided into five parts.

Part 1: The Basics

    Chapter 1, "Getting Started"
    Chapter 2, "Creating SAS Data Sets"
    Chapter 3, "Importing Data"
    Chapter 4, "Summarizing Data"

Part 2: Statistical Background

    Chapter 5, "Understanding Fundamental Statistical Concepts"
    Chapter 6, "Estimating the Mean"

Part 3: Comparing Groups

    Chapter 7, "Comparing Paired Data"
    Chapter 8, "Comparing Two Independent Groups"
    Chapter 9, "Comparing More Than Two Groups"

Part 4: Fitting Lines to Data

    Chapter 10, "Understanding Correlation and Regression"
    Chapter 11, "Performing Basic Regression Diagnostics"

Part 5: Data in Summary Tables

    Chapter 12, "Creating and Analyzing Contingency Tables"

The first part of the book shows how to get your data into SAS and summarize it. The second part gives a statistical foundation for the analyses discussed in parts 3, 4, and 5. Parts 3, 4, and 5 show how to perform various analyses.

Using Part 1

Part 1 is essential for everyone. Chapter 1 gives a brief introduction. Chapters 2 and 3 explain how to get your data into SAS. Chapter 4 shows how to summarize your data with descriptive statistics and plots.

Each chapter ends with three summaries. The "Key Ideas" summary contains the main ideas of the chapter. The "Syntax" summary provides the general form of SAS statements discussed in the chapter. The "Example" summary contains a SAS program that will produce all of the output shown in the chapter.

Using Part 2

Part 2 focuses on statistical concepts and provides a foundation for analyses presented in parts 3, 4, and 5. Topics include populations and samples, the normal distribution, the Empirical Rule, hypothesis testing, the effect of sample size and population variance on estimates of the mean, the Central Limit Theorem, and confidence intervals for the mean. Chapter 5 shows how to test for normality in SAS, and Chapter 6 shows how to get confidence intervals in SAS.

Using Part 3

Part 3 shows how to compare groups of data. The list below gives examples of comparing groups.

- New employees must take a proficiency test before they can operate a dangerous and expensive piece of equipment. Some new employees are experienced with this equipment and some are not. You want to find out if the average test scores for the two groups (experienced and inexperienced) are different.

- Heart rates of students are measured before and after mild exercise on a treadmill. You think the "before" heart rates are likely to be lower than the "after" heart rates. You want to know whether the changes in heart rates are greater than what could happen by chance.

- A study compared the effects of five fertilizers on potted geraniums. At the end of six weeks, you measured the height of the geraniums. Now, you want to compare the average heights for the geraniums that received the five different fertilizers.

- You recorded the number of defective items from an assembly line on the workdays Monday through Friday. You want to determine whether there are differences in the number of defective items made on different workdays.

Chapter 7 shows how to use SAS to summarize data from two independent groups, and how to perform the two-sample *t*-test and Wilcoxon Signed Rank test.

Chapter 8 shows how to use SAS to summarize paired data, and how to perform the paired-difference *t*-test and Wilcoxon Rank Sum test.

Chapter 9 shows how to use SAS to summarize data from more than two groups, and how to perform the analysis of variance, Kruskal-Wallis test, and multiple comparison procedures.

Using Part 4

Part 4 shows how to fit a line or curve to a set of data. The list below gives examples.

- You want to determine the relationship between weight and age in children. You want to produce a plot showing the data and the fitted line or curve. Finally, you want to examine the data to find atypical points far from the fitted line or curve.

- You have varied the temperature of an oven in a manufacturing process. The goal is to estimate the amount of change in hardness of the product that is caused by an amount of change in temperature.

Chapter 10 shows how to use SAS for correlations, straight-line regression, polynomial regression (fitting curves), multiple regression, prediction intervals, and confidence intervals.

Chapter 11 shows how to use SAS to plot residuals and predicted values, residuals and independent variables, and residuals in a time sequence. Chapter 11 discusses investigating outliers and using lack of fit tests.

Using Part 5

Part 5 includes Chapter 12, which discusses how to analyze data in contingency tables. The list below gives examples.

- You want to know whether color preferences for shampoo are the same for men and women.

- You have conducted an opinion poll to evaluate how people in your town feel about the proposal for a new tax to build a sports arena. Do people who live close to the site for the proposed arena feel differently than people who live farther away?

- In conducting a test-market for a new product, your company used a survey that asked shoppers if they had bought the product. You asked other questions about the packaging of the product. Are the opinions on packaging different between people who have bought the new product and people who have not?

Chapter 12 shows how to use SAS for creating contingency tables, measuring association, and performing the statistical tests for independence.

References

Appendix 1, "Further Reading," gives a list of references. Appendix 2 includes a table that identifies the first time the book discusses a SAS procedure, statement, or option.

Appendix 3 gives a brief introduction to the SAS windowing environment. Appendix 4 gives a very brief overview of SAS Enterprise Guide.

Typographic Conventions

The typographic conventions that are used in this book are listed below:

regular   is used for most text.

italic   defines new terms.

bold   highlights points of special importance.

bold   identifies the general form of statements and options in SAS syntax, and identifies menu choices in SAS. This font is also used in lists that explain items in SAS output tables or graphs, to refer to specific aspects of SAS code, and for the first reference to a procedure, statement, or option in a chapter.

plain   is used for all other references to SAS syntax, output, data sets, or variables.

Some chapters include "Technical Details," which give formulas or other details that add to the text, but aren't required for understanding the topic.

How to Use This Book

If you have never used SAS, first, read Part 1. Then, if you aren't familiar with statistics, read Part 2. Otherwise, you can go to other parts of the book.

If you have used SAS, but you are not familiar with statistics, skim Part 1. Then, read Part 2, which explains several statistical concepts. Then, go to other parts of the book that best meet your needs.

If you have used SAS, and you are familiar with statistics, but don't know how to handle your particular problem, first, skim Part 1. Then, go to the chapter that best meets your needs.

Caution: While the book shows how to use SAS to perform analyses and describes the statistical methods and SAS results, it is not a replacement for statistical texts.

Introducing SAS Software

This section describes information that will be useful to know as you begin to use SAS. To run all of the examples in this book, you need Base SAS, SAS/STAT, and SAS/GRAPH software. For many examples, this book provides an alternative that uses low-resolution graphics for people who do not have SAS/GRAPH.

Working with Your Computer

You need to know the basics of using your computer. You should understand single-click, double-click, the CTRL key and click, and so on. The book assumes that you know how to print from your computer. Also, this book assumes that SAS software is already installed on a personal computer (PC). If it is not, you'll need to install it.

You need to know a few things about how to use your computer's operating system to access SAS software. For example, on a mainframe system, you need to know how to submit a SAS program to the computer. For more advanced uses, such as permanently saving a data set or a program, you need to know some file management techniques such as how to create files, list a directory of files, or delete files.

You also need to know how to print from your computer. You might use a printer that is directly connected to your PC. Or, you might use a network printer instead. This book does not discuss installing printers.

Understanding the Ways of Using SAS

You can use SAS software in four ways: batch mode, noninteractive mode, interactive line mode, or the windowing environment mode. In each of these modes, you prepare a SAS program that creates a data set and performs analyses. (In addition to these four modes, SAS includes point-and-click products, such as SAS Enterprise Guide. This book does not discuss the other products.)

In *batch mode*, you submit a batch program by submitting a file that contains the batch program. To submit a file, you need to be familiar with your operating system because this file contains the necessary operating system-specific statements to start SAS software, run the batch program, and route the output. The output prints at either a remote printer or on your terminal, depending on what you specify in the file that is submitted. Some systems refer to batch mode as *background processing*.

In *noninteractive mode*, you prepare a file that contains a SAS program, but no operating system-specific statements. Then, you run SAS software using the **SYSIN** option, which reads the SAS program in your file. Your SAS program runs and the operating system creates a file that contains the output.

In *interactive line mode*, SAS software responds to statements as you enter them. You can enter your data, perform an analysis, look at the results, and perform another analysis without exiting from SAS. In this mode, you see a question mark (?) as a prompt to enter the next statement.

In the *windowing environment mode*, SAS software responds to statements as you enter them, but you see a set of windows on your terminal or PC screen. This mode provides you with all of the capabilities of interactive line mode, as well as a built-in text editor. In this mode, you can select choices from the menu bar. Or, you can use a command line and enter SAS commands directly. As you learn SAS, we recommend that you use the menu choices. On PCs, SAS software automatically starts in the windowing environment mode.

If your site doesn't have interactive capabilities, you need to use batch mode or noninteractive mode. You will find a SAS program in the "Example" summary at the end of each chapter. This program produces all of the output shown in the chapter. For batch mode, you need to add any necessary operating system-specific information to the program.

If your site has interactive capabilities, you can use either interactive line mode or the windowing environment mode. If you are learning to use SAS software for the first time, we recommend the windowing environment mode. Most people prefer this mode once they are familiar with SAS software. See Appendix 3, "Introducing the SAS Windowing Environment," for more detail.

Identifying Your SAS Release

SAS is available for many operating systems, and different releases of SAS software might be available for different operating systems.

This book shows results and reports from SAS 9.2. In most cases, the differences between output for this release and output from other releases are cosmetic. For example, different releases might have different spacing between columns of information in the output. As another example, displays shown in this book might differ slightly for earlier releases. Where needed, this book highlights important differences between releases. Table 1.1 summarizes how to find your release of SAS.

Table 1.1 Identifying Your Release of SAS

Mode	Release Information
Batch	The message on the first line of the log.
Noninteractive	The message on the first line of the log.
Interactive line	The message that appears before the first prompt (1?).
Windowing environment	The message at the top of the Log window.
	Select **Help→About SAS** to display a window that contains release details.

Identifying Your On-Site SAS Support Personnel

At every site where SAS software is licensed, someone is identified as on-site SAS support personnel. This person is usually available to answer questions from other users. Find out who your on-site SAS support personnel is before you begin using SAS software.

In addition, many locations have a Help Desk that answers questions about SAS software. You might know other people who are experienced SAS users and can answer questions if you run into problems. It's always handy to know who to call if you encounter some unexpected situation.

Summarizing the Structure of SAS Software

When working with data, first, you organize it. Second, you perform some analysis on the organized data. For example, you might collect all of your paycheck stubs for the year (organizing), and then add the net salary from each to find your total income for the year (analyzing).

SAS software follows the same approach. First, the *DATA step* helps you organize your data by creating a SAS data set. Second, *PROC steps* use SAS procedures to analyze your data. Once you create a SAS data set, you can use any SAS procedure without re-creating the data set.

Chapter 2 discusses the **DATA** step. Chapter 3 shows how to import data that you have already saved in a format other than SAS. The rest of the book discusses a series of

PROC steps that correspond to the different procedures that you use for different analyses.

The DATA and PROC steps consist of *SAS statements*. Most statements have one or more keywords that must be spelled exactly as shown. Depending on the statement, keywords might be followed by additional information that you supply. When this book shows the general form of SAS statements, keywords and required punctuation appear in bold, and the information that you need to supply appears in italic.

Syntax and Spacing Conventions

The DATA and PROC steps are part of the SAS language. Like any language, it has its own vocabulary and syntax. Some words have special meanings, and there are rules about how words are put together. Fortunately, there are very few restrictions.

Syntax

There is just one basic syntax rule you must always follow. **SAS statements must end with a semicolon.** The semicolon tells the software that you have completed one statement and that the next word starts a new statement. If you complete a statement and forget the semicolon, SAS software continues to read your program as one statement. When the software finds something it doesn't understand, you receive an error message. However, in most cases, the error message won't tell you that you have forgotten a semicolon. Anytime you get an error message, you should first look to see that all statements end with semicolons.

Spacing

With some computer languages, the spacing of your statements and of parts of your statements is important. This isn't true with SAS software. You can put several statements on one line. You can spread one statement over several lines. You can put spaces between statements or not. You can indent statements or not. The key point is that semicolons, not spacing, determine where SAS statements start and end. As an example, three SAS statements are shown below. Each of the methods of spacing in the statements has the same result.

```
data new; input x; datalines;

data new;input
x;
datalines;

data    new;
   input   x;
datalines;
```

```
data new;input x;datalines;

data new;
input x;
datalines;

data new;
   input x;
   datalines;
```

The final example shows the form used in this book. This form is used simply for convenience and readability. You can choose whatever spacing works best for you.

Notice that the sample SAS statements are in lowercase type. You can use either lowercase or uppercase in SAS statements. This book uses lowercase in sample programs and uppercase to show you the general form of SAS statements.

Output Produced by SAS Software

After you run a SAS program, you receive output. The output is in two parts: the log and the procedure output.

- The log shows the program that you submitted.

- The log tells you what the software did with the program.

- The log prints any error or warning messages.

For example, if your program creates a data set, the log tells you that the data set is created, and gives you some information about the data set. The procedure output gives the results of your program. For example, if your program calculates and prints the average of a group of numbers in your data set, the procedure output contains the printed average.

In this book, the log is shown only for the sample program later in this chapter. However, you need to remember that a log exists for every program. If you run into problems, the first step you should take is to look at the log to find error messages and investigate mistakes. Also, as a general strategy, it's a good idea to look at the log to verify that the program ran as expected. For example, checking the notes in the log will help you confirm the number of observations in a data set.

Table 1.2 identifies where your procedure output appears, which depends on how you use SAS.

Table 1.2 Location of SAS Output

Mode	Output Location
Batch	Printed or sent to file, depending on system statements.
Noninteractive	Printed, sent to your terminal, or sent to files on the system.
Interactive line	Sent to your terminal or PC.
Windowing environment	Displayed in Log and Output windows.

Starting SAS

For sites with PCs, you double-click on the **SAS** icon on your desktop. For most other sites, you just type the word "sas" to start SAS. Or, you can select **Start**, then **Programs**, then **SAS**, and then **SAS 9.2**. You need to determine what convention your site uses for starting SAS.

Whether SAS is installed on your PC or you access it over a network, you typically see the Editor, Log, and Explorer windows (assuming that you are in the windowing environment mode). You also typically see a menu bar and toolbars at the top of the SAS window. The details might differ, depending on how SAS was installed on your PC.

Displaying a Simple Example

This section shows you an example of a SAS data set and the results from an analysis.

One indicator of an individual's fitness is the percentage of body fat, which is measured with a special set of calipers. (Calipers look like pliers that have a dial on them.) The calipers are used at three or four different places on the body to measure the amount of skin that can easily be pinched. Women are measured on the triceps, the abdomen, and near the hips. Men are measured on the chest, the abdomen, near the hips, and at the midaxillary line. Skinfold measures are averaged from the different places to provide a measure of the percentage of body fat. Depending on whose guidelines you look at, the normal range for men is 15–20% body fat, and the normal range for women is 20–25% body fat. Table 1.3 shows body fat percentages for several men and women.[1] These men

---

[1] Copyright © 2009, SAS Institute Inc. All rights reserved. Reproduced with permission of SAS Institute Inc., Cary, NC, USA.

and women participated in unsupervised aerobic exercise or weight training (or both) about three times per week for a year. Then, they were measured for their percentages of body fat.

Table 1.3 Body Fat Data

Group	Body Fat Percentage				
Male	13.3	8	20	12	12
	19	18	31	16	24
	20	22	21		
Female	22	16	21.7	21	30
	26	12	23.2	28	23

The SAS statements below create a data set that contains the body fat data, print the data, and summarize it.

```
options nodate nonumber pagesize=50 linesize=80;
data bodyfat;
   input gender $ fatpct @@;
   datalines;
m 13.3 f 22 m 19 f 26 m 20 f 16 m 8 f 12 m 18 f 21.7
m 22 f 23.2 m 20 f 21 m 31 f 28 m 21 f 30 m 12 f 23
m 16 m 12 m 24
;
run;

proc print data=bodyfat;
   title 'Body Fat Data for Fitness Program';
   footnote 'Unsupervised Aerobics and Strength Training';
run;

proc means data=bodyfat;
   class gender;
   var fatpct;
title 'Body Fat for Men and Women in Fitness Program';
run;
```

Figure 1.1 shows the SAS log from the statements above, and Figures 1.2 and 1.3 show the output.

Figure 1.1 SAS Log for Example

```
NOTE: Copyright (c) 2002-2008 by SAS Institute Inc., Cary, NC, USA.
NOTE: SAS (r) Proprietary Software 9.2 (TS1M0)
      Licensed to SANDRA SCHLOTZHAUER-AUTHOR LAPTOP, Site 0070075138.
NOTE: This session is executing on the XP_PRO  platform.


NOTE: SAS initialization used:
      real time           2.15 seconds
      cpu time            1.20 seconds

1    options nodate nonumber ps=50 ls=80;
2    data bodyfat;
3       input gender $ fatpct @@;
4        label fatpct='Body Fat Percentage';
5       datalines;
NOTE: SAS went to a new line when INPUT statement reached past the end of a
      line.
NOTE: The data set WORK.BODYFAT has 23 observations and 2 variables.
NOTE: DATA statement used (Total process time):
      real time           0.03 seconds
      cpu time            0.03 seconds


9    ;
10
11   proc print data=bodyfat;
12       title 'Body Fat Data for Fitness Program';
13       footnote 'Unsupervised Aerobics and Strength Training';
14       run;

NOTE: There were 23 observations read from the data set WORK.BODYFAT.
NOTE: PROCEDURE PRINT used (Total process time):
      real time           0.06 seconds
      cpu time            0.06 seconds


15
16   proc means data=bodyfat;
17      class gender;
18      var fatpct;
19   title 'Body Fat for Men and Women in Fitness Program';
20   run;
NOTE: There were 23 observations read from the data set WORK.BODYFAT.
NOTE: PROCEDURE MEANS used (Total process time):
      real time           0.03 seconds
      cpu time            0.03 seconds
```

The second line of the log identifies the release of SAS.

Figure 1.2 SAS Output for PROC PRINT

```
Body Fat Data for Fitness Program

    Obs      gender      fatpct

     1         m          13.3
     2         f          22.0
     3         m          19.0
     4         f          26.0
     5         m          20.0
     6         f          16.0
     7         m           8.0
     8         f          12.0
     9         m          18.0
    10         f          21.7
    11         m          22.0
    12         f          23.2
    13         m          20.0
    14         f          21.0
    15         m          31.0
    16         f          28.0
    17         m          21.0
    18         f          30.0
    19         m          12.0
    20         f          23.0
    21         m          16.0
    22         m          12.0
    23         m          24.0

Unsupervised Aerobics and Strength Training
```

Figure 1.3 SAS Output for PROC MEANS

```
        Body Fat for Men and Women in Fitness Program

                    The MEANS Procedure

        Analysis Variable : fatpct Body Fat Percentage
```

gender	N Obs	N	Mean	Std Dev	Minimum	Maximum
f	10	10	22.2900000	5.3196596	12.0000000	30.0000000
m	13	13	18.1769231	6.0324337	8.0000000	31.0000000

```
        Unsupervised Aerobics and Strength Training
```

The average body fat for women in the fitness program is 22.29%, and the average body fat for men in the fitness program is 18.18%.

Getting Help

The easiest way to get help is by clicking **Help** in the menu bar. If you do not use SAS in the windowing environment mode, you can use online resources or the printed documentation.

You can find help about SAS at support.sas.com. Visit this site occasionally to view articles about and examples of using SAS. This is where you will learn about upgrades, new releases, and user conferences. In addition, the **Knowledge Base** includes documentation, white papers, and a searchable database of known SAS issues. The **Learning Center** lists SAS courses, provides a bookstore for ordering manuals and other books, and provides details on certification as a SAS programmer.

Exiting SAS

In batch or noninteractive mode, the computer runs your SAS job, and then automatically exits from SAS.

In interactive line mode or the windowing environment mode, the **ENDSAS** statement exits SAS. The general form of this statement is shown below:

> **ENDSAS;**

If you use the automatic menus in the windowing environment mode, selecting **File**, and then **Exit** ends the SAS session. You typically see a message window asking if you are sure you want to exit; just click **OK**. If you have switched from the menus to a command line in the windowing environment mode, you can also type **bye** on any command line to exit SAS.

Introducing Several SAS Statements

This section explains the **OPTIONS, TITLE, FOOTNOTE,** and **RUN** statements. These statements are not exclusively part of the DATA step or of a PROC step, but they can be used in either. You might want to skip this section until you are ready to run your own SAS program.

The TITLE Statement

The TITLE statement provides a title on the top line of each page of output. The TITLE statement is not required. If you don't specify a title, each page of output has the automatic title **The SAS System.**

In the TITLE statement, enclose the text of the title in single quotation marks or double quotation marks. The text between the quotation marks appears as the title for the output. For example, if you use all uppercase, the output title will print in uppercase. If your title contains an apostrophe, enclose the title text in double quotation marks and use a single quotation mark to get an apostrophe. Or, enclose the title text in single quotation marks and use two single quotation marks to get the apostrophe. The code below shows examples of valid TITLE statements:

```
title 'PRODUCTION SUMMARY FOR FY 2009';
title 'Experiment to Test Emotions under Hypnosis';
title 'Children"s Reading Scores';
title "Children's Reading Scores";
```

Each time you use a TITLE statement, the previous title is replaced with the new one. Look again at Figure 1.2 and Figure 1.3. The title changes as a result of the second TITLE statement.

You can specify up to 10 levels of titles. For a second title to appear on a line below the first title, type **TITLE2,** followed by the title text enclosed in single quotation marks. For example:

```
title 'BODY FAT DATA';
title2 'Men and Women in Fitness Program';
title3 'Data Property of SAS Institute Inc.';
```

You might find that using two or three levels of titles is more informative than using one long title.

You can suppress both the automatic title (**The SAS System**) and titles that are specified in previous TITLE statements. To do so, omit the quotation marks and the title text from the TITLE statement. For example, to suppress all three titles shown above, type:

```
title;
```

To suppress the second and third titles in the example statements above, type:

```
title2;
```

Similarly, you can suppress a title and all titles below it in the output by typing TITLE*n*, where *n* is the first title you want to suppress.

The general form of the TITLE statement is shown below:

> **TITLE*n* '*title text*' ;**

where *n* is a number between 1 and 10 for titles 1 through 10. For the first title, you can use either TITLE1 or TITLE. To suppress titles, omit both the quotation marks and the *title text*.

The FOOTNOTE Statement

Text shown in a FOOTNOTE statement appears at the bottom of the page of output. Unlike titles, there is no default footnote.

The syntax of the FOOTNOTE statement is very similar to the TITLE statement. The footnote text is enclosed in single quotation marks, and an apostrophe is indicated by using two single quotation marks together. Or, enclose the footnote text in double quotation marks and use a single quotation mark for an apostrophe. You can specify up to 10 levels of footnotes. A new FOOTNOTE statement replaces a previous one. You can suppress footnotes by simply omitting the quotation marks and footnote text from the FOOTNOTE statement.

Look again at Figure 1.2 and Figure 1.3. The FOOTNOTE statement in the program created the footnote at the bottom of each page. (To save space, several of the blank lines between the output and the footnote have been omitted.)

The general form of the FOOTNOTE statement is shown below:

> **FOOTNOTE*n* '*footnote text*' ;**

where *n* is a number between 1 and 10 for footnotes 1 through 10. For the first footnote, you can use either FOOTNOTE1 or FOOTNOTE. To suppress footnotes, omit both the quotation marks and the footnote text.

The RUN Statement

The RUN statement executes the statement or group of statements immediately above it. If you use batch or noninteractive mode, you don't need to use this statement. However, if you use SAS software interactively, you need to use the RUN statement.

If you use interactive line mode, you should use a RUN statement at the end of each DATA or PROC step. If you use the windowing environment mode, you only need a RUN statement at the end of each group of statements that you submit. However, using a RUN statement at the end of each DATA or PROC step helps to identify where one step ends and another step begins and to find errors in a program that doesn't run.

The general form of the RUN statement is shown below:

RUN;

The OPTIONS Statement

The OPTIONS statement sets certain system-specific options. Although there are many system options, this book uses only four. Your site might have set these four options already, and it might also have set some of the other system options that are available. Although you can override the default settings with an OPTIONS statement, your site probably has these options set in the best way for your system. In the sample program in the previous section, the first SAS statement is:

```
options pagesize=50 linesize=80 nodate nonumber;
```

This statement begins with the keyword OPTIONS. Then, the **PAGESIZE=** keyword is followed by the number **50**, which sets the page size to 50 lines. Next, the **LINESIZE=** keyword is followed by the number **80**, which sets the line size to 80 spaces. You can think of the PAGESIZE= option as giving the number of lines of text on a page, and the LINESIZE= option as giving the number of spaces in each line of text.

Next, the **NODATE** keyword suppresses the date that is normally printed on the same line as the page number.

Next, the **NONUMBER** keyword suppresses the page number that usually appears.

Finally, the statement ends with a semicolon.

Most of the output in this book uses PAGESIZE=60 and LINESIZE=80. When different page sizes or line sizes are used for a program, the SAS code shows the OPTIONS statement. All of the output in this book uses the NODATE and NONUMBER options.

For your own programs, you want to be able to look at the output and see when you ran the program, so you probably don't want to use NODATE. Also, you will find page numbering helpful in your own output, so you probably don't want to use NONUMBER.

The setting of 60 for PAGESIZE= works well for 11-inch-long paper. If your site uses a different paper size, you might want to use another setting. Typical settings for LINESIZE= are 80 for terminal screens, 78 for 8.5-inch-wide paper, and 132 for output printed on 14-inch-wide paper.

Summary

Key Ideas

- The SAS language consists of DATA steps, which organize your data by creating a SAS data set, and PROC steps, which analyze your data. These steps consist of SAS statements, which contain keywords and additional information about your data.

- All SAS statements must end with a semicolon.

- SAS statements can start anywhere on a line, and can continue for several lines. Several SAS statements can be put on one line. Either uppercase or lowercase text can be used.

- SAS output consists of the log and the procedure output. The log displays the program and messages. The procedure output gives the results of the program.

Syntax

- To print a title on your output:
    TITLE*n* **'***title text* **' ;**

    where *n* is a number between 1 and 10 for titles 1 through 10. For the first title, you can use either TITLE1 or TITLE. To suppress titles, omit both the quotation marks and the title text.

- To print a footnote on your output:
    FOOTNOTE*n* **'***footnote text* **' ;**

where *n* is a number between 1 and 10 for footnotes 1 through 10. For the first footnote, you can use either FOOTNOTE1 or FOOTNOTE. To suppress footnotes, omit both the quotation marks and the footnote text.

- To use the same option settings as most of the output in this book:
  OPTIONS PAGESIZE=60 LINESIZE=80 NODATE NONUMBER;

  PAGESIZE= identifies the number of lines of text per page, LINESIZE= identifies the number of spaces in each line of text, NODATE suppresses the automatic date in the output, and NONUMBER suppresses page numbers in the output.

- To execute one or more statements:
  RUN;

- To exit SAS:
  ENDSAS;

Example

The program below produces all of the output shown in this chapter:

```
options pagesize=60 linesize=80 nodate nonumber;
data bodyfat;
   input gender $ fatpct @@;
   datalines;
m 13.3 f 22 m 19 f 26 m 20 f 16 m 8 f 12 m 18 f 21.7
m 22 f 23.2 m 20 f 21 m 31 f 28 m 21 f 30 m 12 f 23
m 16 m 12 m 24
;
run;

proc print data=bodyfat;
title 'Body Fat Data for Men and Women in Fitness Program';
footnote 'Unsupervised Aerobic and Strength Training';
run;

proc means data=bodyfat;
   class gender;
   var fatpct;
title 'Body Fat for Men and Women in Fitness Program';
run;
```

Chapter 2

Creating SAS Data Sets

Your first step in using SAS to summarize and analyze data is to create a SAS data set. If your data is in a spreadsheet or other type of file, see Chapter 3, "Importing Data." This chapter discusses the following topics:

- understanding features of SAS data sets

- creating a SAS data set

- printing a data set

- sorting a data set

The "Special Topics" section discusses the following tasks:

- adding labels to variable names

- adding labels to values of variables

What Is a SAS Data Set?

Most of us work with data every day—in lists or tables or spreadsheets—on paper or on a computer. A data set is how SAS stores data in rows and columns.

Table 2.1 shows the speeding ticket fine amounts for the first speeding offense on a highway, up to 24 miles per hour over the posted speed limit.[1] The next section shows how to create a SAS data set using this data as an example.

For convenience, the speeding ticket data is available in the **tickets** data set in the sample data for this book.

Understanding the SAS DATA Step

A *SAS data set* has some specific characteristics and rules that must be followed so that the data can be analyzed using SAS software. Fortunately, there are only a few simple rules to remember. Recall that the SAS language is divided into **DATA** and **PROC** steps. The process of creating a SAS data set is the **DATA** step. Once you create a SAS data set, it can be used with any SAS procedure.

The speeding ticket data, like most data, consists of pieces of information that are recorded for each of several units. In this case, the pieces of information include the name of the state and the amount of the speeding ticket fine. You could think of a spreadsheet with two columns for the names of the states and the amounts of the speeding ticket fines. Each row contains all of the information for a state. When you use SAS, the columns are called *variables*, and the rows are called *observations*. In general, variables are the specific pieces of information you have, and observations are the units you have measured.

Table 2.1 contains 50 observations (one for each state), and two variables, the name of the state and the speeding ticket fine amount for the state. For example, the first observation has the value "Alabama" for the variable "state," and the value "100" for the variable "amount."

---

[1] Data is adapted from *Summary of State Speed Laws, August 2007*, produced by the National Highway Traffic Safety Administration and available at www.nhtsa.dot.gov. The fines are for situations with no "special circumstances," such as construction, school zones, driving while under the influence, and so on. Some states have varying fines; in these situations, the table shows the maximum fine.

Table 2.1 Speeding Ticket Fines

State	Amount	State	Amount
Alabama	$100	Hawaii	$200
Delaware	$20	Illinois	$1000
Alaska	$300	Connecticut	$50
Arkansas	$100	Iowa	$60
Florida	$250	Kansas	$90
Arizona	$250	Indiana	$500
California	$100	Louisiana	$175
Georgia	$150	Montana	$70
Idaho	$100	Kentucky	$55
Colorado	$100	Maine	$50
Nebraska	$200	Massachusetts	$85
Maryland	$500	Nevada	$1000
Missouri	$500	Michigan	$40
New Mexico	$100	New Jersey	$200
Minnesota	$300	New York	$300
North Carolina	$1000	Mississippi	$100
North Dakota	$55	Ohio	$100
New Hampshire	$1000	Oregon	$600
Oklahoma	$75	South Carolina	$75
Rhode Island	$210	Pennsylvania	$46.50
Tennessee	$50	South Dakota	$200
Texas	$200	Vermont	$1000
Utah	$750	West Virginia	$100
Virginia	$200	Wyoming	$200
Washington	$77	Wisconsin	$300

Summarizing the SAS DATA Step

Table 2.2 uses an example to summarize the parts of a SAS DATA step. Before you begin these tasks, think about names for your data set and variables. See the next section, "Assigning Names," for details.

Table 2.2 Summarizing Parts of a SAS DATA Step

Task	DATA Step Code	Summary
1	`data tickets;`	The **DATA** statement gives a name to the data set.

2	`input state $ 1-2`
	
	`      amount   4-8;`
		The **INPUT** statement tells SAS software what the variable names are, whether the variables have numeric or character values, and where the variables are located.

3	`datalines;`	The **DATALINES** statement tells SAS software that the data lines follow. Do not put a blank line between the DATALINES statement and the start of the data.

4	`AL 100`
	
	`HI 200`
	
	`DE 20`
	
	`AK 300`
		The data lines contain the data. Only four states are shown here, but all states could be entered this way.

5	`;`	After the data lines, a null statement (semicolon alone) ends the data lines.
6	`run;`	The **RUN** statement executes the DATA step.

The only output produced by this DATA step is a note in the log. This note tells you that the data set has been created. It gives the name of the data set, the number of variables, and the number of observations.

The next seven sections give details on the DATA step. The rest of the chapters in the book assume that you understand how to create data sets in SAS.

Assigning Names

Before you can name the data set or name variables, you need to choose names that are acceptable to SAS. Depending on the version of SAS, different rules might apply to names. Table 2.3 summarizes the rules for SAS names for data sets or variables (for all SAS versions and releases after SAS 7).

Table 2.3 Rules for SAS Data Set or Variable Names

- SAS names can contain uppercase letters, lowercase letters, or a mix of the two.
- SAS names must start with a letter or an underscore.
- The remaining characters in a SAS name can be letters, numbers, or underscores.
- SAS names can be up to 32 characters long.
- SAS names cannot contain embedded blanks. For example, DOLAMT is a valid name, but DOL AMT is not.
- SAS reserves some names for internal use. To be completely safe, avoid using an underscore as both the first and last character of a name. This table identifies known names that you should avoid, but SAS might add new names to the list of reserved names.

  For data sets, do not use the name _NULL_, _DATA_, or _LAST_.

  For variables, do not use the name _N_, _ERROR_, _NUMERIC_, _CHARACTER_, or _ALL_.

This book assumes the automatic option of **VALIDVARNAME=V7**, which presumes you are using SAS 7 or later. For your own data, you can use mixed case text, uppercase text, or lowercase text.

You can change the automatic option in the **OPTIONS** statement. However, you might want to use caution. If you need to share your SAS programs with others at your site, and you change the option, they might not be able to run your programs.

If your site uses **VALIDVARNAME=UPCASE**, then the same naming rules apply, with one exception. Variable names must be in uppercase with this setting. This setting is compatible with earlier versions of SAS.

If your site uses **VALIDVARNAME=ANY**, then some of the naming rules are not needed. With this setting, names can contain embedded blanks. Also, with this setting, names can begin with any character, including blanks. If you use the ampersand (&) or percent (%) sign, then review the SAS documentation because special coding is needed for SAS to understand names with these characters.

Task 1: The DATA Statement

The DATA statement starts the DATA step and assigns a name to the data set. The simplest form for the DATA statement is shown below:

DATA *data-set-name*;

data-set-name is the name of the data set.

If you do not provide a *data-set-name*, SAS chooses a name for you. SAS chooses the automatic names DATA1, DATA2, and so on. This book always uses names for data sets.

Task 2: The INPUT Statement

The next statement in the DATA step is the INPUT statement. The statement provides the following information to SAS:

- what to use as variable names
- whether the variables contain numbers or characters
- the columns where each variable is located in the lines of data

SAS can create data sets for virtually any data, and the complete SAS documentation on the INPUT statement discusses many features and options. This chapter discusses the basic features. If your data is more complex, then check the documentation for more features. Table 2.4 shows the basics of writing an INPUT statement, using the data in Table 2.1 as an example.

Table 2.4 Writing an INPUT Statement

Step	Description
1	Begin with the word INPUT.
	`input`
2	Choose a name for the first variable.
	`input state`
3	Look to see whether the values of the variable are character or numeric. If the values are all numbers, the variable is a *numeric variable*, and you can go on to the next step. Otherwise, the variable is a *character variable*, and you need to put a dollar sign ($) after the variable name.
	`input state $`

(continued)

Table 2.4 (*continued*)

Step	Description
4	Enter column numbers for the data values. Enter the beginning column, then a dash, and then the last column. If you have a variable that uses only one column, enter the number for that column. Suppose the two-letter code for the variable **state** is in the first two columns. Here is the INPUT statement:

```
input state $ 1-2
```

If the two-letter code appeared in different columns, then you would specify those columns. For example, suppose the code was indented to appear in columns 5 and 6. Then, you would specify columns **5-6**.

Step	Description
5	Repeat steps 2 through 4 for each variable in your data set. The speeding ticket data has one more variable, **amount**, which is numeric because the values are all numbers. Suppose these values are in columns 4 through 8 of the data. Here is the INPUT statement:

```
input state $ 1-2 amount 4-8
```

Step	Description
6	End the INPUT statement with a semicolon.

```
input state $ 1-2 amount 4-8;
```

Identifying Missing Values

If you do not have the value of a variable for an observation, you can simply leave the data blank. However, this approach often leads to errors in DATA steps. For numeric variables, enter a period for missing values instead. For example, the speeding ticket data has a missing value for Washington, D.C. When providing the data, use a period for this value. For character variables, when you specify the column location, you can continue to leave the data blank. As discussed below, when you omit the column location, or place several observations on a single line, enter a period for a missing character value.

Omitting the Column Location for Variables

In some cases, you can use a simple form of the INPUT statement, and skip the activity of specifying columns for variables. Using the speeding ticket data as an example, here is the simple form:

```
input state $ amount;
```

To use this simple form, the data must satisfy the following conditions:

- Each value on the data line is separated from the next value by at least one blank.

- Any missing values are represented by periods instead of blanks. This is true for both character and numeric variables.

- Values for all character variables are eight characters or less, and they don't contain any embedded blanks.

When these conditions exist, you can do the following:

- List variable names in the INPUT statement in the order in which they appear in the data lines.

- Follow character variable names with a dollar sign.

- End the INPUT statement with a semicolon.

Putting Several Short Observations on One Line

You might have data similar to the speeding ticket data, with only a few variables for each observation. In this situation, you might want to put several observations on a single data line. To use this approach, consider the following:

- Your data must satisfy the three conditions for omitting column locations for variables.

- You can put several observations on a data line and different numbers of observations on different data lines.

- You can use the INPUT statement without column locations and type two at signs (@@) just before the semicolon.

For the speeding ticket data, here is this form of the INPUT statement:

```
input state $ amount @@;
```

SAS documentation refers to @@ as a "double trailing at sign." The @@ tells SAS to continue reading data from the same line, instead of moving to the next line. Without @@, SAS assumes that one observation appears on each data line.

Task 3: The DATALINES Statement

The DATALINES statement is placed immediately before the lines of data. This simple statement has no options.

DATALINES;

In some SAS programs, you might see a CARDS statement instead of the DATALINES statement. The two statements perform the same task.

Existing Data Lines: Using the INFILE Statement

Suppose your data already exists in a text file. You don't have to re-enter the data lines. Instead, you can access the data lines by using the INFILE statement instead of the DATALINES statement. For more information, see Chapter 3.

Task 4: The Data Lines

After the DATALINES statement, enter the data lines according to the INPUT statement. Do not put a blank line between the DATALINES statement and the data. If you specify certain columns for variables, enter the values of those variables in the correct columns. If you don't specify columns in the INPUT statement, put a period in the place where a missing value occurs.

Task 5: The Null Statement

After the last data line, enter a semicolon, alone, on a new line. This is called a *null statement*, and it ends the data lines. The null statement must be alone and on a different line from the data.

Task 6: The RUN Statement

Chapter 1 introduced the RUN statement, and recommended using it at the end of each DATA or PROC step. This approach helps identify where one step ends and another step begins. And, it can find errors in programs. This book follows the approach of ending each DATA step with a RUN statement.

Creating the Speeding Ticket Data Set

This section uses the DATA, INPUT, and DATALINES statements to create a data set for the speeding ticket data. This example uses the simplest form of the INPUT statement, omitting column locations and putting several observations on each data line. The values of **state** are the two-letter codes used by the U.S. Postal Service.

```
data tickets;
input state $ amount @@;
datalines;
AL 100 HI 200 DE 20 IL 1000 AK 300 CT 50 AR 100 IA 60 FL 250
KS 90 AZ 250 IN 500 CA 100 LA 175 GA 150 MT 70 ID 100 KY 55
```

```
CO 100 ME 50 NE 200 MA 85 MD 500 NV 1000 MO 500 MI 40 NM 100
NJ 200 MN 300 NY 300 NC 1000 MS 100 ND 55 OH 100 NH 1000 OR 600
OK 75 SC 75 RI 210 PA 46.50 TN 50 SD 200 TX 200 VT 1000 UT 750
WV 100 VA 200 WY 200 WA 77 WI 300 DC .
;
run;
```

The statements above use the name **tickets** to identify the data set, the name **state** to identify the state, and the name **amount** to identify the dollar amount of the speeding ticket fine.

Look at the last data line of the program above. The missing value for Washington, D.C., is identified using a period.

Printing a Data Set

Chapter 1 describes SAS as consisting of DATA steps and PROC steps. This section shows how to use **PROC PRINT** to print the data. To print the speeding ticket data, add the following statements to the end of the program that creates the data set:

```
proc print data=tickets;
   title 'Speeding Ticket Data';
run;
```

Figure 2.1 shows the results.

Figure 2.1 shows the title at the top that was requested in the **TITLE** statement. The first column labeled **Obs** shows the *observation number*, which you can think of as the order of the observation in the data set. For example, **Alabama** is the first observation in the speeding ticket data set, and the Obs for **AL** is **1**. The second column is labeled **state**. This column gives the values for the variable **state**. The third column is labeled **amount**, and it gives the values for the variable **amount**.

As discussed earlier in "Assigning Names," SAS allows mixed case variable names. When you use all lowercase, SAS displays lowercase in the output.

In general, when you print a data set, the first column shows the observation number and is labeled **Obs**. Other columns are labeled with the variable names provided in the INPUT statement. PROC PRINT automatically prints the variables in the order in which they were listed in the INPUT statement.

Figure 2.1 PROC PRINT Results

```
 Speeding Ticket Data

  Obs     State      Amount

    1      AL        100.0
    2      HI        200.0
    3      DE         20.0
    4      IL       1000.0
    5      AK        300.0
    6      CT         50.0
    7      AR        100.0
    8      IA         60.0
    9      FL        250.0
   10      KS         90.0
   11      AZ        250.0
   12      IN        500.0
   13      CA        100.0
   14      LA        175.0
   15      GA        150.0
   16      MT         70.0
   17      ID        100.0
   18      KY         55.0
   19      CO        100.0
   20      ME         50.0
   21      NE        200.0
   22      MA         85.0
   23      MD        500.0
   24      NV       1000.0
   25      MO        500.0
   26      MI         40.0
   27      NM        100.0
   28      NJ        200.0
   29      MN        300.0
   30      NY        300.0
   31      NC       1000.0
   32      MS        100.0
   33      ND         55.0
   34      OH        100.0
   35      NH       1000.0
   36      OR        600.0
   37      OK         75.0
   38      SC         75.0
   39      RI        210.0
   40      PA         46.5
   41      TN         50.0
   42      SD        200.0
   43      TX        200.0
   44      VT       1000.0
   45      UT        750.0
   46      WV        100.0
   47      VA        200.0
   48      WY        200.0
   49      WA         77.0
   50      WI        300.0
   51      DC          .
```

Printing Only Some of the Variables

If you have a data set with many variables, you might not want to print all of them. Or, you might want to print the variables in a different order than they are in in the data lines. In these situations, use a **VAR** statement with PROC PRINT.

Suppose you conducted a survey that asked opinions about nuclear power plants and collected some demographic information (age, sex, race, and income). Suppose the data set is named **NUCPOWER** and you want to print only the variables **AGE** and **INCOME**.

```
proc print data=NUCPOWER;
   var AGE INCOME;
run;
```

PROC PRINT results show only three columns: Obs, AGE, and INCOME.

Suppressing the Observation Number

Although the observation number is useful in checking your data, you might want to suppress it when printing data in reports. Use the **NOOBS** option in the PROC PRINT statement as shown below:

```
proc print data=tickets noobs;
   title 'Speeding Ticket Variables';
run;
```

Figure 2.2 shows the first few lines of the output.

Figure 2.2 Suppressing the Observation Number

```
Speeding Ticket Variables

    State     Amount

    AL        100.0
    HI        200.0
    DE         20.0
    IL       1000.0
    AK        300.0
    CT         50.0
    AR        100.0
```

Figure 2.2 contains only two columns, one for each variable in the data set.

Adding Blank Lines between Observations

For easier reading, you might want to add a blank line between each observation in the output. Use the **DOUBLE** option in the PROC PRINT statement as shown below:

```
proc print data=tickets double;
    title 'Double-spacing for Speeding Ticket Data';
run;
```

Figure 2.3 shows the first few lines of the output.

Figure 2.3 Adding Double-Spacing to Output

```
Double-spacing for Speeding Ticket Data

        Obs      State      Amount

         1        AL         100

         2        HI         200

         3        DE          20

         4        IL        1000

         5        AK         300

         6        CT          50

         7        AR         100
```

Figure 2.3 shows a blank line between each observation.

Summarizing PROC PRINT

The general form of the statements to print a data set is shown below:

PROC PRINT DATA=*data-set-name options;*

   VAR *variables;*

data-set-name is the name of a SAS data set, and *variables* lists one or more variables in the data set.

The **PROC PRINT** statement *options* can be one or more of the following:

NOOBS   suppresses the observation numbers.

DOUBLE   adds a blank line between each observation.

The VAR statement is optional. SAS automatically prints all the variables in the data set.

Sorting a Data Set

PROC PRINT prints the data in the order in which the values were entered. This might be what you need or it might not. If you want to reorder the data, use **PROC SORT**. This procedure requires a **BY** statement that specifies the variables to use to sort the data.

Suppose you want to see the speeding ticket data sorted by **amount**. In other words, you want to see the observation with the smallest **amount** first, and the observation with the largest **amount** last.

```
proc sort data=tickets;
   by amount;
run;

proc print data=tickets;
   title 'Speeding Ticket Data: Sorted by Amount';
run;
```

Figure 2.4 shows the first few lines of the output.

Figure 2.4 shows the observation for **DC** first. When sorting, missing values are the "lowest" values and appear first. The next several observations show the data sorted by **amount**.

Figure 2.4 Sorting in Ascending Order

```
Speeding Ticket Data: Sorted by Amount

        Obs      State     Amount

          1       DC          .
          2       DE        20.0
          3       MI        40.0
          4       PA        46.5
          5       CT        50.0
          6       ME        50.0
          7       TN        50.0
```

SAS automatically sorts data in English. If you need your data sorted in another language, check the SAS documentation for possible options.

SAS automatically sorts data in ascending order (lowest values first). SAS automatically orders numeric variables from low to high. SAS automatically orders character values in alphabetical order.

In many cases, sorting the data in descending order makes more sense. Use the **DESCENDING** option in the BY statement.

```
proc sort data=tickets;
   by descending amount;
run;

proc print data=tickets;
   title 'Speeding Ticket Data by Descending Amount';
run;
```

Figure 2.5 shows the first few lines of the output.

Figure 2.5 shows the observation for Illinois (**IL**) first. The observation for **DC**, which has a missing value for **amount**, appears last in this output.

Figure 2.5 Sorting in Descending Order

```
Speeding Ticket Data by Descending Amount

         Obs     State     Amount

          1       IL       1000.0
          2       NV       1000.0
          3       NC       1000.0
          4       NH       1000.0
          5       VT       1000.0
          6       UT        750.0
          7       OR        600.0
```

You can sort by multiple variables, sort some in ascending order, and sort others in descending order. Returning to the example of the survey on nuclear power plants, consider the following:

```
proc sort data=NUCPOWER;
   by descending AGE INCOME;
run;

proc print data=NUCPOWER;
   var AGE INCOME Q1;
run;
```

The statements above show results for the first question in the survey. The results are sorted in descending order by AGE, and then in ascending order by INCOME.

The general form of the statements to sort a data set is shown below:

PROC SORT DATA=data-set-name**;**

    BY DESCENDING variables**;**

data-set-name is the name of a SAS data set, and **variables** lists one or more variables in the data set.

The **DESCENDING** option is not required. If it is used, place the option before the variable that you want to sort in descending order.

Summary

Key Ideas

- SAS stores data in a data set, which is created by a DATA step. The rows of the data set are called observations, and the columns are called variables.

- SAS names must follow a few simple rules. Think about appropriate names before you create a data set.

- In a SAS DATA step, use the DATA statement to give the data set a name. Use the INPUT statement to assign variable names and describe the data. Then, use the DATALINES statement to include the data lines, and end with a semicolon on a line by itself.

- Use PROC PRINT to print data sets. Using options, you can suppress observation numbers and add a blank line between each observation.

- Use PROC SORT to sort data sets by one or more variables. Using the DESCENDING option, you can change the automatic behavior of sorting in ascending order.

Syntax

To create a SAS data set using a DATA step

DATA *data-set-name*;

data-set-name	is the name of a SAS data set.

Use one of these INPUT statements:

INPUT *variable* $ *location* . . . ;
INPUT *variable* $ . . . ;
INPUT *variable* $ . . . @@ ;

variable	is a variable name.
$	is used for character variables and omitted for numeric variables.
location	gives the starting and ending columns for the variable.

@@   looks for multiple observations on each line of data.

    DATALINES;

data lines   contain the data.

;   ends the lines of data.

To print a SAS data set

    PROC PRINT DATA=*data-set-name options*;
      VAR *variables*;

data-set-name   is the name of a SAS data set.

variables   lists one or more variables in the data set.

The PROC PRINT statement *options* can be one or more of the following:

NOOBS   suppresses the observation numbers.

DOUBLE   adds a blank line between each observation.

The VAR statement is optional. SAS automatically prints all the variables in the data set.

To sort a SAS data set

    PROC SORT DATA=*data-set-name*;
      BY DESCENDING *variables*;

data-set-name   is the name of a SAS data set.

variables   lists one or more variables in the data set.

The DESCENDING option is not required. If it is used, place the option before the variable that you want to sort in descending order.

Example

The program below produces all of the output shown in this chapter:

```
options nodate nonumber ps=60 ls=80;
data tickets;
   input state $ amount @@;
datalines;
AL 100 HI 200 DE 20 IL 1000 AK 300 CT 50 AR 100 IA 60 FL 250
KS 90 AZ 250 IN 500 CA 100 LA 175 GA 150 MT 70 ID 100 KY 55
CO 100 ME 50 NE 200 MA 85 MD 500 NV 1000 MO 500 MI 40 NM 100
NJ 200 MN 300 NY 300 NC 1000 MS 100 ND 55 OH 100 NH 1000 OR 600
OK 75 SC 75 RI 210 PA 46.50 TN 50 SD 200 TX 200 VT 1000 UT 750
WV 100 VA 200 WY 200 WA 77 WI 300 DC .
;
run;

proc print data=tickets;
   title 'Speeding Ticket Data';
   run;

proc print data=tickets noobs;
   title 'Speeding Ticket Variables';
   run;

proc print data=tickets double;
   title 'Double-spacing for Speeding Ticket Data';
   run;

proc sort data=tickets;
   by amount;
run;

proc print data=tickets;
   title 'Speeding Ticket Data: Sorted by Amount';
run;

proc sort data=tickets;
   by descending amount;
run;

proc print data=tickets;
   title 'Speeding Ticket Data by Descending Amount';
run;
```

Special Topics

This section shows how to add labels to variables and to values of variables. SAS refers to adding labels to values of variables as *formatting the variable* or *adding formats*.

SAS provides ways to add labels and formats in the PROC step. However, the simplest approach is to add labels and formats in the DATA step. Then, SAS automatically uses the labels and formats in most procedures.

Labeling Variables

Many programmers prefer short variable names. However, these short names might not be descriptive enough for printed results. Use a **LABEL** statement in the DATA step to add labels. Place the LABEL statement between the INPUT and DATALINES statements.

```
data tickets1;
   input state $ amount @@;
   label state='State Where Ticket Received'
              amount='Cost of Ticket';
   datalines;
```

Follow these statements with the lines of data, a null statement, and a RUN statement.

The LABEL statement is similar to the TITLE and FOOTNOTE statements, which Chapter 1 discusses. Enclose the text for the label in quotation marks. If the label itself contains a single quotation mark, enclose the text for the label in double quotation marks. For example, the statements below are valid:

```
label cost09 = 'Cost from Producer in 2009';
label month1 = "Current Month's Results";
```

To see the effect of adding the LABEL statement, use PROC PRINT. Many SAS procedures automatically add labels. For PROC PRINT, use the **LABEL** option in the PROC PRINT statement. Here is what you would type for the speeding ticket data:

```
proc print data=tickets1 label;
   title 'Speeding Ticket Data with Labels';
run;
```

Figure ST2.1 shows the first few lines of the output.

Figure ST2.1 Adding Labels to Variables

```
Speeding Ticket Data with Labels

                State
                Where
                Ticket        Cost of
        Obs    Received        Ticket

         1        AL           100.0
         2        HI           200.0
         3        DE            20.0
         4        IL          1000.0
         5        AK           300.0
         6        CT            50.0
         7        AR           100.0
```

Figure ST2.1 shows the labels instead of the variable names.

Formatting Values of Variables

SAS provides multiple ways to add formats to variables. Here are three approaches:

- Assign an existing SAS format to the variable. SAS includes dozens of formats, so you can avoid the effort of creating your own format in many cases.

- Use a SAS *function* to assign a format. This approach creates a new variable that contains the values you want. SAS includes hundreds of functions, so you can save time and effort in many cases.

- Create your own SAS format, and assign it to the variable.

The next three topics discuss these approaches.

Using an Existing SAS Format

SAS includes many existing formats. If you want to add a format to a variable, check first to see whether an existing format meets your needs. For more information about existing formats, see the SAS documentation or review the online Help.

For the speeding ticket data, suppose you want to format **amount** to show dollars and cents. With the **DOLLAR** format, you specify the total length of the variable and the number of places after the decimal point. For **amount**, the **FORMAT** statement below specifies a length of 8 with 2 characters after the decimal point:

```
format amount dollar8.2;
```

You can assign existing SAS formats in a DATA step, which formats the values for all procedures. Or, you can assign existing SAS formats in a PROC step, which formats the values for only that procedure. The statements below format **amount** for all procedures:

```
data ticketsf1;
    input state $ amount @@;
    format amount dollar8.2;
    datalines;
```

Follow these statements with the lines of data, a null statement, and a RUN statement to complete the DATA step.

You can assign formats to several variables in one FORMAT statement. For example, the statement below assigns the **DOLLAR10.2** format to three variables:

```
format salary dollar10.2 bonus dollar10.2 sales dollar10.2;
```

The statement can be used either in a DATA step or a PROC step.

The statements below format **amount** for PROC PRINT only, and use the data set created earlier in this chapter:

```
proc print data=tickets;
    format amount dollar8.2;
    title 'Speeding Ticket Data Using DOLLAR Format';
run;
```

Figure ST2.2 shows the first few lines of the output from the PROC PRINT step.

Figure ST2.2 Using Existing SAS Formats

```
Speeding Ticket Data Using DOLLAR Format

        Obs      state       amount

         1        AL        $100.00
         2        HI        $200.00
         3        DE         $20.00
         4        IL       $1000.00
         5        AK        $300.00
         6        CT         $50.00
         7        AR        $100.00
```

Figure ST2.2 shows the formatted values for the speeding ticket fines.

Table ST2.1 summarizes a few existing SAS formats for numeric variables.

Table ST2.1 Commonly Used SAS Formats

Format Name	Description
COMMAw.d	Formats numbers with commas and decimal points. The total width is w, and d is 0 for no decimal points, or 2 for decimal values. Useful for dollar amounts.
DOLLARw.d	Formats numbers with dollar signs, commas, and decimal points. The total width is w, and d is 0 for no decimal points, or 2 for decimal values. Useful for dollar amounts.
Ew	Formats numbers in scientific notation. The total width is w.
FRACTw	Formats numbers as fractions. The total width is w. For example, 0.333333 formats as 1/3.
PERCENTw.d	Formats numbers as percentages. The total width is w, and the number of decimal places is d. For example, 0.15 formats as 15%.

Using a SAS Function

SAS includes many functions, which you can use to create new variables that contain the formatted values you want. As with existing SAS formats, you might want to check the list of SAS functions before creating your own format.

For the speeding ticket data, suppose you want to use the full state names instead of the two-letter codes. After checking SAS Help, you find that the **STNAMEL** function meets your needs.

```
data ticketsf2;
   input state $ amount @@;
   statetext = stnamel(state);
   datalines;
```

Follow these statements with the lines of data, a null statement, and a RUN statement to complete the DATA step.

The statements below print the resulting data:

```
proc print data=ticketsf2;
   title 'Speeding Ticket Data with STNAMEL Function';
run;
```

Figure ST2.3 shows the first few lines of the output from the PROC PRINT step.

Figure ST2.3 Using Existing SAS Functions

```
     Speeding Ticket Data with STNAMEL Function

  Obs      state     amount     statetext

   1        AL        100.0     Alabama
   2        HI        200.0     Hawaii
   3        DE         20.0     Delaware
   4        IL       1000.0     Illinois
   5        AK        300.0     Alaska
   6        CT         50.0     Connecticut
   7        AR        100.0     Arkansas
```

Figure ST2.3 shows all three variables. The INPUT statement creates the **state** and **amount** variables. The STNAMEL function creates the **statetext** variable.

Table ST2.2 summarizes a few SAS functions.

Table ST2.2 Commonly Used SAS Formats

Function	Description
ABS(*variable*)	Takes the absolute value of the numeric *variable*.
LOG(*variable*)	Takes the natural logarithm of the numeric *variable*.
ROUND(*variable*, *unit*)	Rounds the numeric *variable* according to the *unit*. For example, round(amount,1) rounds amount to the nearest whole number.
LOWCASE(*variable*)	Converts mixed case text of the character *variable* to all lowercase.
STNAMEL(*variable*)	Converts the postal code to state name, where *variable* contains the 2-letter code used by the U.S. Postal Service.
UPCASE(*variable*)	Converts mixed case text of the character *variable* to all uppercase.

Creating Your Own Formats

Sometimes, neither an existing SAS format nor SAS function meets your needs. Chapter 1 introduced the body fat data, where **gender** has the values of **m** and **f** for males and females. Suppose you want to use full text values for this variable. To do so, you can use **PROC FORMAT** to create your own format. After creating the format, you can assign it to a variable in the DATA step. Here is what you would type for the body fat data:

```
proc format;
    value $sex 'm'='Male' 'f'='Female';
run;

data bodyfat2;
    input gender $ fatpct @@;
    format gender $sex.;
    datalines;
m 13.3 f 22 m 19 f 26 m 20 f 16 m 8 f 12 m 18 f 21.7
m 22 f 23.2 m 20 f 21 m 31 f 28 m 21 f 30 m 12 f 23
m 16 m 12 m 24
;
run;

proc print data=bodyfat2;
    title 'Body Fat Data for Fitness Program';
run;
```

PROC FORMAT creates a new format.

The **VALUE** statement gives a name to the format. For the example above, the new format is named **$SEX**. Formats for character values must begin with a dollar sign.

The VALUE statement also specifies the *format* you want for each data *value*. For character values, enclose both the *format* and the *value* in quotation marks. Both the *format* and the *value* are case sensitive. In the example above, suppose the specified value was **M**. When SAS applied the $SEX format to the data, no matches would be found.

When you use the format in the DATA step, follow the name of the format with a period. For the example above, the FORMAT statement in the DATA step specifies $SEX. for the GENDER variable.

Figure ST2.4 shows the results. Compare Figure ST2.4 and Figure 1.2. The values for GENDER use the new format.

Figure ST2.4 Creating a New Format

```
Body Fat Data for Fitness Program

    Obs      gender      fatpct

      1      Male         13.3
      2      Female       22.0
      3      Male         19.0
      4      Female       26.0
      5      Male         20.0
      6      Female       16.0
      7      Male          8.0
      8      Female       12.0
      9      Male         18.0
     10      Female       21.7
     11      Male         22.0
     12      Female       23.2
     13      Male         20.0
     14      Female       21.0
     15      Male         31.0
     16      Female       28.0
     17      Male         21.0
     18      Female       30.0
     19      Male         12.0
     20      Female       23.0
     21      Male         16.0
     22      Male         12.0
     23      Male         24.0
```

The steps for creating your own format are summarized below:

1. Confirm that you cannot use an existing SAS format or SAS function.

2. Use PROC FORMAT to create a new format. The VALUE statement gives a name to the format and specifies formats for values. You can create multiple formats in one PROC FORMAT step. Use a VALUE statement to define each new format.

3. After using PROC FORMAT, assign the format to a variable in the DATA step by using the FORMAT statement. All SAS procedures use the format whenever possible. Or, you can specify the format with each procedure.

Combining Labeling and Formatting

You can combine labeling and formatting. You can use the LABEL and FORMAT statements in any order, as long as they both appear before the DATALINES statement. Similarly, you can create a new variable with a SAS function as long as this statement appears before the DATALINES statement. The list below summarizes possible approaches:

- Assign labels in the DATA step with the LABEL statement. Many SAS procedures automatically use the labels, but PROC PRINT requires the LABEL option to use the labels.

- If you use an existing SAS format in a DATA step, then SAS procedures automatically use the format.

- If you use an existing SAS format for a specific procedure, then only that procedure uses the format.

- If you use a SAS function in a DATA step, then SAS procedures display the values of the new variable. You can assign a label to the new variable in the LABEL statement.

- If you create your own format, and assign the format to a variable in a DATA step, then SAS procedures automatically use the format.

- If you create your own format, you can assign it for a specific procedure, and then only that procedure uses the format.

For the speeding ticket data, the statements below combine labeling, using a SAS function, and using an existing SAS format:

```
data tickets2;
    input state $ amount @@;
    format amount dollar8.2;
    label state='State Code'
          statetext='State Where Ticket Received'
          amount='Cost of Ticket';
    statetext = stnamel(state);
    datalines;
AL 100 HI 200 DE 20 IL 1000 AK 300 CT 50 AR 100 IA 60 FL 250
KS 90 AZ 250 IN 500 CA 100 LA 175 GA 150 MT 70 ID 100 KY 55
CO 100 ME 50 NE 200 MA 85 MD 500 NV 1000 MO 500 MI 40 NM 100
NJ 200 MN 300 NY 300 NC 1000 MS 100 ND 55 OH 100 NH 1000 OR 600
OK 75 SC 75 RI 210 PA 46.50 TN 50 SD 200 TX 200 VT 1000 UT 750
WV 100 VA 200 WY 200 WA 77 WI 300 DC .
;
run;
```

```
proc print data=tickets2 label;
   title 'Speeding Ticket Data with Labels and Formats';
run;
```

Figure ST2.5 shows the first few lines of the output.

Figure ST2.5 Combining Labeling and Formatting

```
   Speeding Ticket Data with Labels and Formats

          State       Cost of    State Where Ticket
   Obs    Code        Ticket     Received

    1      AL          $100.00    Alabama
    2      HI          $200.00    Hawaii
    3      DE           $20.00    Delaware
    4      IL         $1000.00    Illinois
    5      AK          $300.00    Alaska
    6      CT           $50.00    Connecticut
    7      AR          $100.00    Arkansas
```

Syntax

To label variables in a SAS data set

> **DATA** *statement that specifies a data set name*
>   **INPUT** *statement appropriate for your data*
>   **LABEL** *variable='label'*;
>   **DATALINES;**
>   *data lines*
>   **;**
>   **RUN;**
>   **PROC PRINT DATA=***data-set-name* **LABEL;**

data-set-name   is the name of a SAS data set.

variable   is the variable you want to label.

label   is the label you want for the variable. The label must be enclosed in single quotation marks and can be up to 256 characters long. (Blanks count as characters.) You can associate labels with several variables in one LABEL statement.

For your data, use DATA and INPUT statements that describe the data lines, and end each statement with a semicolon.

To use existing formats in a SAS data set

DATA *statement that specifies a data set name*
INPUT *statement appropriate for your data*
FORMAT *variable format-name*;
DATALINES;
data lines
;
RUN;

variable   is the variable you want to format.

format-name   is an existing SAS format. You can associate formats with several variables in one FORMAT statement.

To use existing formats in a SAS procedure

Many SAS procedures can use a FORMAT statement and apply a format to a variable for only that procedure. As an example, to use an existing SAS format and print the data set,

PROC PRINT DATA=*data-set-name*;
FORMAT *variable format-name*;

To create new variables with SAS functions in a SAS data set

Use functions in a DATA step to create new variables that contain the values you want.

DATA *statement that specifies a data set name*
INPUT *statement appropriate for your data*
new-variable=function(existing-variable);
DATALINES;
data lines
;
RUN;

new-variable   is the new variable.

function   is a SAS function.

existing-variable   is an appropriate variable in the INPUT statement.

The parentheses and the semicolon are required.

To create a new format and apply it in a SAS data set

```
PROC FORMAT;
  VALUE format-name value=format
                    value=format
                         .
                         .
                         .
                    value=format;
RUN;

DATA
  INPUT
  FORMAT variable format-name. ;
  DATALINES;
  data lines
;
RUN;
```

format-name	is a SAS name you assign to the list of formats.
	If the variable has character values, *format-name* must begin with a dollar sign (**$**). Also, if the variable has character values, the *format-name* can be up to 31 characters long, and cannot end in a number. (Blanks count as characters.)
	If the variable has numeric values, *format-name* can be up to 32 characters long, and can end in a number.
	The *format-name* cannot be the name of an existing SAS format.
value	is the value of the variable you want to format. If the *value* contains letters or blanks, enclose it in single quotation marks.
format	is the format you want to attach to the value of the variable. If the *value* contains letters or blanks, enclose it in single quotation marks.
	Some SAS procedures display only the first 8 or the first 16 characters of the *format*. The *format* can be up to 32,767 characters long, but much shorter text is more practical. (Blanks count as characters.)
variable	is the name of the variable that you want to format.

In the FORMAT statement in the DATA step, the period immediately following the *format-name* is required.

You can create multiple formats in one PROC FORMAT step. Use a VALUE statement to define each new format.

To create a new format and apply it in a SAS procedure

```
PROC FORMAT;
  VALUE format-name value=format
                    value=format
                        .
                        .
                        .
                    value=format;

DATA
  INPUT
  FORMAT variable format-name.;
  DATALINES;
  data lines
  ;
  RUN;

  PROC PRINT DATA=data-set-name;
    FORMAT variable format-name. ;
```

In the FORMAT statement in the PROC step, the period immediately following the *format-name* is required.

C h a p t e r 3

Importing Data

Your first step in using SAS to summarize and analyze data is to create a SAS data set. If your data is available electronically, SAS can almost certainly import it.

SAS provides many features for reading external data files, some of which differ based on your operating system. This chapter discusses a few basic features. If your data cannot be imported using the techniques in this chapter, then see SAS documentation or other references for more information. (See Appendix 1, "Further Reading," for suggestions.)

If your data is in an electronic form, it's worth the extra time to investigate ways to import it, instead of entering the data all over again. Importing data improves the overall data quality because quality control checks that were previously performed on the data do not need to be performed again.

This chapter discusses:

- opening an existing SAS data set
- reading data from an existing text file
- importing Microsoft Excel spreadsheets
- using the Import Wizard on a PC

Opening an Existing SAS Data Set

When you create a data set using the programs in this book, SAS creates a *temporary data set* and stores it in the **Work** library. To access this data set in a future SAS session, you would need to recreate it.

For your own data, you probably want to create a *permanent data set* that you can access in future SAS sessions. SAS stores your data set in a *SAS library*. This section discusses the simplest approach for accessing saved data sets.

When your data set is stored in a SAS library, you need to tell SAS where this library is located. This requirement is true for all operating systems. The easiest approach is to assign a SAS library reference (*libref*), and then associate the libref with the physical file location for the existing data set. The existing data set must be located in an existing directory. Also, you cannot use the **LIBNAME** statement to create file directories or SAS data sets. After assigning the libref, use the LIBNAME statement to access the existing SAS data set for any procedure.

The following statements access the saved **bodyfat** data, and then print it:

```
libname mydata 'c:\My Documents\SASdata';

proc print data=mydata.bodyfat;
run;
```

When you assign a libref, SAS prints a note to the log, but it does not create any printed output. When you access a permanent data set, you use a two-level name. The first part of the name provides the libref, and the second part of the name identifies the data set. SAS requires that the two parts of the name be separated with a period. Figure 3.1 shows the first few lines of the printed **bodyfat** data. (See Figure 1.2 for the full data set.)

Figure 3.1 Printing Saved Data

```
          The SAS System

  Obs       gender       fatpct

   1           m          13.3
   2           f          22.0
   3           m          19.0
   4           f          26.0
   5           m          20.0
```

The general form of the LIBNAME statement is shown below:

LIBNAME *libref* **'**SAS-data-library-location**';**

libref is a nickname (library reference) for the SAS library. The libref can be up to eight characters long. SAS reserves some librefs, so do not use **Sasuser, Work, Sasmsg,** or **Sashelp.**

SAS-data-library-location identifies the physical file location for the saved SAS data sets. Enclose the *SAS-data-library-location* in single quotation marks. The text for the physical file location depends on your operating system.

To use the *libref* to access a data set with a SAS procedure:

PROC PRINT DATA=*libref.data-set-name***;**

libref is defined above.

data-set-name is the name of a SAS data set.

The period between *libref* and *data-set-name* is required.

While this example uses PROC PRINT, you can use the same approach with every procedure in this book.

Similar to accessing an existing data set with a LIBNAME statement, you use a LIBNAME statement and a two-level name in a **DATA** step to create a permanent SAS data set. For example, suppose you use the LIBNAME statement previously shown. Then, the DATA statement below creates a data set named **tickets** in the libref **mydata:**

```
data mydata.tickets;
```

After completing the statements for the DATA step, SAS saves the **tickets** data set as a permanent data set.

Reading Data from an Existing Text File

Suppose your data lines—and only the data lines—are already stored in a plain text file. This type of file is sometimes referred to as an ASCII text file. On a PC, this type of file typically has a file extension of **.txt**. You can use the SAS **INFILE** statement to read the data.

If your text file also includes the variable names, then you can use **PROC IMPORT** (discussed in the next section). On a PC, you can also use the SAS Import Wizard (discussed later in this chapter).

The INFILE statement identifies the file that contains the data lines. Place the INFILE statement between the **DATA** statement (which names the data set) and the **INPUT** statement (which explains how to read the information in the file). Suppose the speeding ticket data is stored in the **ticket.txt** file in the **c:\example** directory on your PC. Here is the DATA step:

```
data tickets;
    infile 'c:\example\ticket.txt';
    input state $ amount @@ ;
run;
```

When you use the INFILE statement, then neither the **DATALINES** statement nor the null statement (which is a semicolon on a line by itself) is needed.

You can use the various forms of the INPUT statement with the INFILE statement. Be sure the information in the INPUT statement matches the appearance of the data in the external text file.

The list below shows sample INFILE statements for several operating systems:

PC   INFILE 'c:\example\ticket.txt';

UNIX   INFILE 'usr/*yourid*/example/ticket.dat';

OpenVMS   INFILE '[*yourid.data*]ticket.dat';

z/OS   INFILE '*yourid*.ticket.dat';

To complete the statements, the italic text should be replaced with your user ID.

The general form of the INFILE statement is shown below:

INFILE '*file-name*';

file-name is the name of the file that contains the data. Enclose *file-name* in single quotation marks.

Importing Microsoft Excel Spreadsheets

SAS provides multiple features for importing external data such as Microsoft Excel spreadsheets. The simplest approach is to use PROC IMPORT, where you specify the location of the existing Excel file and the name of the SAS data set that you want to create. This procedure is available for Windows, UNIX, and OpenVMS systems. Suppose the speeding ticket data is stored in an Excel file. The statements below import the Excel file and create a temporary SAS data set:

```
proc import datafile="c:\example\ticket.xls"
            dbms=xls
            out=tickets;
run;
```

The **DATAFILE=** option specifies the complete physical location for the Excel spreadsheet, which must be enclosed in double quotation marks. The **DBMS=XLS** option specifies the file type as an Excel file. The **OUT=** option specifies the name of the SAS data set. SAS creates the temporary data set and prints a note in the log. SAS automatically reads the variable names from the first row in the Excel spreadsheet.

To create a permanent SAS data set, you can use similar statements. First, use a LIBNAME statement to specify the location for the saved SAS data set. Second, use the two-level SAS name in the OUT= option in PROC IMPORT. The statements below show an example of creating a permanent SAS data set from an imported Excel file. SAS creates a permanent data set and prints a note in the log.

```
libname mydata 'c:\My Documents\SASdata';
proc import datafile="c:\example\ticket.xls"
            dbms=xls
            out=mydata.tickets;
run;
```

When you create a permanent SAS data set, you use a two-level name. See "Opening an Existing SAS Data Set" earlier in the chapter for more discussion on data set names and the LIBNAME statement.

SAS provides many choices for the DBMS= option. With these other choices, you can import many other types of files using the same approach as shown above. The list below gives a partial set of values for the DBMS= option. See SAS documentation for a complete list.

CSV   Comma-separated-values files with a file extension of .**csv**.

TAB   Tab-delimited files with a file extension of .**txt**.

SAV   SPSS files.

JMP   JMP files.

XLS   Excel files with a file extension of .**xls**.

EXCEL   Most types of Excel files, including older versions of Excel and .**xlsx** files. Reading .**xlsx** files requires the second maintenance release after SAS 9.2, or later.

The general form of importing from Excel and creating a temporary data set is shown below:

PROC IMPORT DATAFILE="*data-location***"**
            DBMS=XLS
            OUT=*data-set-name***;**

data-location identifies the location of the Excel file.

data-set-name is the name of the data set that you want to create.

To create a permanent SAS data set, use a **LIBNAME** statement, and then use a two-level name for the *data-set-name*.

Using the Import Wizard on a PC

On the PC, SAS includes an Import Wizard that guides you through the tasks for importing external files. You can use this wizard when you use SAS in the windowing environment. This section shows an example of importing the tickets data, which is saved in an Excel file.

1. Select **File→Import Data** from the menu bar. SAS shows the first page of the Import Wizard as shown below:

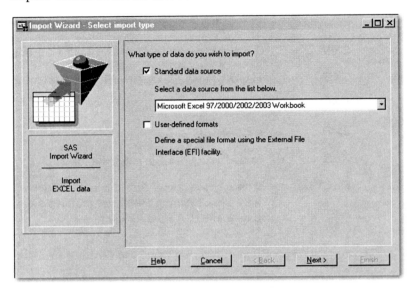

The wizard automatically selects **Standard data source**, and automatically chooses **Microsoft Excel** files. Select the drop-down menu to view other standard data sources as shown below:

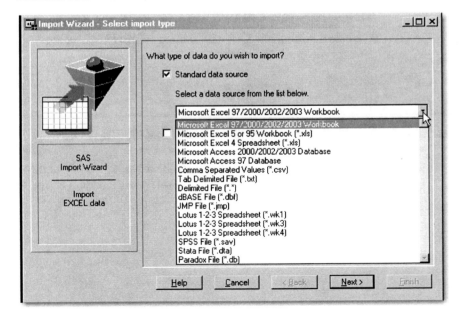

If none of these data sources meets your needs, you can select **User-defined formats**. This selection starts the SAS External File Interface (EFI) feature, which this book does not discuss.

2. Click **Next**. On the next page, click **Browse**, and then select the location of your external data file. After navigating to the location, click **Open**. Or, you can simply type the location into the blank field. After you provide the location, SAS displays it in the field as shown below. Click **OK**.

The next page automatically selects the first sheet in the Excel workbook, shown in the display below as **Sheet1$**. If you named the spreadsheet in Excel, use the drop-down menu and select the name you want.

Click **Options** to see how SAS will import your data. The following display shows the options that are automatically selected by SAS. If you have a variable that contains characters for some observations and numbers for other observations, you should select the option for conversion. (This option is not selected below.) Otherwise, the observations with numeric values will convert to missing variables.

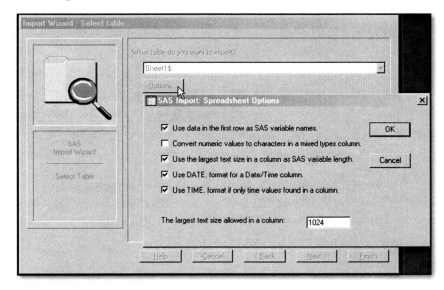

In general, accept the options that SAS automatically selects unless you know that these options will not work for your data. Click **OK** to close the **Spreadsheet Options** window.

3. Click **Next**.

4. The next page displays the SAS library and data set name for your new data set. SAS automatically chooses the **WORK** library. (See "Opening an Existing SAS Data Set" earlier in this chapter.) If you want to permanently save the new data set, then select the drop-down menu in the **Library** field and select a library. (To do this, you must have already defined the library with a LIBNAME statement.)

    SAS does not automatically give your data set a name. Enter the name in the **Member** field. Although the drop-down menu lists available data sets (either in the WORK library or another library, whichever library is selected), using the menu to select a data set name will overwrite the existing data set. The safest approach is to enter a new name, such as **TICKETS2**, as shown below:

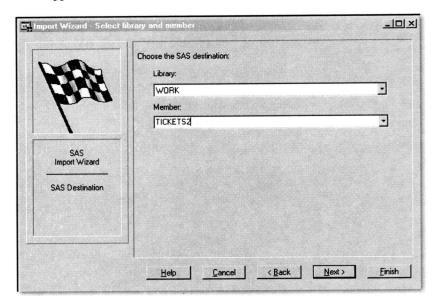

5. Click **Next**.

6. The next page provides you with an option to save the PROC IMPORT program that SAS creates behind the scenes. You can simply click **Finish**, and SAS imports the data.

However, if you do want to save the program, click **Browse** and navigate to the location where you want to save the program. Click **Save**. SAS displays the location as shown below:

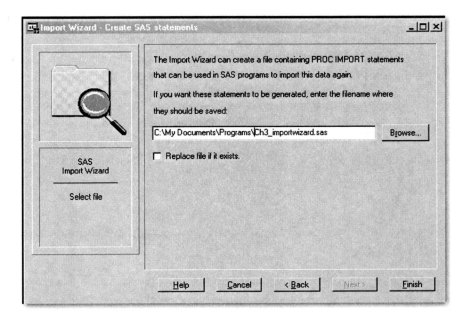

Optionally, you can select the check box for **Replace file if it exists**. This approach can be useful if you previously tried to import the data and were unsuccessful. For example, you might not have selected the option for conversion (discussed in step 2), and you discover that you do need to select this option. By replacing the program if it exists, you do not need to keep the old, incorrect version of the program.

7. After importing has completed, check the SAS log. The log displays a note informing you that the data set was successfully created. You might want to print your data set to check it before performing any analyses.

While using the Import Wizard, you can click **Back** to return to the previous page, or you can click **Cancel** to exit the wizard.

Introducing Advanced Features

As mentioned at the beginning of this chapter, SAS can read almost any data file from almost any source. This chapter discusses only the basics. Providing full details for every operating system would require a book by itself.

If the tools in this chapter do not help you read your data, see Appendix 1, "Further Reading," for references. If you use a PC, you might want to start with the SAS documentation for DDE (Dynamic Data Exchange). If you use SAS in the windowing environment mode, you might want to start with documentation for EFI. The simplest way to access EFI is to select **User-defined formats** in the SAS Import Wizard. Then, you identify the external file location and the location of your new data set. SAS guides you through the tasks of identifying the variables, and specifying the type (character or numeric) and location for each variable.

Summary

Key Ideas

- If your data exists in an electronic format, investigate ways to import it into SAS instead of entering the data all over again.

- To import an existing SAS data set, use a LIBNAME statement to identify the physical file location of the data set. Then, use the two-level name in SAS procedures.

- To read data from an existing text file, the simplest approach is to use an INFILE statement to identify the file location. Then, the INFILE statement replaces the DATALINES and null statements in the DATA step. You can also use PROC IMPORT if you use SAS on Windows, UNIX, or OpenVMS.

- To import a Microsoft Excel spreadsheet into SAS on Windows, UNIX, and OpenVMS, use PROC IMPORT.

- On a PC, you can use the SAS Import Wizard, which guides you through the tasks for importing external files.

Syntax

To assign a library reference

LIBNAME *libref* '*SAS-data-library-location*' ;

libref	is a nickname (library reference) for the SAS library. The libref can be up to eight characters long. SAS reserves some librefs, so do not use **Sasuser, Work, Sasmsg,** or **Sashelp.**
SAS-data-library-location	
	identifies the physical file location for the saved SAS data sets. Enclose the *SAS-data-library-location* in single quotation marks. The text for the physical file location depends on your operating system.

To use the libref to access a data set with a SAS procedure

PROC PRINT DATA=*libref.data-set-name*;

libref	is defined above.
data-set-name	is the name of a SAS data set.

The period between *libref* and *data-set-name* is required.

While this example uses **PROC PRINT**, you can use the same approach with every procedure in this book.

To create a permanent SAS data set

First, use a **LIBNAME** statement to identify the SAS library. Then, use a two-level name in the **DATA** statement as shown below:

DATA *libref.data-set-name*;

After completing the statements for the **DATA** step, SAS saves the data set as a permanent data set.

To identify an existing text file

INFILE '*file-name*';

file-name	is the name of the file that contains the data. Enclose the *file-name* in single quotation marks.

To import from Excel and create a temporary data set

> **PROC IMPORT DATAFILE="***data-location***"**
> **DBMS=XLS**
> **OUT=***data-set-name***;**

data-location   identifies the location of the Excel file.

data-set-name   is the name of the data set that you want to create.

To create a permanent SAS data set, use a **LIBNAME** statement, and then use a two-level name for the *data-set-name*.

Chapter **4**

Summarizing Data

After you know how to create a data set, the next step is to learn how to summarize it. This chapter explains the concepts of levels of measurement and types of response scales. Depending on the variable, appropriate methods to summarize it include the following:

- descriptive statistics, such as the average

- frequencies or counts of values for a variable

- graphs that are appropriate for the variable, such as bar charts or histograms, box plots, or stem-and-leaf plots

SAS provides several procedures for summarizing data. This chapter discusses the **UNIVARIATE, MEANS, FREQ, CHART**, and **GCHART** procedures.

This chapter uses the body fat data from Chapter 2, and the speeding ticket data from Chapter 3. This chapter introduces the dog walk data set.

Understanding Types of Variables

Chapter 2 discussed character and numeric variables. Character variables have values containing characters or numbers, and numeric variables have values containing numbers only.

Another way to classify variables is based on the level of measurement. The level of measurement is important because some methods of summarizing or analyzing data are based on the level of measurement. Levels of measurement can be *nominal*, *ordinal*, or *continuous*. In addition, variables can be measured on a *discrete* or a *continuous* scale.

Levels of Measurement

The list below summarizes levels of measurement:

Nominal
Values of the variable provide names. These values have no implied order. Values can be character (Male, Female) or numeric (1, 2). If the value is numeric, then an interim value does not make sense. If gender is coded as 1 and 2, then a value of 1.5 does not make sense.

When you use numbers as values of a nominal variable, the numbers have no meaning, other than to provide names.

SAS uses numeric values to order numeric nominal variables. SAS orders character nominal variables based on their sorting sequence.

Ordinal
Values of the variable provide names and also have an implied order. Examples include High-Medium-Low, Hot-Warm-Cool-Cold, and opinion scales (Strongly Disagree to Strongly Agree). Values can be character or numeric. The distance between values is not important, but their order is. High-Medium-Low could be coded as 1-2-3 or 10-20-30, with the same result.

SAS uses numeric values to order numeric ordinal variables. SAS orders character ordinal variables based on their sorting sequence. A variable with values of High-Medium-Low is sorted alphabetically as High-Low-Medium.

Continuous   Values of the variable contain numbers, where the distance between values is important. Many statistics books further classify continuous variables into interval and ratio variables. See the "Technical Details" box for an explanation.

Technical Details: Interval and Ratio

Many statistical texts refer to interval and ratio variables. Here's a brief explanation.

Interval variables are numeric and have an inherent order. Differences between values are important. Think about temperature in degrees Fahrenheit (F). A temperature of 120°F is 20° warmer than 100°F. Similarly, a temperature of 170°F is 20° warmer than 150°F. Contrast this with an ordinal variable, where the numeric difference between values is not meaningful. For an ordinal variable, the difference between 2 and 1 does not necessarily have the same meaning as the difference between 3 and 2.

Ratio variables are numeric and have an inherent order. Differences between values are important and the value of 0 is meaningful. For example, the weight of gold is a ratio variable. The number of calories in a meal is a ratio variable. A meal with 2,000 calories has twice as many calories as a meal with 1,000 calories. Contrast this with temperature. It does not make sense to say that a temperature of 100°F is twice as hot as 50°F, because 0° on a Fahrenheit scale is just an arbitrary reference point.

For the analyses in this book, SAS does not consider interval and ratio variables separately. From a practical view, this makes sense. You would perform the same analyses on interval or ratio variables. The difference is how you would summarize the results in a written report. For an interval variable, it does not make sense to say that the average for one group is twice as large as the average for another group. For a ratio variable, it does.

Types of Response Scales

The list below summarizes response scales:

Discrete   Variables measured on a *discrete scale* have a limited number of numeric values in any given range. The number of times you walk your dog each day is a variable measured on a discrete scale. You can choose not to walk your dog at all, which gives a value of 0. You can walk your dog 1, 2, or 7 times, which gives values of 1, 2, and 7, respectively. You can't walk your dog 3.5 times in one day, however. The discrete values for this variable are the numbers 0, 1, 2, 3, and so on. Test scores are another example of a variable measured on a discrete scale. With a range of scores from 0 to 100, the possible values are 0, 1, 2, and so on.

Continuous   Variables measured on a *continuous scale* have conceptually an unlimited number of values. As an example, your exact body temperature is measured on a continuous scale. A thermometer doesn't measure your exact body temperature, but measures your approximate body temperature. Body temperature conceptually can take on any value within a specific range, say 95° to 105° Fahrenheit. In this range, conceptually there are an unlimited number of values for your exact body temperature (98.6001, 98.6002, 98.60021, and so on). But, in reality, you can measure only a limited number of values (98.6, 98.7, 98.8, and so on). Contrast this with the number of times you walk your dog. Conceptually and in reality, there are a limited number of values to represent the number of times you walk your dog each day.

The important idea that underlies a continuous scale is that the variable can theoretically take on an unlimited number of values. In reality, the number of values for the variable is limited only by your ability to measure.

Relating Levels of Measurement and Response Scales

Nominal and ordinal variables are usually measured on a discrete scale. However, for ordinal variables, the discrete scale is sometimes one that separates a continuous scale into categories. For example, a person's age is often classified according to categories (0–21, 22–45, 46–65, and over 65, for example).

Interval and ratio variables are measured on discrete and continuous scales. The number of times you walk your dog each day is a ratio measurement on a discrete scale. Conceptually, your weight is a ratio measurement on a continuous scale.

Table 4.1 shows the relationships between the levels of measurement and the summary methods in this chapter. For example, box plots are appropriate for continuous variables, but not for nominal variables.

Table 4.1 Levels of Measurement and Basic Summaries

Level of Measurement	Methods for Summarizing Data			
	Descriptive Statistics	Frequency Tables	Bar Charts	Histograms Box Plots
Nominal		✓	✓	
Ordinal	✓	✓	✓	
Continuous	✓	✓	✓	✓

Summarizing a Continuous Variable

This section shows two SAS procedures for summarizing a continuous variable. PROC UNIVARIATE gives an extensive summary, and PROC MEANS gives a brief summary. To get an extensive summary of the speeding ticket data, type the following:

```
options ps=55;
proc univariate data=tickets;
    var amount;
    id state;
    title 'Summary of Speeding Ticket Data';
run;
```

The PROC UNIVARIATE statement requests a summary for the tickets data set, the **VAR** statement lists variables to be summarized, and the **ID** statement names a variable that identifies observations in one part of the output. These statements produce Figure 4.1.

The output includes five tables with headings **Moments, Basic Statistical Measures, Quantiles, Extreme Observations**, and **Missing Values**. This chapter discusses these five tables. The procedure also prints the **Tests for Location** table, which this book discusses in Chapter 7.

Figure 4.1 PROC UNIVARIATE Results for Tickets Data Set

```
                    Summary of Speeding Ticket Data

                        The UNIVARIATE Procedure
                          Variable:  amount

                                Moments

N                          50      Sum Weights                    50
Mean                   265.67      Sum Observations          13283.5
Std Deviation      291.176845      Variance               84783.9552
Skewness            1.71408532     Kurtosis               1.83191198
Uncorrected SS     7683441.25      Corrected SS           4154413.81
Coeff Variation     109.600951     Std Error Mean         41.1786244


                      Basic Statistical Measures

          Location                        Variability

     Mean      265.6700      Std Deviation          291.17685
     Median    162.5000      Variance                   84784
     Mode      100.0000      Range                  980.00000
                             Interquartile Range    223.00000


                    Tests for Location: Mu0=0

        Test             -Statistic-       -----p Value------

     Student's t     t   6.451648     Pr > |t|      <.0001
     Sign            M         25      Pr >= |M|      <.0001
     Signed Rank     S      637.5     Pr >= |S|      <.0001


                    Quantiles (Definition 5)

                    Quantile        Estimate

                    100% Max         1000.0
                    99%              1000.0
                    95%              1000.0
                    90%               875.0
                    75% Q3            300.0
                    50% Median        162.5
                    25% Q1             77.0
                    10%                50.0
                    5%                 46.5
                    1%                 20.0
                    0% Min             20.0
```

Figure 4.1 PROC UNIVARIATE Results for Tickets Data Set (continued)

```
              Summary of Speeding Ticket Data

                    The UNIVARIATE Procedure
                    Variable:   amount

                    Extreme Observations

    ---------Lowest---------        --------Highest---------

  Value    state       Obs        Value    state       Obs

   20.0    DE           3          1000    IL           4
   40.0    MI          26          1000    NV          24
   46.5    PA          40          1000    NC          31
   50.0    TN          41          1000    NH          35
   50.0    ME          20          1000    VT          44


                    Missing Values

                             -----Percent Of-----
            Missing                          Missing
            Value        Count    All Obs        Obs

              .            1        1.96      100.00
```

From Figure 4.1, the average speeding ticket fine is $265.67. The median is $162.50 and the mode is $100. The standard deviation is $291.18. The smallest fine is $20, and the largest fine is $1000. The data set has one missing value, which is for Washington, D.C. Although Figure 4.1 shows many other statistics, these few summary statistics (mean, median, mode, standard deviation, minimum, and maximum) provide an overall summary of the distribution of speeding ticket fines.

The next five topics discuss the tables of results.

Reviewing the Moments Table

Figure 4.1 shows the Moments table that PROC UNIVARIATE prints automatically. The list below describes the items in this table.

N	Number of rows with nonmissing values for the variable. This number might not be the same as the number of rows in the data. The tickets data set has 51 rows, but N is **50** because the value for Washington, D.C. is missing.
	SAS uses N in calculating other statistics, such as the mean.

Mean	Average of the data. It is calculated as the sum of values, divided by the number of nonmissing values. The mean is the most common measure used to describe the center of a distribution of values. The mean for **amount** is **265.67**, so the average speeding ticket fine for the 50 states is $265.67.
Std Deviation	Standard deviation, which is the square root of the variance. For **amount**, this is about 291.18 (rounded off from the output shown). The standard deviation is the most common measure used to describe dispersion or variability around the mean. When all values are close to the mean, the standard deviation is small. When the values are scattered widely from the mean, the standard deviation is large. Like the variance, the standard deviation is a measure of dispersion about the mean, but it is easier to interpret because the units of measurement for the standard deviation are the same as the units of measurement for the data.
Skewness	Measure of the tendency for the distribution of values to be more spread out on one side than the other side. Positive skewness indicates that values to the right of the mean are more spread out than values to the left. Negative skewness indicates the opposite. For **amount**, the skewness of about 1.714 is caused by the group of states with speeding ticket fines above $300, which pulls the distribution to the right of the mean, and spreads out the right side more than the left side.
Uncorrected SS	Uncorrected sum of squares. Chapter 9 discusses this statistic.

Coeff Variation Coefficient of variation, sometimes abbreviated as CV. It is calculated as the standard deviation, divided by the mean, and multiplied by 100. Or, (Std Dev/Mean)*100.

Sum Weights Sum of weights. When you don't use a Weight column, Sum Weights is the same as N. When you do use a Weight column, Sum Weights is the sum of that column, and SAS uses Sum Weights instead of N in calculations. This book does not use a Weight column.

Sum Observations Sum of values for all rows. It is used in calculating the average and other statistics.

Variance Square of the standard deviation. The Variance is never less than 0, and is 0 only if all rows have the same value.

Kurtosis Measure of the shape of the distribution of values. Think of kurtosis as measuring how peaked or flat the shape of the distribution of values is. Large values for kurtosis indicate that the distribution has heavy tails or is flat. This means that the data contains some values that are very distant from the mean, compared to most of the other values in the data.

For amount, the right tail of the distribution of values is heavy because of the group of states with speeding ticket fines above $300.

Corrected SS Corrected sum of squares. Chapter 9 discusses this statistic.

Std Error Mean Standard error of the mean. It is calculated as the standard deviation, divided by the square root of the sample size. It is used in calculating confidence intervals. Chapter 6 gives more detail.

Reviewing the Basic Statistical Measures Table

Figure 4.1 shows the Basic Statistical Measures table that PROC UNIVARIATE prints automatically. The **Location** column shows statistics that summarize the center of the distribution of data values. The **Variability** column shows statistics that summarize the breadth or spread of the distribution of data values. This table repeats many of the statistics in the Moments table because you can suppress the Moments table and display only these few key statistics. See "Special Topic: Using ODS to Control Output Tables" at the end of this chapter. The list below describes the items in this table.

Mean	Average of the data, as described for the Moments table.
Median	50th percentile. Half of the data values are above the median, and half are below it. The median is less sensitive than the mean to skewness, and is often used with skewed data.
Mode	Data value with the largest associated number of observations. The mode of **100** indicates that more states give speeding ticket fines of $100 than any other amount.
	Sometimes the data set contains more than one mode. In that situation, SAS displays the smallest mode and prints a note below the table that explains that the printed mode is the smallest.
Std Deviation	Standard deviation, as described for the Moments table.
Variance	Square of the standard deviation, as described for the Moments table.
Range	Difference between highest and lowest data values.
Interquartile Range	Difference between the 25th percentile and the 75th percentile.

Reviewing the Quantiles Table

Figure 4.1 shows the Quantiles table that PROC UNIVARIATE prints automatically. This table gives more detail about the distribution of data values. The 0th percentile is the lowest value and is labeled as **Min**. The 100th percentile is the highest value and is labeled as **Max**. The difference between the highest and lowest values is the range.

The 25th percentile is the *first quartile* and is greater than 25% of the values. The 75th percentile is the *third quartile* and is greater than 75% of the values. The difference between these two percentiles is the interquartile range, which also appears in the Basic Statistical Measures table. The 50th percentile is the median, which also appears in the Basic Statistical Measures table.

SAS shows several other quantiles, which are interpreted in a similar way. For example, the 90th percentile is **875**, which means that 90% of the values are less than 875.

The heading for the Quantiles table identifies the definition that SAS uses for percentiles in the table. SAS uses **Definition 5** automatically. For more detail on these definitions, see the SAS documentation on the **PCTLDEF=** option.

Reviewing the Extreme Observations Table

Figure 4.1 shows the Extreme Observations table that PROC UNIVARIATE prints automatically. This table lists the five lowest and five highest observations, in the **Lowest** and **Highest** columns. The lowest value is the *minimum*, and the highest value is the *maximum*. For this data, the minimum is **20** and the maximum is **1000**.

Figure 4.1 shows two observations with the same lowest value in this table. This table doesn't identify the five lowest distinct values. For example, two states have an amount of **50**, and two of the lowest observations show this. The five highest observations for amount are all **1000**.

The Extreme Observations table shows the impact of using the ID statement. The table identifies the observations with the value of **state**. Without an ID statement, the procedure identifies observations with their observation numbers (**Obs** in this table in Figure 4.1).

Reviewing the Missing Values Table

Figure 4.1 shows the Missing Values table that PROC UNIVARIATE prints automatically.

Missing Value shows how missing values are coded, and **Count** gives the number of missing values. The right side of the table shows percentages. The **All Obs** column shows the percentage of all observations with a given missing value, and the **Missing Obs** column shows the percentage of missing observations with a given value. For the speeding ticket data, **1.96%** of all observations are missing, and **100%** of the missing values are coded with a period.

Reviewing Other Tables

Figure 4.1 shows the Tests for Location table that PROC UNIVARIATE prints automatically. Chapter 7 discusses this table.

> The general form of the statements to summarize data is shown below:
>
> **PROC UNIVARIATE DATA**=*data-set-name*;
>
>   **VAR** *variables*;
>
>   **ID** *variable*;
>
> *data-set-name* is the name of a SAS data set, and *variables* lists one or more variables in the data set. If you omit the variables from the VAR statement, then the procedure uses all numeric variables. The ID statement is optional. If you omit the ID statement, then SAS identifies the extreme observations using their observation numbers.

Adding a Summary of Distinct Extreme Values

The Extreme Observations table prints the five lowest and five highest observations in the data set. These might be the five lowest distinct values and the five highest distinct values. Especially with large data sets, the more likely situation is that this is not the case.

To see the lowest and highest distinct values, you can use the **NEXTRVAL=** option. You can suppress the Extreme Observations table with the **NEXTROBS=0** option.

```
proc univariate data=tickets nextrval=5 nextrobs=0;
   var amount;
   id state;
   title 'Summary of Speeding Ticket Data';
run;
```

These statements produce most of the tables in Figure 4.1. The Extreme Observations table does not appear. Figure 4.2 shows the **Extreme Values** table, printed as a result of the NEXTRVAL= option.

Figure 4.2 Distinct Extreme Values for Speeding Ticket Data

```
                    Extreme Values

--------- Lowest ---------       -------- Highest --------

 Order    Value    Freq       Order    Value    Freq

   1      20.0      1           18      300       4
   2      40.0      1           19      500       3
   3      46.5      1           20      600       1
   4      50.0      3           21      750       1
   5      55.0      2           22     1000       5
```

Compare the results in Figures 4.1 and 4.2. Figure 4.2 shows the five lowest distinct values of **20**, **40**, **46.5**, **50**, and **55**. Compare these with Figure 4.1, which shows the five lowest observations and does not show the value of **55**. Similarly, Figure 4.2 shows the five highest distinct values, and Figure 4.1 shows the five highest observations. The five highest observations are all **1000**, but the five highest distinct values are **300**, **500**, **600**, **750**, and **1000**.

For some data, the two sets of observations are the same—the five highest observations are also the five highest distinct values. Or, the five lowest observations are also the five lowest distinct values.

The general form of the statements to suppress the lowest and highest observations and to print the lowest and highest distinct values is:

PROC UNIVARIATE DATA=*data-set-name*
                    NEXTRVAL=5 NEXTROBS=0;

   VAR *variables*;

data-set-name is the name of a SAS data set, and *variables* lists one or more variables in the data set.

The **NEXTRVAL=5** option prints the five lowest distinct values and the five highest distinct values. This option creates the **Extreme Values** table. PROC UNIVARIATE automatically uses NEXTRVAL=0. You can use values from 0 to half the number of observations. For example, for the speeding ticket data, NEXTRVAL= has a maximum value of 25.

The **NEXTROBS=0** option suppresses the **Extreme Observations** table, which shows the five lowest and five highest observations in the data. PROC UNIVARIATE automatically uses NEXTROBS=5. You can use values from 0 to half the number of observations. For example, for the speeding ticket data, NEXTROBS= has a maximum value of 25.

You can use the NEXTRVAL= and NEXTROBS= options together or separately to control the tables that you want to see in the output.

Using PROC MEANS for a Brief Summary

PROC UNIVARIATE prints detailed summaries of variables, and prints the information for each variable in a separate table. If you want a concise summary, you might want to use PROC MEANS. To get a brief summary of the speeding ticket data, type the following:

```
proc means data=tickets;
   var amount;
   title 'Brief Summary of Speeding Ticket Data';
run;
```

These statements create the output shown in Figure 4.3.

This summary shows the number of observations (**N**), average value (**Mean**), standard deviation (**Std Dev**), minimum (**Minimum**), and maximum (**Maximum**) for the amount variable. The definitions of these statistics were discussed earlier for PROC UNIVARIATE.

Figure 4.3 PROC MEANS Results for Speeding Ticket Data

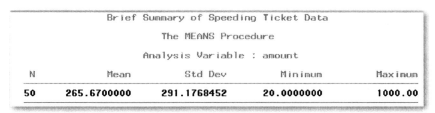

		Brief Summary of Speeding Ticket Data		
		The MEANS Procedure		
		Analysis Variable : amount		
N	Mean	Std Dev	Minimum	Maximum
50	265.6700000	291.1768452	20.0000000	1000.00

The PROC MEANS output is much briefer than the PROC UNIVARIATE output. If there were several variables in the tickets data set, and you wanted to get only these few descriptive statistics for each variable, PROC MEANS would produce a line of output for each variable. PROC UNIVARIATE would produce at least one page of output for each variable.

In addition, you can add many statistical keywords to the PROC MEANS statement, and create a summary that is still brief, but provides more statistics.

```
proc means data=tickets
      n nmiss mean median stddev range qrange min max;
   var amount;
   title 'Single-line Summary of Speeding Ticket Data';
run;
```

These statements create the output shown in Figure 4.4.

The summary condenses information that PROC UNIVARIATE displays in multiple tables into a single table. With a **LINESIZE=** value greater than 80, the information appears on a single line. This approach can be very useful for data sets with many variables.

Figure 4.4 PROC MEANS Results with Specified Statistics

```
            Single-line Summary of Speeding Ticket Data

                       The MEANS Procedure

                    Analysis Variable : amount

                                                               Quartile
      N                                                           Range
 N   Miss        Mean        Median       Std Dev        Range
 50     1   265.6700000   162.5000000   291.1768452   980.0000000   223.0000000


                    Analysis Variable : amount

                      Minimum          Maximum
                     20.0000000         1000.00
```

The general form for PROC MEANS is shown below:

PROC MEANS DATA=*data-set-name*
        N NMISS MEAN MEDIAN STDDEV RANGE QRANGE MIN MAX;
    VAR *variables*;

data-set-name is the name of a SAS data set, and *variables* lists one or more variables in the data set. If you omit the variables from the VAR statement, then the procedure uses all continuous variables.

All of the statistical keywords are optional, and the procedure includes other keywords that this book does not discuss. The keywords create statistics as listed below:

N	number of observations with nonmissing values
NMISS	number of missing values
MEAN	mean
MEDIAN	median
STDDEV	standard deviation
RANGE	range
QRANGE	interquartile range
MIN	minimum
MAX	maximum

The discussion for PROC UNIVARIATE earlier in this chapter defines the statistics above. (PROC UNIVARIATE shows NMISS as Count in the Missing Values table.) If you omit all of the statistical keywords, then PROC MEANS prints N, MEAN, STDDEV, MIN, and MAX.

Creating Line Printer Plots for Continuous Variables

PROC UNIVARIATE can create line printer plots and high-resolution graphs. To create the high-resolution graphs, you must have SAS/GRAPH software licensed. This section shows the line printer box plot and line printer stem-and-leaf plot. For the speeding ticket data, type the following:

```
proc univariate data=tickets plot;
    var amount;
    title 'Line Printer Plots for Speeding Ticket Data';
run;
```

The first part of your output is identical to the output in Figure 4.1. Figure 4.5 shows partial results from the **PLOT** option.

Your output will show complete results, which are three plots labeled **Stem Leaf**, **Boxplot**, and **Normal Probability Plot**. Chapter 5 discusses the normal probability plot. The next two topics discuss the box plot and stem-and-leaf plot.

Figure 4.5 Line Printer Plots from PROC UNIVARIATE

```
          Line Printer Plots for Speeding Ticket Data

                    The UNIVARIATE Procedure
                     Variable:   amount

    Stem Leaf                            #            Boxplot
      10 00000                           5               *
       9
       9
       8
       8
       7 5                               1               0
       7
       6
       6 0                               1
       5                                                 |
       5 000                             3               |
       4                                                 |
       4                                                 |
       3                                                 |
       3 0000                            4           +-----+
       2 55                              2           |  +  |
       2 00000001                        8           |     |
       1 58                              2           *-----*
       1 000000000                       9           |     |
       0 5555666788889                  13           +-----+
       0 24                              2               |
        ----+----+----+----+
    Multiply Stem.Leaf by 10**+2
```

Understanding the Stem-and-Leaf Plot

Figure 4.5 shows the stem-and-leaf plot on the left.[1] The stem-and-leaf plot shows the value of the variable for each observation.

The stem is on the left side under the **Stem** heading, and the leaves are on the right under the **Leaf** heading. The instructions at the bottom tell you how to interpret the stem-and-leaf plot. In Figure 4.5, the stems are the 100s and the leaves are 10s. Look at the lowest stem. Take 0.2 (**Stem.Leaf**) and multiply it by $10**+2$ (=100) to get 20 (0.2*100=20).

---

[1] If your data contains more than 49 observations in a single stem, PROC UNIVARIATE produces a horizontal bar chart instead of a stem-and-leaf plot.

For your data, follow the instructions to determine the values of the variable.[2] SAS prints instructions except when you do not need to multiply **Stem.Leaf** by anything.

For the speeding ticket data, the lowest stem is for the values from 0 to 49, the second is for the values from 50 to 99, the third is for the values from 100 to 149, and so on. The plot is roughly mound shaped. The highest part of the mound is for the data values from 50 to 149, which makes sense, given the mean, median, and mode.

SAS might round data values in order to display all values in the plot. For example, SAS rounds the value of 46.5 to 50.

The plot shows a lot of skewness or sidedness. The values above 300 are more spread out than the values near 0. The data is skewed to the right—in the direction where more values trail at the end of the distribution of values. This skewness is seen in the box plot and **Moments** table.

The plot shows five values of 1000. You can see how much farther away these values are from the rest of the data. These extremely high values help cause the positive kurtosis of 1.83.

The column labeled with the pound sign (#) identifies how many values appear in each stem. For example, there are 2 values in the stem labeled 0.

Understanding the Box Plot

Figure 4.5 shows the box plot on the right. This plot uses the scale from the stem-and-leaf plot on the left.

The bottom and top of the box are the 25th percentile and the 75th percentile. The length of the box is the interquartile range.

The line inside the box with an asterisk at each end is the median. The plus sign (+) inside the box indicates the mean, which might be the same as the median, but usually is not. If the median and the mean are close to each other, the plus sign (+) falls on the median line inside the box.

The dotted vertical lines that extend from the box are *whiskers*. Each whisker extends up to 1.5 interquartile ranges from the end of the box. Values outside the whiskers are marked with a 0 or an asterisk (*). A 0 is used if the value is between 1.5 and 3 interquartile ranges of the box, and an asterisk (*) is used if the value is farther away. Values that are far away from the rest of the data are called *outliers*.

---

[2] In SAS, double asterisks (**) means "raised to the power of," so 10**2 is 10^2, or 100.

The box plot for the speeding ticket data shows the median closer to the bottom of the box, which indicates skewness or sidedness to the data values. The mean and the median are different enough so that the plus sign (+) does not appear on the line for the median. The plot shows one potential outlier point at 1000. This point represents the five data values at 1000 (all of which are correct in the data set). The plot also shows one point outside the whiskers with a value of 750. This example shows how using the stem-and-leaf plot and box plot together can be helpful. With the box plot alone, you do not know that the potential outlier point at 1000 actually represents five data values.

The general form of the statements to create line printer plots is shown below:

PROC UNIVARIATE DATA=*data-set-name* **PLOT;**

   VAR *variables*;

data-set-name is the name of a SAS data set, and *variables* lists one or more variables in the data set.

You can use the PLOT option with the NEXTRVAL= and NEXTROBS= options.

Creating Histograms for Continuous Variables

PROC UNIVARIATE can create histograms, which show the shape of the distribution of data values. Histograms are appropriate for continuous variables. For the speeding ticket data, type the following:

```
proc univariate data=tickets noprint;
   var amount;
   histogram amount;
   title 'Histogram for Speeding Ticket Data';
run;
```

The **NOPRINT** option suppresses the printed output tables from PROC UNIVARIATE. Without this option, the statements would create the output shown in Figure 4.1. The **HISTOGRAM** statement creates the plot shown in Figure 4.6.

Figure 4.6 Histogram for Speeding Ticket Data

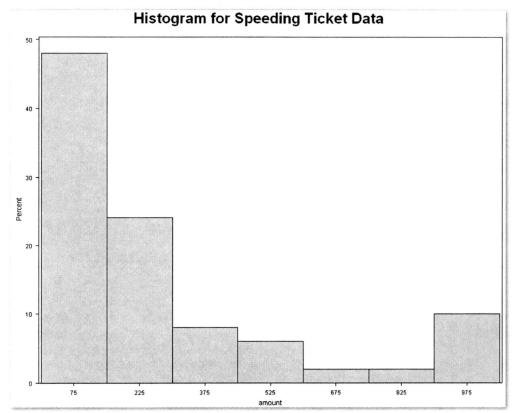

The **amount** variable has a somewhat mound-shaped distribution at the lower end of the values. Then, several bars appear to the right. These bars represent the speeding ticket fines above $300. The final bar represents the five states with speeding ticket fines of $1000. Like the box plot and stem-and-leaf plot, the histogram shows the skewness in the data.

SAS labels each bar with the midpoint of the range of values. The next table summarizes the midpoints and range of values for each bar in Figure 4.6.

Midpoint	Range of Values
75	$0 \leq$ amount < 150
225	$150 \leq$ amount < 300
375	$300 \leq$ amount < 450
525	$450 \leq$ amount < 600
675	$600 \leq$ amount < 750
825	$750 \leq$ amount < 900
975	$900 \leq$ amount < 1050

The general form of the statements to create histograms and suppress other printed output is shown below:

PROC UNIVARIATE DATA=*data-set-name* **NOPRINT**;

  VAR *variables*;

    HISTOGRAM *variables*;

Items in italic were defined earlier in the chapter. If you omit the variables from the HISTOGRAM statement, then the procedure uses all numeric variables in the VAR statement. If you omit the variables from the HISTOGRAM and VAR statements, then the procedure uses all numeric variables. You cannot specify variables in the HISTOGRAM statement unless the variables are also specified in the VAR statement.

You can use the NOPRINT option with other options.

Creating Frequency Tables for All Variables

For nominal and ordinal variables, you might want to know the count for each value of the variable. Also, for continuous variables measured on a discrete scale, you might want to know the count for each value.

The data in Table 4.2 shows the number of times people in a neighborhood walk their dogs each day.

Table 4.2 Dog Walk Data

Number of Times Dogs Walked per Day				
3	1	2	0	1
2	3	1	1	2
1	2	2	2	1
3	2	4	2	1

This data is available in the dogwalk data set in the sample data for this book.

Using PROC UNIVARIATE

The number of daily walks is a numeric variable, and you can use PROC UNIVARIATE to summarize it.

```
data dogwalk;
   input walks @@;
   label walks='Number of Daily Walks';
   datalines;
 3 1 2 0 1 2 3 1 1 2 1 2 2 2 1 3 2 4 2 1
;
run;

options ps=50;
proc univariate data=dogwalk freq;
   var walks;
title 'Summary of Dog Walk Data';
run;
```

Figure 4.7 shows the PROC UNIVARIATE results.

Figure 4.7 PROC UNIVARIATE Results for Dog Walk Data

```
                    Summary of Dog Walk Data

                    The UNIVARIATE Procedure
             Variable:  walks  (Number of Daily Walks)

                            Moments

N                           20    Sum Weights               20
Mean                       1.8    Sum Observations          36
Std Deviation       0.95145318    Variance          0.90526316
Skewness            0.43996348    Kurtosis           0.2533696
Uncorrected SS              82    Corrected SS            17.2
Coeff Variation     52.8585101    Std Error Mean     0.2127514


                  Basic Statistical Measures

        Location                        Variability

    Mean      1.800000    Std Deviation          0.95145
    Median    2.000000    Variance               0.90526
    Mode      2.000000    Range                  4.00000
                          Interquartile Range    1.00000


                 Tests for Location: Mu0=0

     Test              -Statistic-        -----p Value------

     Student's t    t  8.460579    Pr > |t|      <.0001
     Sign           M       9.5    Pr >= |M|     <.0001
     Signed Rank    S        95    Pr >= |S|     <.0001


                  Quantiles (Definition 5)

                  Quantile        Estimate

                  100% Max          4.0
                  99%               4.0
                  95%               3.5
                  90%               3.0
                  75% Q3            2.0
                  50% Median        2.0
                  25% Q1            1.0
                  10%               1.0
                  5%                0.5
                  1%                0.0
                  0% Min            0.0
```

Figure 4.7 PROC UNIVARIATE Results for Dog Walk Data (continued)

```
                        Summary of Dog Walk Data

                          The UNIVARIATE Procedure
                  Variable:  walks  (Number of Daily Walks)

                          Extreme Observations

                 ----Lowest----              ----Highest---

                 Value       Obs             Value      Obs

                   0          4                2         19
                   1         20                3          1
                   1         15                3          7
                   1         11                3         16
                   1          9                4         18


                            Frequency Counts
```

Value	Count	Percents Cell	Percents Cum	Value	Count	Percents Cell	Percents Cum	Value	Count	Percents Cell	Percents Cum
0	1	5.0	5.0	2	8	40.0	80.0	4	1	5.0	100.0
1	7	35.0	40.0	3	3	15.0	95.0				

This chapter discussed most of the tables in Figure 4.7 for the speeding ticket data. The last table in Figure 4.7 is the **Frequency Counts** table, which is created by the **FREQ** option in the PROC UNIVARIATE statement. Most people in the neighborhood walk their dogs one or two times a day. One person walks the dog four times a day, and one person does not take the dog on walks. The list below gives more detail on the items in the table:

Value
Value of the variable. When the data set has missing values, the procedure prints a row that shows the missing values.

Count
Number of rows with a given value.

Cell Percent
Percentage of rows with a given value. For example, 35% of people in the neighborhood walk their dogs once each day.

Cum Percent
Cumulative percentage of the row and all rows with lower values. For example, 80% of people in the neighborhood walk their dogs two or fewer times each day.

Use caution when adding the FREQ option for numeric variables with many values. PROC UNIVARIATE creates a line in the frequency table for each value of the variable. Consequently, for a numeric variable with many values, the resulting frequency table has many rows.

> The general form of the statements to create a frequency table with PROC UNIVARIATE is shown below:
>
> **PROC UNIVARIATE DATA**=*data-set-name* **FREQ;**
>
> **VAR** *variables*;
>
> Items in italic were defined earlier in the chapter. You can use the FREQ option with other options and statements.

Using PROC FREQ

For the **dogwalk** data set, the descriptive statistics might be less useful to you than the frequency table. For example, you might be less interested in knowing the average number of times these dogs are walked is 1.8 times per day. You might be more interested in knowing the percentage of dogs walked once a day, twice a day, and so on. You can use PROC FREQ instead and create only the frequency table. (See "Special Topic: Using ODS to Control Output Tables" for another approach.)

For other data sets, you cannot use PROC UNIVARIATE to create the frequency table because your variables are character variables. With those data sets, PROC FREQ is your only choice.

The statements below create frequency tables for the **dogwalk** and **bodyfat** data sets. (The program at the end of the chapter also includes the statements to create the **bodyfat** data set and format the **gender** variable.)

```
proc freq data=dogwalk;
   tables walks;
   title 'Frequency Table for Dog Walk Data';
run;

proc freq data=bodyfat;
   tables gender;
   title 'Frequency Table for Body Fat Data';
run;
```

Figure 4.8 shows the PROC FREQ results.

Figure 4.8 PROC FREQ Results for Dog Walk and Body Fat Data

```
              Frequency Table for Dog Walk Data

                    The FREQ Procedure

                  Number of Daily Walks

                                   Cumulative    Cumulative
   walks    Frequency    Percent    Frequency      Percent

     0          1         5.00          1           5.00
     1          7        35.00          8          40.00
     2          8        40.00         16          80.00
     3          3        15.00         19          95.00
     4          1         5.00         20         100.00
```

```
              Frequency Table for Body Fat Data

                    The FREQ Procedure

                                   Cumulative    Cumulative
   gender    Frequency    Percent    Frequency      Percent

   Female       10        43.48         10          43.48
   Male         13        56.52         23         100.00
```

The results for the **dogwalk** data match the results for PROC UNIVARIATE in Figure 4.7. The results for the **bodyfat** data summarize the counts and percentages for men and women in the fitness program. Because **gender** is a character variable, you cannot use PROC UNIVARIATE to create the frequency table. The list below gives more detail on the items in the table:

Variable Name	Values of the variable.
Frequency	Number of rows with a given value. The counts for each value match the **Count** column for PROC UNIVARIATE results. However, when the data set has missing values, the procedure does not print a row that shows the missing values. The procedure prints a note at the bottom of the table instead. See the next topic for more detail.

Percent	Percentage of rows with a given value. The value matches the **Cell Percent** column for PROC UNIVARIATE results.
Cumulative Frequency	Cumulative frequency of the row and all rows with lower values. For example, 16 of the neighbors walk their dogs two or fewer times each day. The PROC UNIVARIATE results do not display this information.
Cumulative Percent	Cumulative percentage of the row and all rows with lower values. The value matches the **Cum Percent** column for PROC UNIVARIATE results.

The general form of the statements to create a frequency table with PROC FREQ is shown below:

PROC FREQ DATA=*data-set-name***;**

 TABLES *variables***;**

Items in italic were defined earlier in the chapter. If you omit the **TABLES** statement, then the procedure prints a frequency table for every variable in the data set.

Missing Values in PROC FREQ

PROC UNIVARIATE results show you the number of missing values. You can add this information to PROC FREQ results. Suppose that when collecting data for the dogwalk data set, three neighbors were not at home. These cases were recorded as missing values in the data set. The statements below create results that show three ways PROC FREQ handles missing values:

```
data dogwalk2;
   input walks @@;
   label walks='Number of Daily Walks';
   datalines;
 3 1 2 . 0 1 2 3 1 . 1 2 1 2 . 2 2 1 3 2 4 2 1
;
run;

proc freq data=dogwalk2;
   tables walks;
   title 'Dog Walk with Missing Values';
   title2 'Automatic Results';
run;
```

```
proc freq data=dogwalk2;
   tables walks / missprint;
   title 'Dog Walk with Missing Values';
   title2 'MISSPRINT Option';
run;

proc freq data=dogwalk2;
   tables walks / missing;
   title 'Dog Walk with Missing Values';
   title2 'MISSING Option';
run;
```

Figure 4.9 shows the PROC FREQ results.

Figure 4.9 PROC FREQ and Missing Value Approaches

```
                 Dog Walk with Missing Values
                      Automatic Results

                      The FREQ Procedure

                    Number of Daily Walks

                                    Cumulative     Cumulative
    walks     Frequency     Percent  Frequency      Percent

      0           1          5.00        1           5.00
      1           7         35.00        8          40.00
      2           8         40.00       16          80.00
      3           3         15.00       19          95.00
      4           1          5.00       20         100.00

                   Frequency Missing = 3
```

```
                 Dog Walk with Missing Values
                       MISSPRINT Option

                      The FREQ Procedure

                    Number of Daily Walks

                                    Cumulative     Cumulative
    walks     Frequency     Percent  Frequency      Percent

      .           3            .          .            .
      0           1          5.00        1           5.00
      1           7         35.00        8          40.00
      2           8         40.00       16          80.00
      3           3         15.00       19          95.00
      4           1          5.00       20         100.00

                   Frequency Missing = 3
```

Figure 4.9 PROC FREQ and Missing Value Approaches (continued)

```
                     Dog Walk with Missing Values
                            MISSING Option

                          The FREQ Procedure

                        Number of Daily Walks

                                          Cumulative      Cumulative
     walks      Frequency      Percent     Frequency        Percent
     ─────────────────────────────────────────────────────────────────
        .            3          13.04          3            13.04
        0            1           4.35          4            17.39
        1            7          30.43         11            47.83
        2            8          34.78         19            82.61
        3            3          13.04         22            95.65
        4            1           4.35         23           100.00
```

The first table shows how PROC FREQ automatically handles missing values. The note at the bottom of the table identifies the number of missing values.

The second table shows the effect of using the **MISSPRINT** option. PROC FREQ adds a row in the frequency table to show the count of missing values. However, the procedure omits the missing values from calculations for percents, cumulative frequency, and cumulative percents.

The third table shows the effect of using the **MISSING** option. PROC FREQ adds a row in the frequency table to show the count of missing values. The procedure includes the missing values in calculations for percents, cumulative frequency, and cumulative percents.

Depending on your situation, select the approach that best meets your needs.

The general form of the statements to handle missing values with PROC FREQ is shown below:

PROC FREQ DATA=*data-set-name***;**

TABLES *variables* **/ MISSPRINT MISSING;**

Items in italic were defined earlier in the chapter.

The MISSPRINT option prints the count of missing values in the table. The MISSING option prints the count of missing values and also uses the missing values in calculating statistics. Use either the MISSPRINT or MISSING option; do not use both options. If you use an option, the slash is required.

Creating Bar Charts for All Variables

When your data has character variables, or when you want to create a bar chart that shows each value of a variable, you can use PROC GCHART. This procedure creates high-resolution graphs and requires SAS/GRAPH software. If you do not have SAS/GRAPH, then you can use PROC CHART to create line printer bar charts.

For variables with character values, either PROC GCHART or PROC CHART creates a bar chart that has one bar for each value. For variables with numeric values, both procedures group values together, similar to a histogram. In some situations, this approach is what you want to do. Especially for numeric variables that represent nominal or ordinal data, you might want to create a bar chart with one bar for each value.

The topics below start with simple bar charts, and then discuss adding options to create bar charts that best meet your needs.

Specifying Graphics Options

Just as you specify options for line printer results with the **OPTIONS** statement, you specify options for graphs with the **GOPTIONS** statement. For bar charts, you can control the appearance of the bars in the chart with a **PATTERN** statement. Both of these statements have many options. This book uses as few options as possible. Depending on your situation, you might want to look at SAS/GRAPH documentation to see whether other options would be useful.

The GOPTIONS statement sets general options for graphics output. This book uses only the option that specifies the device. All graphs in this chapter use the following:

```
goptions device=win;
```

This option identifies the graphics device as a monitor.

The PATTERN statement controls the appearance of bars in bar charts. This book specifies a simple appearance of filling the bars with solid gray shading. All bar charts in this chapter use the following:

```
pattern v=solid color=gray;
```

The programs for each graph show these statements for completeness. In your own programs, you can use the statements once in each SAS session.

Simple Vertical Bar Charts with PROC GCHART

In vertical bar charts, the horizontal axis shows the values of a variable, and the vertical axis shows the frequency counts for the variable. The bars rise vertically from the horizontal axis. The statements below create a vertical bar chart to summarize the men and women in the fitness program:

```
goptions device=win;
pattern v=solid color=gray;
proc gchart data=bodyfat;
    vbar gender;
    title 'Bar Chart for Men and Women in Fitness Program';
run;
```

Figure 4.10 shows the bar chart.

Compare the results in Figure 4.8 and Figure 4.10. The **FREQUENCY** label on the vertical axis explains that the bar chart displays the frequency counts. The **gender** label on the horizontal axis identifies the variable.

Figure 4.10 Vertical Bar Chart for Body Fat Data

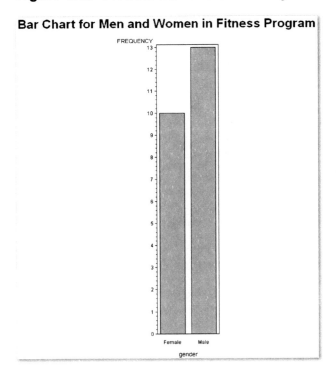

Bar Chart for Men and Women in Fitness Program

The general form of the statements to create a simple vertical bar chart with PROC GCHART is shown below:

PROC GCHART DATA=*data-set-name*;

  VBAR *variables*;

Items in italic were defined earlier in the chapter. You must use the VBAR statement to identify the variables that you want to chart.

Simple Vertical Bar Charts with PROC CHART

The syntax for PROC CHART is almost identical to PROC GCHART. The statements below create a line printer vertical bar chart to summarize the men and women in the fitness program:

```
options ps=50;
proc chart data=bodyfat;
   vbar gender;
 title 'Line Printer Bar Chart for Fitness Program';
run;
```

Figure 4.11 shows the bar chart.

Figure 4.11 Line Printer Vertical Bar Chart for Body Fat Data

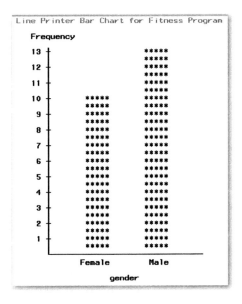

Figures 4.10 and 4.11 display the same results. The rest of this chapter creates high-resolution bar charts, and identifies the corresponding SAS procedure for line printer charts where possible.

> The general form of the statements to create a simple line printer vertical bar chart with PROC CHART is shown below:
>
> **PROC CHART DATA**=*data-set-name*;
>
> **VBAR** *variables*;
>
> Items in italic were defined earlier in the chapter. You must use the VBAR statement to identify the variables that you want to chart.

Discrete Vertical Bar Charts with PROC GCHART

For character variables like gender for the fitness program data, PROC GCHART creates a bar chart that displays the unique values of the variable. This can also be useful for numeric variables, especially when the numeric variables have many values but are measured on a discrete scale. The statements below use the **DISCRETE** option for the **dogwalk** data:

```
goptions device=win;
pattern v=solid color=gray;
proc gchart data=dogwalk;
   vbar walks / discrete;
   title 'Number of Daily Dog Walks';
run;
```

Figure 4.12 shows the bar chart.

Figure 4.12 Bar Chart for Dog Walk Data

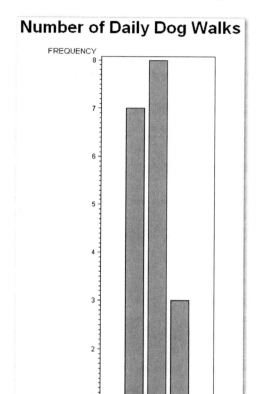

The chart has one bar for each numeric value for the **walks** variable. Because this variable has only a few values, PROC GCHART would have created the same chart without the DISCRETE option.

Use caution with the DISCRETE option. For numeric variables with many values, this option displays a bar for each value, but the bars are not proportionally spaced on the horizontal line. For example, the next statements create Figure 4.13 for the speeding ticket data:

```
goptions device=win;
pattern v=solid color=gray;
proc gchart data=tickets;
    vbar amount / discrete;
    title 'Bar Chart with DISCRETE Option';
run;
```

Figure 4.13 Discrete Bar Chart for Speeding Ticket Data

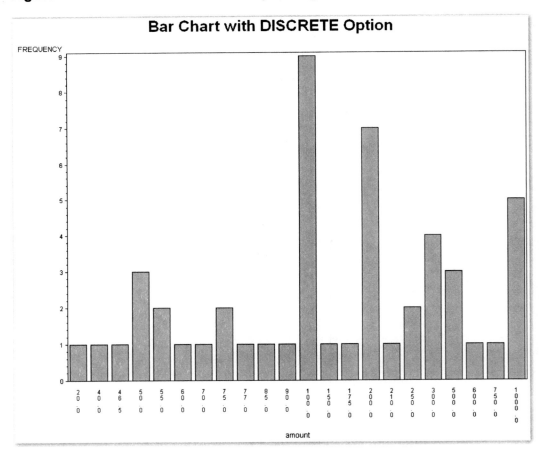

The chart has one bar for each distinct numeric value in the data set. However, the values are not proportionally spaced on the horizontal line. The bars for 750 and 1000 are as close together as the bars for 20 and 40. This spacing might cause confusion when interpreting the chart. To help identify this situation, the SAS log contains a warning message. The message informs you that the intervals on the axis are not evenly spaced.

You can use PROC CHART with the VBAR statement to create a line printer vertical bar chart. The syntax is the same as for PROC GCHART.

> The general form of the statements to create a vertical bar chart with a bar for every value of a numeric variable using PROC GCHART is shown below:
>
> **PROC GCHART DATA=***data-set-name***;**
>
> **VBAR** *variables* **/ DISCRETE;**
>
> Items in italic were defined earlier in the chapter. You must use the VBAR statement to identify the variables that you want to chart.

Simple Horizontal Bar Charts with PROC GCHART

In horizontal bar charts, the vertical axis shows the values of a variable, and the horizontal axis shows the frequency counts for the variable. Sometimes, horizontal bar charts are preferred for display purposes. For example, a horizontal bar chart of **gender** in the **bodyfat** data requires much less room to display than the vertical bar chart. The statements below create a horizontal bar chart to summarize the number of dog walks each day:

```
goptions device=win;
pattern v=solid color=gray;
proc gchart data=dogwalk;
   hbar walks;
   title 'Horizontal Bar Chart for Dog Walk Data';
run;
```

Figure 4.14 shows the bar chart.

Figure 4.14 Horizontal Bar Chart for Dog Walk Data

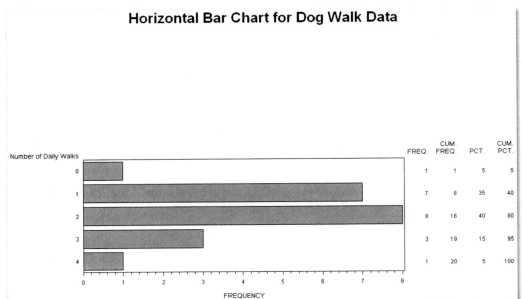

Figure 4.14 shows the horizontal bar chart and frequency table statistics for each bar. You can omit these statistics from the bar chart.

You can use PROC CHART with the **HBAR** statement to create a line printer horizontal bar chart. The syntax is the same as for PROC GCHART.

The general form of the statements to create a simple horizontal bar chart with PROC GCHART is shown below:

PROC GCHART DATA=*data-set-name*;

    HBAR *variables*;

Items in italic were defined earlier in the chapter. You must use the HBAR statement to identify the variables that you want to chart. You can use the DISCRETE option with the HBAR statement.

Omitting Statistics in Horizontal Bar Charts

To suppress the frequency table statistics, use the **NOSTAT** option. The statements below create the horizontal bar chart in Figure 4.15:

```
goptions device=win;
pattern v=solid color=gray;
proc gchart data=bodyfat;
    hbar gender / nostat;
run;
```

Figure 4.15 Horizontal Bar Chart with Statistics Omitted

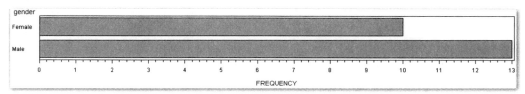

When you use the NOSTAT option, the scale for frequencies might be expanded from the automatic scale. Because the statistics have been omitted, there is more room on the page for the horizontal bar chart. Compare Figures 4.15 and 4.11 to understand why the horizontal bar chart might be preferred for display purposes in a report or presentation. The horizontal bar chart requires much less space.

You can use PROC CHART with the HBAR statement and the NOSTAT option. The syntax is the same as for PROC GCHART.

The general form of the statements to create a horizontal bar chart with statistics omitted with PROC GCHART is shown below:

PROC GCHART DATA=*data-set-name***;**

HBAR *variables* **/ NOSTAT;**

Items in italic were defined earlier in the chapter. You must use the HBAR statement to identify the variables that you want to chart. You can use the DISCRETE option with the NOSTAT option.

Creating Ordered Bar Charts

Sometimes you want a customized bar chart for presenting results, and neither a histogram nor a simple horizontal or vertical bar chart meets your needs. For example, for the speeding ticket data, suppose you want a bar chart of the speeding ticket fine amounts with each bar labeled based on the state. Also, you want the bars to be in descending order so that the highest fine amounts appear at the top, and the lowest at the bottom. You can use the NOSTAT, **DESCENDING**, and **SUMVAR=** options to create the chart you want.

```
goptions device=win;
pattern v=solid color=gray;
proc gchart data=tickets;
    hbar state / nostat descending sumvar=amount;
run;
```

Figure 4.16 shows the bar chart.

Figure 4.16 Ordered Bar Chart for Speeding Ticket Data

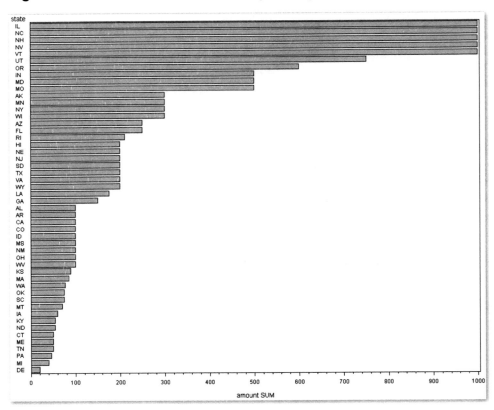

The HBAR statement identifies **state** as the variable to chart. The SUMVAR= option identifies **amount** as the second variable. The length of the bars represents the sum of this second variable, instead of the frequency counts for the first variable. Without the SUMVAR= option, the bar for each state would have a length of 1 because the data set has 1 observation for each state. With SUMVAR=amount, the bar for each state has a length that corresponds to the speeding ticket fine for that state.

The DESCENDING option orders the bars on the chart from the maximum to the minimum value. Without the DESCENDING option, the bars on the chart would be ordered by **state**. While this chart might make it easier to find a particular state, it is not as useful when trying to compare the speeding ticket fine amounts for different states.

Using the SUMVAR= and DESCENDING options together produces an ordered chart for a variable, where the lengths of the bars give information about the numeric variable. For example, in the speeding ticket data, the bar for each state shows the value of the numeric variable for the amount of the speeding ticket fine. The NOSTAT option suppresses the summary statistics from the bar chart.

You can use PROC CHART with the HBAR statement and these options. The syntax is the same as for PROC GCHART.

The general form of the statements to create an ordered bar chart with statistics omitted with PROC GCHART is shown below:

PROC GCHART DATA=*data-set-name***;**

HBAR *variable* **/ NOSTAT DESCENDING SUMVAR=***numeric-variable***;**

variable is the variable whose values you want to see on the chart axis. *numeric-variable* is the variable whose values appear as the lengths of the bars. You must use the HBAR statement to identify the variables that you want to chart.

You can use the options independently. For your data, choose the combination of options that creates the chart you want.

Checking Data for Errors

As you summarize your data, you can also check it for errors.

For small data sets, you can print the data set, and then compare it with the original data source.

This process is difficult for larger data sets, where it's easier to miss errors when checking manually. Here are some other ways to check:

- For continuous variables, look at the maximum and minimum values. Are there any values that seem odd or wrong? For example, you would not expect to see a value of 150° for a person's temperature.

- For continuous variables, look at the descriptive statistics and see if they make sense. For example, if your data consists of scores on a test, you should have an idea of what the average should be. If it's very different from what you expect, investigate the data for outlier points or for mistakes in data values.

- For nominal and ordinal variables, are there duplicate values that are misspellings? For example, do you have both "test" and "gest" as values for a variable? Do you have both "TRue" and "True" as values for a variable? (Because SAS is case sensitive, these two values are different.)

- For nominal and ordinal variables, are there too many values? If you have coded answers to an opinion poll with numbers 1 through 5, all of your values should be 1, 2, 3, 4, or 5. If you have a 6, you know it's an error.

The next two topics give details on checking data for errors using SAS procedures.

Checking for Errors in Continuous Variables

You can use PROC UNIVARIATE to check for errors.

When checking your data for errors, investigate points that are separated from the main group of data—either above it or below it—and confirm that the points are valid. Use the PROC UNIVARIATE line printer plots and graphs to perform this investigation. In the histogram, investigate bars that are separated from other bars. In the box plot, investigate the outlier points. Also, use stem-and-leaf plots to identify outlier points and to check the minimum and maximum values.

Use the **Extreme Observations** table to check the minimum and maximum values. See these values make sense for your data. Suppose you measured the weights of dogs and you find that your **Highest** value is listed as 200 pounds. Although this value is potentially valid, it is unusual enough that you should check it before you use it. Perhaps the dog really weighs 20 pounds, and an extra 0 has been added. For the speeding ticket data, a minimum value of 0 is expected. For other types of data, negative values might make sense.

Use the **Basic Statistical Measures** table to check the data against what you expect. Does the average make sense given the scale for your data?

Look at the number of missing values in the **Missing Values** table. If you have many missing values, you might need to check your data for errors. For example, in clinical trials, some data (such as age) must be recorded for all patients. Missing values are not allowed. If the value of **Count** is higher than you expect, perhaps your **INPUT** statement is incorrect, so the values of the variables are wrong. For example, you might have used incorrect columns for a variable in the **INPUT** statement, so several values of your variable are listed as missing. For the speeding ticket data, the missing value for Washington, D.C., occurred because the researcher didn't collect this data.

Use the plots to check for outliers. The stem-and-leaf plot and box plot both help to show potential outliers.

Checking for Errors in Nominal or Ordinal Variables

You can use the frequency tables in **PROC UNIVARIATE** or **PROC FREQ** to check for errors. You can use the bar charts from **PROC CHART** or **PROC GCHART** to check for errors.

The **Frequency** table lists all of the values for a variable. Use this table to check for unusual or unexpected values.

Similarly, bar charts show all of the values and are another option for checking for errors.

You can use other tables in **PROC UNIVARIATE** to check for errors as described above for continuous variables. These approaches require numeric variables, and you will find the following specific checks to be useful:

- Look at the maximum and minimum values and confirm that these values make sense for your data.
- Check the number of missing values to see whether you might have an error in the **INPUT** statement.

Summary

Key Ideas

- The levels of measurement for variables are nominal, ordinal, and continuous. The choice of an appropriate analysis for a variable is often based on the level of measurement for the variable.

- The table below summarizes SAS procedures discussed in this chapter, and relates the procedure to the level of measurement for a variable.

Level of Measurement	SAS Procedures for Summarizing Data			
	Descriptive Statistics	Frequency Tables	Bar Charts	Histograms / Box Plots
Nominal	UNIVARIATE or MEANS	FREQ or UNIVARIATE	GCHART or CHART	Not applicable
Ordinal				
Continuous				UNIVARIATE

- Use PROC UNIVARIATE for a complete summary of numeric variables, and PROC MEANS for a brief summary of numeric variables. Some of the statistics make sense only for continuous variables. Both procedures require numeric variables.

- Use PROC FREQ for frequency tables for any variable, and PROC UNIVARIATE for frequency tables for numeric variables. Use PROC GCHART or PROC CHART for bar charts for any variable. Either PROC GCHART or PROC CHART has several options to customize the bar chart. For a continuous variable, histograms from PROC UNIVARIATE are typically more appropriate.

- Histograms show the distribution of data values, and help you see the center, spread (dispersion or variability), and shape of your data. SAS produces histograms for numeric variables.

- Box plots show the distribution of continuous data. Like histograms, they show the center and spread of your data. Box plots highlight outlier points, and display skewness (sidedness) by showing the mean and median in relationship to the dispersion of data values. SAS provides box plots only for numeric variables.

- Stem-and-leaf plots are similar to histograms because they show the shape of your data. Stem-and-leaf plots show the individual data values in each bar, which provides more detail. SAS provides stem-and-leaf plots only for numeric variables.

- The mean (or average) is the most common measure of the center of a distribution of values for a variable. The median (or 50th percentile) is another measure. The median is less sensitive than the mean to skewness, and is often preferred for ordinal variables.

- Standard deviation is the most common measure used to describe dispersion or variability of values for a variable. Standard deviation is the square root of the variance. The interquartile range, which is the difference between the 25th percentile and the 75th percentile, is another measure. The length of the box in a box plot is the interquartile range.

- For nominal and ordinal variables, frequency counts of the values are the most common way to summarize the data.

- As you summarize your data, check it for errors. For continuous variables, look at the maximum and minimum values, and look at potential outlier points. Also, check the data against your expectations—do the values make sense? For nominal and ordinal variables, check for unusual or unexpected values. For all types of variables, confirm that the number of missing values is reasonable.

Syntax

To get a brief summary of a numeric variable

PROC MEANS DATA=*data-set-name*
      N NMISS MEAN MEDIAN STDDEV RANGE QRANGE MIN MAX;
VAR *measurement-variables***;**

data-set-name   is the name of the data set.

measurement- variables
      are the variables that you want to summarize. If you omit the variables from the VAR statement, then SAS uses all numeric variables in the data set.

statistical-keywords	are optional. The procedure includes other keywords that this book does not discuss. If you omit the statistical keywords, then PROC MEANS prints N, MEAN, STDDEV, MIN, and MAX.

To get an extensive summary of a numeric variable

PROC UNIVARIATE DATA=*data-set-name*
            FREQ PLOT NEXTROBS=*value* **NEXTRVAL=***value* **NOPRINT;**
    VAR *measurement-variables*;
    HISTOGRAM *measurement-variables*;
    ID *identification-variable*;

data-set-name	was defined earlier.
measurement- variables	
	was defined earlier.
identification-variable	is the variable that you want to use to identify extreme observations.

If you omit the variables from the VAR statement, then SAS uses all numeric variables in the data set. If you omit the ID statement, then SAS identifies the extreme observations using their observation numbers. The optional HISTOGRAM statement creates a high-resolution histogram.

FREQ	creates a frequency table.
PLOT	creates three line printer plots: a box plot, a stem-and-leaf plot, and a normal probability plot. This chapter discusses the first two plots, and Chapter 5 discusses the third plot.
NEXTROBS=	defines the number of extreme observations to print. SAS automatically uses NEXTROBS=5. Use NEXTROBS=0 to suppress the Extreme Observations table. You can use values from 0 to half the number of observations in the data set.
NEXTRVAL=	defines the number of distinct extreme observations to print. SAS automatically uses NEXTRVAL=0. You can use values from 0 to half the number of observations in the data set.
NOPRINT	suppresses the printed output tables.

To create a frequency table for any variable

> **PROC FREQ DATA=**_data-set-name_;
>   **TABLES** _variables_ **/ MISSPRINT MISSING;**

data-set-name   was defined earlier.

variables   are the variables that you want to summarize.

If you omit the TABLES statement, then the procedure prints a frequency table for every variable in the data set.

The MISSPRINT option prints the count of missing values in the table. The MISSING option prints the count of missing values and uses the missing values in calculating statistics. Use either the MISSPRINT or MISSING option; do not use both options. If you use an option, the slash is required.

To create a high-resolution vertical bar chart

> **PROC GCHART DATA=**_data-set-name_;
>   **VBAR** _variables_ **/ DISCRETE;**

data-set-name   was defined earlier.

variables   are the variables that you want to summarize.

You must use the VBAR statement to identify the variables that you want to chart. The DISCRETE option prints a vertical bar for every value of a numeric variable. If you use the option, the slash is required.

To create a high-resolution horizontal bar chart

> **PROC GCHART DATA=**_data-set-name_;
>   **HBAR** _variables_ **/ DISCRETE NOSTAT;**

data-set-name   was defined earlier.

variables   are the variables that you want to summarize.

You must use the HBAR statement to identify the variables that you want to chart. The DISCRETE option prints a horizontal bar for every value of a numeric variable. The NOSTAT option suppresses summary statistics. You can use the options independently. If you use an option, the slash is required.

To create a high-resolution ordered horizontal bar chart

> **PROC GCHART DATA=***data-set-name***;**
> **HBAR** *variables*
> **/ NOSTAT DESCENDING SUMVAR=***numeric-variable***;**

data-set-name   was defined earlier.

variables   are the variables that you want to summarize.

numeric-variable is the variable whose values appear as the lengths of the bars.

You must use the HBAR statement to identify the variables that you want to chart.

You can omit the NOSTAT option to add summary statistics to the ordered bar chart.

To create a line printer vertical bar chart

> **PROC CHART DATA=***data-set-name***;**
> **VBAR** *variables***;**

data-set-name   was defined earlier.

variables   are the variables that you want to summarize.

You must use the VBAR statement to identify the variables that you want to chart.

You can use the DISCRETE option. You can use an HBAR statement instead to create a low-resolution horizontal bar chart. With the HBAR statement, you can use the DISCRETE option, the NOSTAT option, or both. If you use an option, the slash is required.

Example

The program below produces all of the output shown in this chapter:

```
options nodate nonumber ls=80;
data tickets;
   input state $ amount @@;
   datalines;
AL 100 HI 200 DE 20 IL 1000 AK 300 CT 50 AR 100 IA 60 FL 250
KS 90 AZ 250 IN 500 CA 100 LA 175 GA 150 MT 70 ID 100 KY 55
CO 100 ME 50 NE 200 MA 85 MD 500 NV 1000 MO 500 MI 40 NM 100
NJ 200 MN 300 NY 300 NC 1000 MS 100 ND 55 OH 100 NH 1000 OR 600
OK 75 SC 75 RI 210 PA 46.50 TN 50 SD 200 TX 200 VT 1000 UT 750
WV 100 VA 200 WY 200 WA 77 WI 300 DC .
;
run;

options ps=55;
proc univariate data=tickets ;
   var amount;
   id state;
title 'Summary of Speeding Ticket Data' ;
run;

options ps=60;
proc univariate data=tickets nextrval=5 nextrobs=0;
   var amount;
   id state;
title 'Summary of Speeding Ticket Data' ;
run;

proc means data=tickets;
   var amount;
title 'Brief Summary of Speeding Ticket Data' ;
run;

proc means data=tickets
      n nmiss mean median stddev range qrange min max;
   var amount;
title 'Single-line Summary of Speeding Ticket Data' ;
run;

ods select plots;
proc univariate data=tickets  plot;
   var amount;
   id state;
```

```
title 'Line Printer Plots for Speeding Ticket Data' ;
run;

proc univariate data=tickets noprint;
   var amount;
   histogram amount;
title 'Histogram for Speeding Ticket Data' ;
run;

data dogwalk;
   input walks @@;
   label walks='Number of Daily Walks';
   datalines;
3 1 2 0 1 2 3 1 1 2 1 2 2 2 1 3 2 4 2 1
;
run;

options ps=50;
proc univariate data=dogwalk freq;
   var walks;
title 'Summary of Dog Walk Data' ;
run;

options ps=60;
proc freq data=dogwalk;
   tables walks;
title 'Frequency Table for Dog Walk Data' ;
run;

proc format;
value $gentext 'm' = 'Male'
               'f' = 'Female';
run;

data bodyfat;
   input gender $ fatpct @@;
   format gender $gentext.;
   label fatpct='Body Fat Percentage';
   datalines;
m 13.3 f 22 m 19 f 26 m 20 f 16 m 8 f 12 m 18 f 21.7
m 22 f 23.2 m 20 f 21 m 31 f 28 m 21 f 30 m 12 f 23
m 16 m 12 m 24
;
run;
```

```
proc freq data=bodyfat;
   tables gender;
title 'Frequency Table for Body Fat Data' ;
run;

data dogwalk2;
   input walks @@;
   label walks='Number of Daily Walks' ;
   datalines;
3 1 2 . 0 1 2 3 1 . 1 2 1 2 . 2 2 1 3 2 4 2 1
;
run;

proc freq data=dogwalk2;
   tables walks;
title 'Dog Walk with Missing Values' ;
title2 'Automatic Results' ;
run;

proc freq data=dogwalk2;
   tables walks / missprint;
title 'Dog Walk with Missing Values' ;
title2 'MISSPRINT Option' ;
run;

proc freq data=dogwalk2;
   tables walks / missing;
title 'Dog Walk with Missing Values' ;
title2 'MISSING Option' ;
run;

goptions device=win; pattern v=solid color=gray;

proc gchart data=bodyfat;
   vbar gender;
title 'Bar Chart for Men and Women in Fitness Program' ;
run;

options ps=50;
proc chart data=bodyfat;
   vbar gender;
title 'Line Printer Bar Chart for Fitness Program' ;
run;
```

```
proc gchart data=dogwalk;
   vbar walks / discrete;
title 'Number of Daily Dog Walks' ;
run;

proc gchart data=tickets;
   vbar amount / discrete;
title 'Bar Chart with DISCRETE Option' ;
run;

proc gchart data=dogwalk;
   hbar walks;
title 'Horizontal Bar Chart for Dog Walk Data' ;
run;

title;
proc gchart data=bodyfat;
   hbar gender / nostat;
run;

proc gchart data=tickets;
   hbar state / nostat descending sumvar=amount;
title;
run;
```

Special Topic: Using ODS to Control Output Tables

Each SAS procedure automatically produces output tables. Each procedure also includes options that produce additional output tables. Although it is not discussed in this book, most SAS procedures can also create output data sets.

With the Output Delivery System (ODS), you can control which output tables appear. ODS is a versatile and complex system that provides many features. This topic discusses the basics of using ODS to control the output tables.

The simplest way to use ODS is to continue to send SAS output to the Listing (or Output) window and select the output tables that you want to view. Here is the simplified form of the **ODS** statement:

ODS SELECT *output-table-names*;

output-table-names identifies output tables for the procedure. Use the **ODS** statement before the procedure. For example, the statements below print only the **Moments** table:

```
ods select moments;
proc univariate data=tickets;
   var amount;
run;
```

This book uses the simplified **ODS** statement for many examples. Starting with the next chapter, the chapter identifies the ODS table names for the output that SAS automatically produces, and the output produced by the options that are discussed.

Table ST4.1 identifies the output tables for the procedures in this chapter:

Table ST4.1 ODS Table Names for Chapter 4

SAS Procedure	ODS Table Name	Output Table Name
UNIVARIATE	Moments	Moments
	BasicMeasures	Basic Statistical Measures
	Quantiles	Quantiles
	ExtremeObs	Extreme Observations
	ExtremeValues	Extreme Values
	MissingValues	Missing Values
	Frequencies	Frequency Counts
	Plots	stem-and-leaf plot, box plot, and normal probability plot
MEANS	Summary	summary statistics
FREQ	Table$n$	frequency table, where n is 1 for the first table, 2 for the second table, and so on
CHART	VBAR or HBAR	vertical or horizontal bar charts

ODS focuses on output tables and does not apply to high-resolution graphics output.

Most of the ODS table names are very similar to the output table names. However, the ODS table names do not always exactly match the output table names. The next topic shows how to find ODS table names.

Finding ODS Table Names

If you use SAS in the windowing environment mode, you can find ODS table names by looking at the Output window. Right-click on the item in the output, and then select **Properties**. Figure ST4.1 shows an example of right-clicking on the Moments table name, and then selecting Properties. The **Name** row in the **Moments Properties** dialog box identifies the output table name as **Moments**.

Figure ST4.1 Finding ODS Table Names from SAS Output

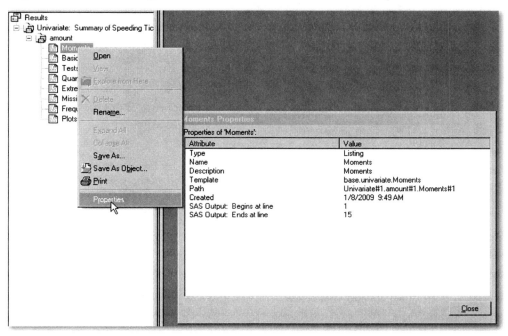

SAS documentation for each procedure also identifies ODS table names. As a third approach, you can activate the ODS **TRACE** function. The procedure prints details on the output tables in the SAS log. To activate the function, type the following:

ODS TRACE ON;

To deactivate the function, type the following:

ODS TRACE OFF;

Introducing Other Features

ODS includes features for creating the following:

- HTML output that you can view in a Web browser
- PostScript output for high-resolution printers
- RTF output that you can view in Microsoft Word or similar programs
- Output data sets

You can use a combination of these features for any procedure. For example, you can specify that some tables be formatted in HTML, others in RTF, and others sent to the Output window. For some types of output, such as HTML and RTF, you can control the appearance of the table border, spacing, and the colors and fonts for text in the table.

For more advanced features, you can create *ODS templates*, which you can use repeatedly. You can create *style definitions*, which control the appearance of the output for all SAS procedures in your program.

For an introduction to these topics, you can type **Help ODS** in the SAS windowing environment mode. For full details, see Appendix 1 for references and SAS documentation.

Part 2

Statistical Background

Chapter 5

Understanding Fundamental Statistical Concepts

This chapter focuses more on statistical concepts than on using SAS. It discusses using **PROC UNIVARIATE** to test for normality. It also discusses hypothesis testing, which is the foundation of many statistical analyses. The major topics are the following:

- understanding populations and samples
- understanding the normal distribution
- defining parametric and nonparametric statistical methods
- testing for normality
- building hypothesis tests
- understanding statistical and practical significance

Testing for normality is appropriate for continuous variables. The other statistical concepts in this chapter are appropriate for all types of variables.

Populations and Samples

Definitions

A *population* is a collection of values that has one value for every member in the group of interest. For example, consider the speeding ticket data. If you consider only the 50 states (and ignore Washington, D.C.), the data set contains the speeding ticket fine amounts for the entire group of 50 states. This collection contains one value for every member in the group of interest (the 50 states), so it is the entire population of speeding ticket fine amounts.

This definition of a population as a collection of values might be different from the definition you are more familiar with. You might think of a population as a group of people, as in the population of Detroit. In statistics, think of a population as a collection of measurements on people or things.

A *sample* is also a collection of values, but it does not represent the entire group of interest. For the speeding ticket data, a sample could be the collection of values for speeding ticket fine amounts for the states located in the Southeast. Notice how this sample differs from the population. Both are collections of values, but the population represents the entire group of interest (all 50 states), and the sample represents only a subgroup (states located in the Southeast).

Consider another example. Suppose you are interested in estimating the average price of new homes sold in the United States during 2008, and you have the prices for a few new homes in several cities. This is only a sample of the values. To have the entire population of values, you would need the price for every new home sold in the United States in 2008.

Figure 5.1 shows the relationship between a population and a sample. Each value in the population appears as a small circle. Filled-in circles are values that have been selected for a sample.

Figure 5.1 Relationship between a Population and a Sample

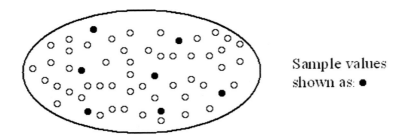

Sample values
shown as: •

Consider a final example. Think about opinion polls. You often see the results of opinion polls in the news, yet most people have never been asked to participate in an opinion poll. Rather than ask every person in the country (collect a population of values for the group of interest), the companies that conduct these polls ask a small number of people. They collect a sample of values for the group of interest.

Random Samples

Most of the time, you cannot collect the entire population of values. Instead, you collect a sample. To make a valid decision about the population based on a sample, the sample must be representative of the population. This is usually accomplished by collecting a random sample.

A sample is a *simple random sample* if the process that is used to collect the sample ensures that any one sample is as likely to be selected as any other sample. For example, the companies that conduct opinion polls often collect random samples. Any group of people (and their associated sample values) is as likely to be selected as any other group of people. In contrast, the collection of speeding ticket fine amounts for the states located in the Southeast is not a random sample of the entire population of speeding ticket fine amounts for the 50 states. This is because the Southeastern states were deliberately chosen to be in the sample.

In many cases, the process of collecting a random sample is complicated and requires the help of a statistician. Simple random samples are only one type of sampling scheme. Other sampling schemes are often preferable. Suppose you conduct a survey to find out students' opinions of their professors. At first, you decide to randomly sample students on campus, but you then realize that seniors might have different opinions from freshmen, sophomores from juniors, and so on. Over the years, seniors have developed opinions about what they expect in professors, while freshmen haven't yet had the chance to do so. With a simple random sample, the results could be affected by the differences between the students in the different classes. Thus, you decide to group students according to class, and to get a random sample from each class (seniors, juniors, sophomores, and freshmen). This *stratified random sample* is a better representation of the students' opinions and examines the differences between the students in the different classes. Before conducting a survey, consult a statistician to develop the best sampling scheme.

Describing Parameters and Statistics

Another difference between populations and samples is the way summary measures are described. Summary measures for a population are called *parameters*. Summary measures for a sample are called *statistics*. As an example, suppose you calculate the average for a set of data. If the data set is a population of values, the average is a parameter, which is called the *population mean*. If the data set is a sample of values, the average is a statistic, which is called the *sample average* (or the *average*, for short). The rest of this book uses the word "mean" to indicate the population mean, and the word "average" to indicate the sample average.

To help distinguish between summary measures for populations and samples, different statistical notation is used for each. In general, population parameters are denoted by letters of the Greek alphabet. For now, consider only three summary measures: the mean, the variance, and the standard deviation.

Recall that the mean is a measure that describes the center of a distribution of values. The variance is a measure that describes the dispersion around the mean. The standard deviation is the square root of the variance. The sample average is denoted with $\overline{X}$ and is called *x bar*. The population mean is denoted with μ and is called *Mu*. The sample variance is denoted with s^2 and is called *s-squared*. The population variance is denoted with σ^2 and is called *sigma-squared*. Because the standard deviation is the square root of the variance, it is denoted with s for a sample and σ for a population. Table 5.1 summarizes the differences in notation.

Table 5.1 Symbols Used for Populations and Samples

	Average	Variance	Standard Deviation
Population	μ	σ^2	σ
Sample	$\overline{X}$	s^2	s

For the average or mean, the same formula applies to calculations for the population parameter and the sample statistic. This is not the case for other summary measures, such as the variance. See the box labeled "Technical Details: Sample Variance" for formulas and details. SAS automatically calculates the sample variance, not the population variance. Because you almost never have measurements on the entire population, SAS gives summary measures, such as the variance, for a sample.

Technical Details: Sample Variance

To calculate the sample variance for a variable, perform the following steps:

1. Find the average.
2. For each value, calculate the difference between the value and the average.
3. Square each difference.
4. Sum the squares.
5. Divide by $n-1$, where n is the number of differences.

For example, suppose your sample values are 10, 11, 12, and 15. The sample size is 4, and the average is 12. The variance is calculated as:

$$s^2 = \frac{(10-12)^2 + (11-12)^2 + (12-12)^2 + (15-12)^2}{4-1}$$

$$= 14/3 = 4.67$$

More generally, the formula is:

$$s^2 = \frac{\sum (X_i - \overline{X})^2}{(n-1)}$$

where Σ stands for sum, X_i represents each sample value, $\overline{X}$ represents the sample average, and n represents the sample size.

The difference between computing the sample variance and the population variance is in the denominator of the formula. The population variance uses n instead of $n-1$.

The Normal Distribution

Definition and Properties

Many methods of statistical analysis assume that the data is a sample from a population with a normal distribution. The *normal distribution* is a theoretical distribution of values for a population. The normal distribution has a precise mathematical definition. Rather than describing the complex mathematical definition, this book describes some of the properties and characteristics of the normal distribution.

Figure 5.2 shows several normal distributions. Notice that μ (the population mean) and σ (the standard deviation of the population) are different. The two graphs with μ=100 have the same scaling on both axes, as do the three graphs with μ=30.

Figure 5.2 Graphs of Several Normal Distributions

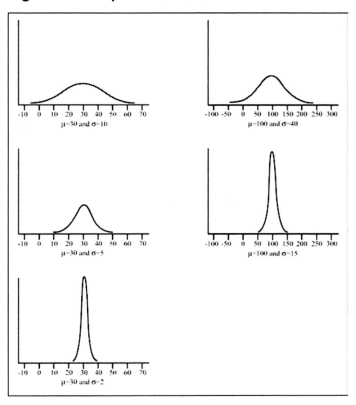

The list below summarizes properties of the normal distribution.

- **The normal distribution is completely defined by its mean and standard deviation**. For a given mean and standard deviation, there is only one normal distribution whose graph you can draw.

- **The normal distribution has mean=mode=median**. The mode is the most frequently occurring value, and the median is greater than half of the values, and less than half of the values. If you draw a graph of the normal distribution, and then fold the graph in half, the center of the distribution (at the fold) is the mean, the mode, and the median of the distribution.

- **The normal distribution is symmetric**. If you draw a graph of the normal distribution, and then fold the graph in half, each side of the distribution looks the same. A symmetric distribution has a skewness of 0, so a normal distribution has a skewness of 0. The distribution does not lean to one side or the other. It is even on both sides.

- **The normal distribution is smooth**. From the highest point at the center of the distribution, out to the ends of the distribution (the tails), there are no irregular bumps.

- **The normal distribution has a kurtosis of 0**, as calculated in SAS. (There are different formulas for kurtosis. In some formulas, a normal distribution has a kurtosis of 3.) Kurtosis describes the heaviness of the tails of a distribution. Extremely non-normal distributions can have high positive or high negative kurtosis values, while nearly normal distributions have kurtosis values close to 0. Kurtosis is positive if the tails are heavier than they are for a normal distribution, and negative if the tails are lighter than they are for a normal distribution.

Because the plots of the normal distribution are smooth and symmetric, and the normal distribution resembles the outline of a bell, the normal distribution is sometimes said to have a bell-shaped curve.

You might be wondering about the use of the word "normal." Does it mean that data from a non-normal distribution is abnormal? The answer is no. Normal distribution is one of many distributions that can occur. For example, the time to failure for computer chips does not have a normal distribution. Experience has shown that computer chips fail more often early (often on their first use), and then the time to failure slowly decreases over the length of use. This distribution is not symmetrical.

The Empirical Rule

If data is from a normal distribution, the Empirical Rule gives a quick and easy way to summarize the data. The Empirical Rule says the following:

- About 68% of the values are within one standard deviation of the mean.

- About 95% of the values are within two standard deviations of the mean.

- More than 99% of the values are within three standard deviations of the mean.

Figure 5.3 shows a normal distribution. About 68% of the values occur between μ–σ and μ+σ, corresponding to the Empirical Rule.

Figure 5.3 The Normal Distribution and the Empirical Rule

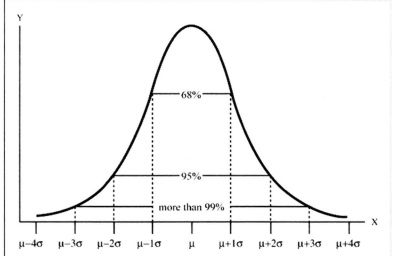

To better understand the Empirical Rule, consider an example. Suppose the population of the individual weights for 12-year-old girls is normally distributed with a mean of 86 pounds and a standard deviation of 10 pounds. Using the Empirical Rule, you find the following:

- About 68% of the weights are between 76 and 96 pounds.

- About 95% of the weights are between 66 and 106 pounds.

- More than 99% of the weights are between 56 and 116 pounds.

Parametric and Nonparametric Statistical Methods

Many statistical methods rely on the assumption that the data values are a sample from a normal distribution. Other statistical methods rely on an assumption of some other distribution of the data. Statistical methods that rely on assumptions about distributions are called *parametric methods*.

There are statistical methods that do not assume a particular distribution for the data. Statistical methods that don't rely on assumptions about distributions are called *nonparametric methods*.

This distinction is important in later chapters, which explain both parametric and nonparametric methods for solving problems. Use a parametric method if your data meets the assumptions, and use a nonparametric method if it doesn't. These later chapters provide details about the assumptions.

Testing for Normality

Recall that a normal distribution is the theoretical distribution of values for a population. Many statistical methods assume that the data values are a sample from a normal distribution. For a given sample, you need to decide whether this assumption is reasonable. Because you have only a sample, you can never be absolutely sure that the assumption is correct. What you can do is test the assumption, and, based on the results of this test, decide whether the assumption is reasonable. This testing and decision process is called *testing for normality*.

Statistical Test for Normality

When testing for normality, you start with the idea that the sample is from a normal distribution. Then, you verify whether the data agrees or disagrees with this idea. Using the sample, you calculate a statistic and use this statistic to try to verify the idea. Because this statistic tests the idea, it is called a *test statistic*. The test statistic compares the shape of the sample distribution with the shape of a normal distribution.

The result of this comparison is a number called a *p*-value, which describes how doubtful the idea is in terms of probability. A *p*-value can range from 0 to 1. A *p*-value close to 0 means that the idea is very doubtful, and provides evidence against the idea. If you find enough evidence to reject the idea, you decide that the data is not a sample from a normal distribution. If you cannot find enough evidence to reject the idea, you proceed with the analysis based on the assumption that the data is a sample from a normal distribution.

SAS provides the formal test for normality in PROC UNIVARIATE with the **NORMAL** option. To illustrate the test, Table 5.2 shows data values for the heights of aeries, or nests, for prairie falcons in North Dakota.[1]

Table 5.2 Prairie Falcons Data

Aerie Height in Meters				
15.00	3.50	3.50	7.00	1.00
7.00	5.75	27.00	15.00	8.00
4.75	7.50	4.25	6.25	5.75
5.00	8.50	9.00	6.25	5.50
4.00	7.50	8.75	6.50	4.00
5.25	3.00	12.00	3.75	4.75
6.25	3.25	2.50		

This data is available in the **falcons** data set in the sample data for this book. To see the test for normality in SAS, use the code below:

```
data falcons;
   input aerieht @@;
   label aerieht='Aerie Height in Meters';
   datalines;
15.00 3.50 3.50 7.00 1.00 7.00 5.75 27.00 15.00 8.00
4.75 7.50 4.25 6.25 5.75 5.00 8.50 9.00 6.25 5.50
4.00 7.50 8.75 6.50 4.00 5.25 3.00 12.00 3.75 4.75
6.25 3.25 2.50
;
run;
```

---

[1] Data is from Dr. George T. Allen, U.S. Fish & Wildlife Service, Arlington, Virginia. Used with permission.

```
proc univariate data=falcons normal;
   var aerieht;
   title 'Normality Test for Prairie Falcon Data';
run;
```

Figure 5.4 shows the portion of the results from the NORMAL option.

Figure 5.4 Testing for Normality of Prairie Falcons Data

```
                       Tests for Normality

   Test                    --Statistic---      -----p Value------

   Shapiro-Wilk           W    0.744634     Pr < W      <0.0001
   Kolmogorov-Smirnov     D    0.207788     Pr > D      <0.0100
   Cramer-von Mises       W-Sq 0.411609     Pr > W-Sq   <0.0050
   Anderson-Darling       A-Sq 2.349306     Pr > A-Sq   <0.0050
```

In Figure 5.4, the **Tests for Normality** table shows the results and details of four formal tests for normality.

Test	Identifies the test. SAS provides the Shapiro-Wilk test only for sample sizes less than or equal to 2000.
Statistic	Identifies the test statistic and the value of the statistic.
p Value	Specifies the direction of the test and the *p*-value.

Depending on the test, values of the test statistic that are either too small or too large indicate that the data is not a sample from a normal distribution. Focus on the *p*-values, which describe how doubtful the idea of normality is. Probability values (*p*-values) can range from 0 to 1 ($0 \le p \le 1$). Values very close to 0 indicate that the data is not a sample from a normal distribution, and produce the most doubt in the idea. For the prairie falcons data, you conclude that the aerie heights are not normally distributed.

For the Shapiro-Wilk test, the test statistic, *W*, has values between 0 and 1 ($0 < W \le 1$). This test depends directly on the sample size, unlike the three other tests. SAS calculates this test statistic when the sample size is less than or equal to 2000.

For the other three tests, values of the test statistic that are too large indicate that the data is not a sample from a normal distribution. PROC UNIVARIATE automatically uses the sample average and sample standard deviation as estimates of the population mean and population standard deviation. In this case, PROC UNIVARIATE automatically performs all three of these tests.

In general, when the Shapiro-Wilk test appears, use the *p*-value to make conclusions. Otherwise, use the *p*-value from the Anderson-Darling test. Statisticians have different opinions about which test is best, and no single test is best in every situation. Most statisticians recommend against the Kolmogorov-Smirnov test because the other tests are better able to detect non-normal distributions. See Appendix 1, "Further Reading," for more information. See the SAS documentation on PROC UNIVARIATE for formulas and information about each test.

The general form of the statements to test for normality is shown below:

> PROC UNIVARIATE DATA=*data-set-name* NORMAL;
>
> > VAR *variables*;

data-set-name is the name of a SAS data set, and *variables* lists one or more variables in the data set. For valid normality tests, use only continuous variables in the VAR statement.

Other Methods of Checking for Normality

In addition to the formal test for normality, there are other methods for checking for normality. These methods improve your understanding about the distribution of sample values. These methods include checking the values of skewness and kurtosis, looking at a stem-and-leaf plot, looking at a box plot, looking at a histogram of the data, and looking at a normal probability plot. The next five topics discuss these methods.

Skewness and Kurtosis

Recall that for a normal distribution, the skewness is 0. As calculated in SAS, the kurtosis for a normal distribution is 0. PROC UNIVARIATE automatically prints these two statistics.

Figure 5.5 shows the **Moments** table for the prairie falcons data. This table contains the skewness and kurtosis. For the prairie falcons data, neither of these values is close to 0. This fact reinforces the results of the formal test for normality that led you to reject the idea of normality.

Figure 5.5 Moments Table for Prairie Falcons Data

```
            Normality Test for Prairie Falcon Data

                   The UNIVARIATE Procedure
          Variable:  aerieht  (Aerie Height in Meters)

                            Moments

N                            33    Sum Weights              33
Mean                  6.87878788   Sum Observations        227
Std Deviation         4.79180665   Variance           22.961411
Skewness              2.65178222   Kurtosis           9.27630676
Uncorrected SS           2296.25   Corrected SS      734.765152
Coeff Variation       69.6606253   Std Error Mean    0.83414647
```

Stem-and-Leaf Plot

Chapter 4 discussed the stem-and-leaf plot as a way to explore data. Stem-and-leaf plots are useful for visually checking data for normality. If the data is normal, then there will be a single bell-shaped curve of data. You can create a stem-and-leaf plot for the prairie falcons data by adding the **PLOT** option to the PROC UNIVARIATE statement. Figure 5.6 shows the stem-and-leaf plot for the falcons data.

Figure 5.6 Stem-and-Leaf and Box Plots for Prairie Falcons Data

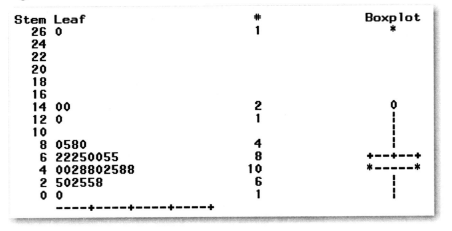

```
Stem Leaf                            #        Boxplot
 26 0                                1           *
 24
 22
 20
 18
 16
 14 00                               2           0
 12 0                                1           |
 10                                              |
  8 0580                             4           |
  6 22250055                         8        +--+--+
  4 0028802588                      10        *-----*
  2 502558                          6           |
  0 0                               1           |
    ----+----+----+----+
```

Most of the data forms a single group that is roughly bell-shaped. Depending on how you look at the data, there are three or four outlier points. These are the data points for the aerie heights of 12, 15, 15, and 27 meters. (You might see the bell shape more clearly if you rotate the output sideways.)

Box Plot

Chapter 4 discussed the box plot as a way to explore the data. Box plots are useful when checking for normality because they help identify potential outlier points. You can create a line printer box plot for the prairie falcons data by adding the PLOT option to the PROC UNIVARIATE statement. Figure 5.6 also shows the box plot for the falcons data.

This box plot highlights the potential outlier points that also appear in the stem-and-leaf plot.

Histogram

Chapter 4 discussed the histogram as a way to explore the data. Histograms are useful when checking for normality. You can create a histogram for the prairie falcons data by using the **HISTOGRAM** statement in PROC UNIVARIATE.[2] By adding the NORMAL option to the HISTOGRAM statement, you can add a fitted normal curve. The statements below create the results shown in Figure 5.7.

```
proc univariate data=falcons;
   var aerieht;
   histogram aerieht / normal(color=red);
   title 'Normality Test for Prairie Falcon Data';
run;
```

Figure 5.7 shows a histogram of the aerie heights. The histogram does not have the bell shape of a normal distribution. Also, the histogram is not symmetric. The histogram shape indicates that the sample is not from a normal distribution.

As a result of the NORMAL option, SAS adds a fitted normal curve to the histogram. The curve uses the sample average and sample standard deviation from the data. The **COLOR=** option specifies a red curve. You can use other colors. If the data were normal, the smooth red curve would more closely match the bars of the histogram. For the prairie falcons data, the curve helps show that the sample is not from a normal distribution.

---

[2] The HISTOGRAM statement is available if you have licensed SAS/GRAPH software.

Figure 5.7 Histogram with Fitted Normal Curve for Prairie Falcons Data

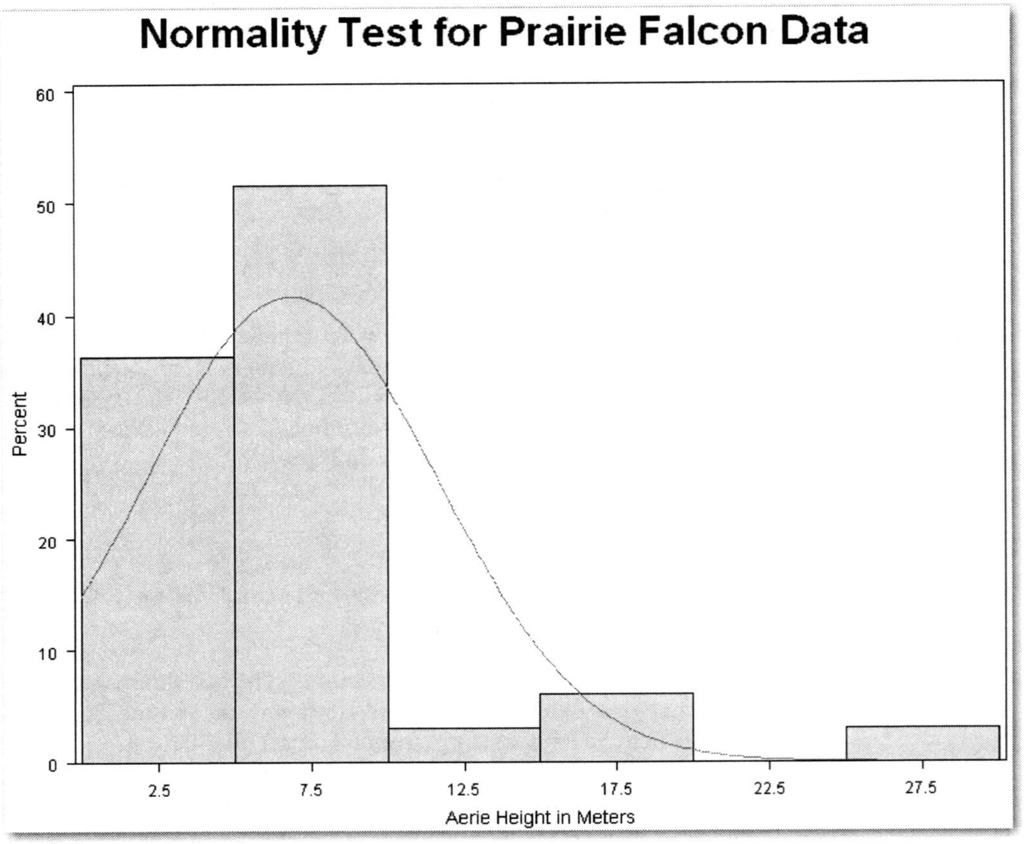

Also as a result of the NORMAL option, PROC UNIVARIATE prints three tables shown in Figure 5.8. The list below summarizes these tables.

- The **Parameters for Normal Distribution** table shows the estimated mean and standard deviation of the fitted normal curve. PROC UNIVARIATE automatically uses the sample mean and standard deviation. As a result, the estimates in Figure 5.8 are the same as the estimates in the **Moments** table. The **Parameters for Normal Distribution** table contains the **Symbol** column, which specifies the symbols used for the mean and standard deviation as **Mu** and **Sigma**. PROC UNIVARIATE can also fit many other distributions, and it identifies the symbols appropriate for each distribution.

Figure 5.8 Histogram Tables for Prairie Falcons Data

```
              Normality Test for Prairie Falcon Data

                      The UNIVARIATE Procedure
              Fitted Normal Distribution for aerieht

                  Parameters for Normal Distribution

                   Parameter    Symbol    Estimate

                   Mean         Mu        6.878788
                   Std Dev      Sigma     4.791807


          Goodness-of-Fit Tests for Normal Distribution

   Test                    ----Statistic-----   ------p Value------

   Kolmogorov-Smirnov    D       0.20778836    Pr > D       <0.010
   Cramer-von Mises      W-Sq    0.41160863    Pr > W-Sq    <0.005
   Anderson-Darling      A-Sq    2.34930625    Pr > A-Sq    <0.005


             Quantiles for Normal Distribution

                      -------Quantile------
              Percent     Observed    Estimated

                  1.0      1.00000     -4.26862
                  5.0      2.50000     -1.00303
                 10.0      3.25000      0.73784
                 25.0      4.00000      3.64676
                 50.0      5.75000      6.87879
                 75.0      7.50000     10.11081
                 90.0     12.00000     13.01974
                 95.0     15.00000     14.76061
                 99.0     27.00000     18.02620
```

- The **Goodness-of-Fit Tests for Normal Distribution** table shows the results of three of the tests for normality. PROC UNIVARIATE does not create the results for the Shapiro-Wilk test when you use the HISTOGRAM statement. Otherwise, the results shown in this table are the same as the results described earlier for Figure 5.4.

- The **Quantiles for Normal Distribution** table shows the observed quantiles for several percentiles, and the estimated quantiles that would occur for data from a normal distribution. For example, the observed median for the prairie falcons data is 5.75, and the estimated median for a normal distribution is about 6.88. In general, the closer the observed and estimated values are, the more likely the data is from a normal distribution.

You can suppress all three of these tables by adding the **NOPRINT** option to the HISTOGRAM statement.

The general form of the statements to add a histogram and a fitted normal curve is shown below:

PROC UNIVARIATE DATA=*data-set-name*;

   VAR *variables*;

     HISTOGRAM *variables* **/ NORMAL(COLOR=**_color_ **NOPRINT);**

data-set-name is the name of a SAS data set, and *variables* lists one or more variables in the data set. For valid normality tests, use only continuous variables in the VAR and HISTOGRAM statements. The COLOR= and NOPRINT options are not required.

Normal Probability Plot

A normal probability plot of the data shows both the sample data and a line for a normal distribution. Because of the way the plot is constructed, the normal distribution appears as a straight line, instead of as the bell-shaped curve seen in the histogram. If the sample data is from a normal distribution, then the points for the sample data are very close to the normal distribution line. The PLOT option in PROC UNIVARIATE creates a line printer normal probability plot. Figure 5.9 shows this plot for the prairie falcons data.

The normal probability plot shows both asterisks (*) and plus signs (+). The asterisks show the data values. The plus signs form a straight line that shows the normal distribution based on the sample mean and standard deviation. If the sample is from a normal distribution, the asterisks also form a straight line. Thus, they cover most of the plus signs. When using the plot, first, look to see whether the asterisks form a straight line. This indicates a normal distribution. Then, look to see whether there are few visible plus signs. Because the asterisks in the plot for the prairie falcons data don't form a straight line or cover most of the plus signs, you, again, conclude that the data isn't a sample from a normal distribution.

Figure 5.9 Line Printer Normal Probability Plot for Prairie Falcons Data

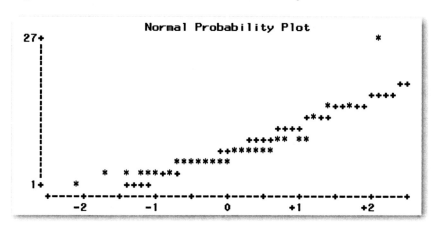

You can create a high-resolution normal probability plot for the prairie falcons data by using the **PROBPLOT** statement in PROC UNIVARIATE.[3] By adding the **NORMAL** option to this statement, you can also add a fitted normal curve. The statements below create the results shown in Figure 5.10.

```
proc univariate data=falcons;
   var aerieht;
   probplot aerieht / normal(mu=est sigma=est color=red);
   title 'Normality Test for Prairie Falcon Data';
run;
```

This plot shows a diagonal reference line that represents the normal distribution based on the sample mean and standard deviation. The **MU=EST** and **SIGMA=EST** options create this reference line. The plot shows the data as plus signs. If the falcons data were normal, then the plus signs would be close to the red (diagonal) line. For the falcons data, the data points do not follow the straight line. Once again, this fact reinforces the conclusion that the data isn't a sample from a normal distribution.

The normal probability plot in Figure 5.10 shows that there are three or four data points that are separated from most of the data points. As discussed in Chapter 4, points that are separated from the main group of data points should be investigated as potential outliers.

---

[3] The PROBPLOT statement is available if you have licensed SAS/GRAPH software.

Figure 5.10 Normal Probability Plot for Prairie Falcons Data

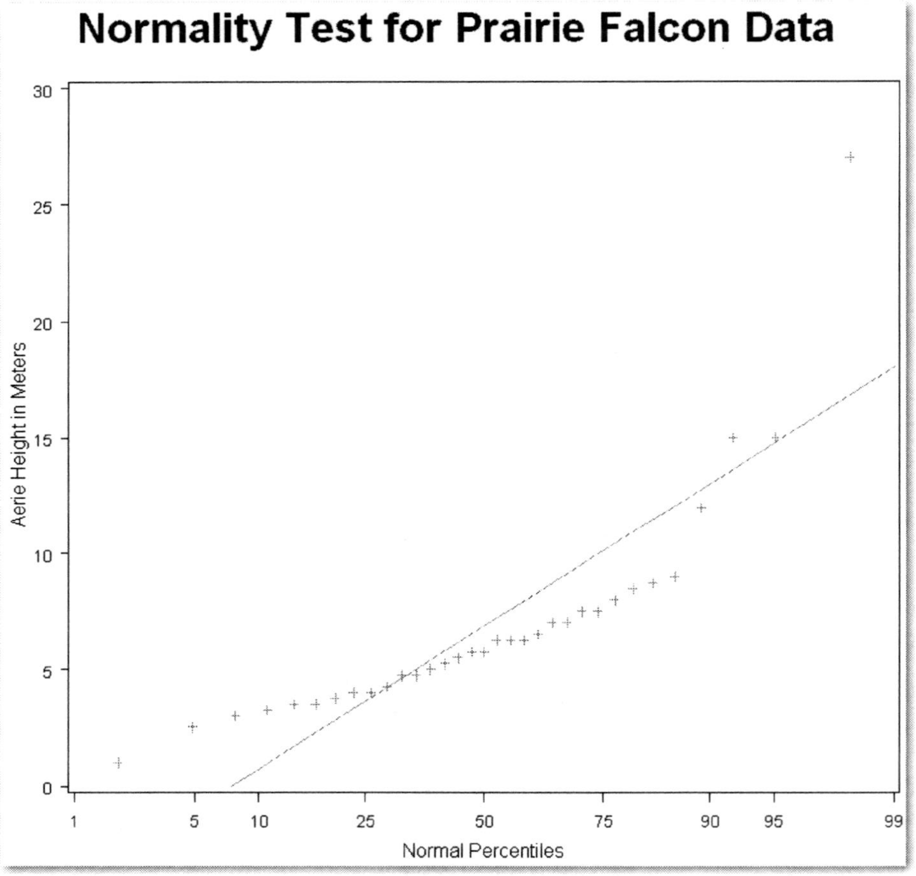

The general form of the statements to add a normal probability plot with a diagonal
reference line based on the estimated mean and standard deviation is shown below:

PROC UNIVARIATE DATA=*data-set-name*;

   VAR *variables*;

     PROBPLOT *variables* **/ NORMAL(MU=EST SIGMA=EST)**;

data-set-name is the name of a SAS data set, and *variables* lists one or more variables
in the data set. For valid normality tests, use only continuous variables in the VAR and
PROBPLOT statements. You can also add the COLOR= option.

Figure 5.11 shows several patterns that can occur in normal quantile plots, and gives the interpretations for these patterns. Compare the patterns in Figure 5.11 to the normal probability plots in Figures 5.9 and 5.10. The closest match is the pattern for a distribution that is skewed to the right. (Remember that the positive skewness measure indicates that the data is skewed to the right.)

Figure 5.11 How to Interpret Patterns in Normal Quantile Plots

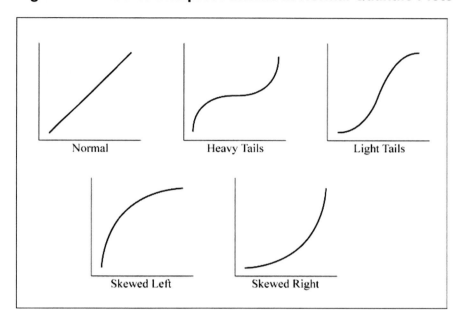

Summarizing Conclusions

The formal tests for normality led to the conclusion that the prairie falcons data set is not normally distributed. The informal methods all support this conclusion. Also, the histogram, box plot, normal quantile plot, and stem-and-leaf plot show outlier points that should be investigated. Perhaps these outlier points represent an unusual situation. For example, prairie falcons might have been using a nest built by another type of bird. If your data has outlier points, you should carefully investigate them to avoid incorrectly concluding that the data isn't a sample from a normal distribution. Suppose during an investigation you find that the three highest points appear to be eagle nests. See "Rechecking the Falcons Data," which shows how to recheck the data for normality.

Identifying ODS Tables

PROC UNIVARIATE creates several tables and graphs that are useful when checking for normality. These include the formal normality tests and the informal methods. Table 5.3 identifies the tables.

Table 5.3 ODS Table Names Useful for Normality Testing

ODS Table Name	Output Table
Moments	Moments table
Plots	Stem-and-leaf plot, box plot, normal probability plot
TestsForNormality	Tests for Normality table
ParameterEstimates	Parameter Estimates table, created by adding the NORMAL option to the HISTOGRAM statement
GoodnessOfFit	Goodness of Fit table, created by adding the NORMAL option to the HISTOGRAM statement
FitQuantiles	Quantiles for Normal Distribution table, created by adding the NORMAL option to the HISTOGRAM statement

Rechecking the Falcons Data

The full **falcons** data set has four potential outlier points. Suppose you investigate and find that the three highest aerie heights are actually eagle nests that have been adopted by prairie falcons. You want to omit these three data points from the check for normality. In SAS, this involves defining a subset of the data, and then using PROC UNIVARIATE with the additional tables and plots to check for normality. In SAS, the following code checks for normality:

```
proc univariate data=falcons normal plot;
   where aerieht<15;
   var aerieht;
   histogram aerieht / normal(color=red) noprint;
   probplot aerieht / normal(mu=est sigma=est color=red);
   title 'Normality Test for Subset of Prairie Falcon Data';
run;
```

The code above uses a **WHERE** statement to omit the outlier points. See "Understanding the WHERE Statement" for more detail.

Figure 5.12 shows selected output tables and line printer plots from the code above. Figure 5.13 shows the graphs created by the HISTOGRAM and PROBPLOT statements above.

In Figure 5.12, the **Tests for Normality** table shows the formal tests for normality. Reviewing the *p*-values, you conclude that the subset of the original **falcons** data is a sample from a normal distribution. The other methods of checking for normality support this conclusion.

The **Moments** table shows that skewness and kurtosis are both close to 0.

Figure 5.12 Selected PROC UNIVARIATE Results for Subset of Falcons Data

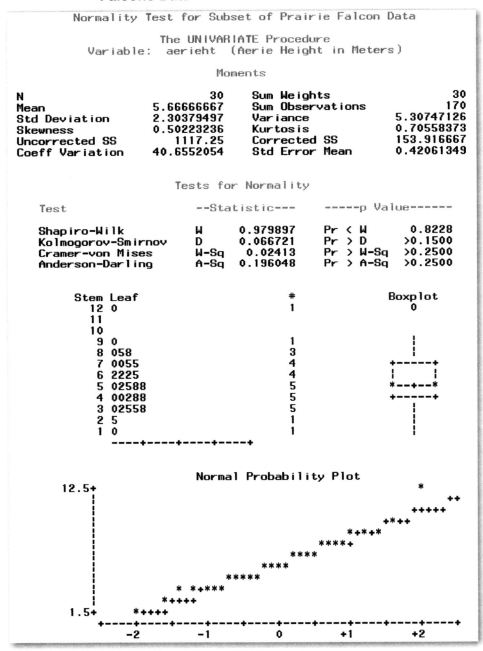

In Figure 5.12, the stem-and-leaf plot shows a single group of data that is roughly bell-shaped. The line printer box plot shows one potential outlier point for the aerie height of 12 meters. However, earlier investigation found this data point to be valid. In the line printer normal probability plot, the asterisks form a straight line and cover most of the plus signs.

Figure 5.13 PROC UNIVARIATE Graphs for Subset of Falcons Data

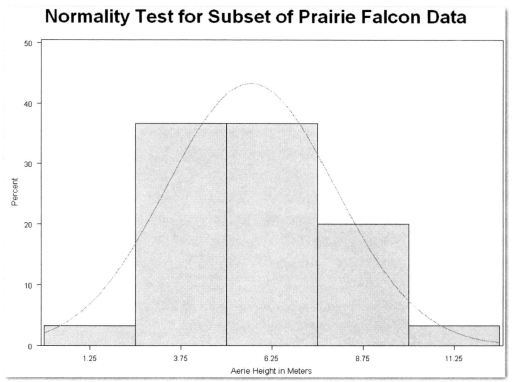

The histogram shows a single bell-shaped distribution, and the normal curve overlaying the histogram closely follows the bars of the histogram. The histogram is also mostly symmetric.

**Figure 5.13 PROC UNIVARIATE Graphs for Subset of Falcons Data
(continued)**

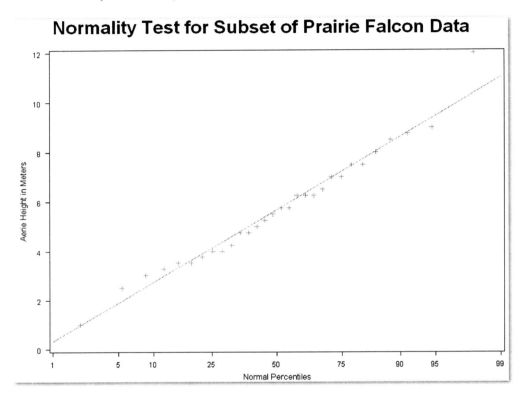

In the normal probability plot, most of the data points are near the red (diagonal) line for the normal distribution. The plot shows one potential outlier point for the aerie height of 12 meters. However, earlier investigation found this data point to be valid.

A word of caution is important here. When working with your own data, do not assume that outlier points can be omitted from the data. If you find potential outlier points, you must investigate them. If your investigation does not find a valid reason for omitting the data points, then the outlier points must remain in the data.

Understanding the WHERE Statement

The WHERE statement is the simplest way to subset data. This statement specifies the observations to include in the analysis. SAS excludes observations that do not meet the criteria in the WHERE statement. SAS documentation provides details on using the

WHERE statement with procedures. The list below gives a brief summary of the WHERE statement.

- The WHERE statement can help your programs run faster because SAS uses only the observations that meet the criteria in the WHERE statement. This advantage can be very helpful with large data sets.

- The WHERE statement uses unformatted values of variables, so you need to specify unformatted values in the statement.

- You can use the WHERE statement with any SAS procedure.

- The WHERE statement allows all of the usual arithmetic and comparison operators. For example, you can add, subtract, multiply, and divide in the statement. You can specify comparison operators like equal to, not equal to, greater than, and so on.

- The WHERE statement has special operators that are available only in the statement. Three operators that are especially useful are **IS MISSING**, **BETWEEN-AND**, and **CONTAINS**.

The general form of the statement to specify data to include in a procedure is shown below:

WHERE *expression*;

expression is a combination of variables, comparison operators, and values that specify the observations to include in the procedure. The list below shows sample expressions:

```
where gender='m' ;
where age >=65;
where country='US' and region='southeast' ;
where country='China' or country='India' ;
where state is missing ;
where name contains 'Smith' ;
where income between 5000 and 10000;
```

See Appendix 1, "Further Reading," for papers and books about the WHERE statement. See SAS documentation for complete details.

Building a Hypothesis Test

Earlier in this chapter, you learned how to test for normality to decide whether your data is from a normal distribution. The process used for testing for normality is one example of a statistical method that is used in many types of analyses. This method is also known as *performing a hypothesis test*, usually abbreviated to *hypothesis testing*. This section describes the general method of hypothesis testing. Later chapters discuss the hypotheses that are tested, in relation to this general method.

Recall the statistical test for normality. You start with the idea that the sample is from a normal distribution. Then, you verify whether the data set agrees or disagrees with the idea. This concept is basic to hypothesis testing.

In building a hypothesis test, you work with the *null hypothesis*, which describes one idea about the population. The null hypothesis is contrasted with the *alternative hypothesis*, which describes a different idea about the population. When testing for normality, the null hypothesis is that the data set is a sample from a normal distribution. The alternative hypothesis is that the data set is not a sample from a normal distribution.

Null and alternative hypotheses can be described with words. In statistics texts and journals, they are usually described with special notation. Suppose you want to test to determine whether the population mean is equal to a certain number. You want to know whether the average price of hamburger is the same as last year's average price of \$3.29 per pound. You collect prices of hamburger from several grocery stores, and you want to use this sample to determine whether the population mean is different from \$3.29. The null and alternative hypotheses are written as the following:

H_o: $\mu = 3.29$

H_a: $\mu \neq 3.29$

H_o represents the null hypothesis that the population mean (μ) equals \$3.29, and H_a represents the alternative hypothesis that the population mean (μ) does not equal \$3.29. Combined, the null and alternative hypotheses describe all possibilities. In this example, the possibilities are equal to (=) and not equal to ($\neq$). In other examples, the possibilities might be less than or equal to ($\leq$) and greater than (>).

Once you have the null and alternative hypotheses, you use the data to calculate a statistic to test the null hypothesis. Then, you compare the calculated value of this *test statistic* to the value that could occur if the null hypothesis were true. The result of this comparison is a probability value, or *p*-value, which tells you if the null hypothesis should be believed. The *p*-value is the probability that the value of your test statistic or one more extreme could have occurred if the null hypothesis were true. Figure 5.13 shows the general process for hypothesis testing.

Figure 5.13 Performing a Hypothesis Test

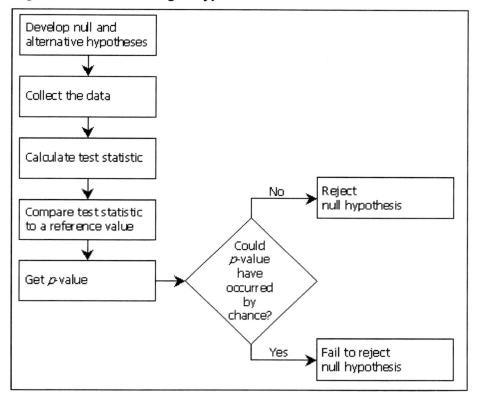

A *p*-value close to 0 indicates that the value of the test statistic could not reasonably have occurred by chance. You reject the null hypothesis and conclude that the null hypothesis is not true. However, if the *p*-value indicates that the value of the test statistic could have occurred by chance, you fail to reject the null hypothesis. You never accept the null hypothesis. Instead, you conclude that you do not have enough evidence to reject it. Unless you measure an entire population, you never have enough evidence to conclude that the population is exactly as described by the null hypothesis. For example, you could not conclude that the true mean price of hamburger is $3.29 per pound, unless you collected prices from every place that sold hamburger.

For the example of testing for normality, you test the null hypothesis that the data is from a normal distribution. You calculate a test statistic that summarizes the shape of the distribution. Then, you compare the test statistic with a reference value, and you get a *p*-value. Based on the *p*-value, you can reject the null hypothesis that the data is normally distributed. Or you can fail to reject the null hypothesis and proceed with the assumption that the data is normally distributed.

The next section explains how to decide whether a *p*-value indicates that a difference is larger than what would be expected by chance. It also discusses the concept of practical significance.

Statistical and Practical Significance

When you perform a statistical test and get a *p*-value, you usually want to know whether the result is significant. In other words, is the value of the test statistic larger than what you would expect to find by chance? To answer this question, you need to understand statistical significance and practical significance.

Statistical Significance

Statistical significance is based on *p*-values. Typically, you decide, in advance, what level of significance to use for the test. Choosing the significance level is a way of limiting the chance of being wrong. What chance are you willing to take that you are wrong in your conclusions?

For example, a significance level of 5% means that if you collected 100 samples and performed 100 hypothesis tests, you would make the wrong conclusion about 5 times (5/100=0.05). In the previous sentence, "wrong" means concluding that the alternative hypothesis is true when it is not. For the hamburger example, "wrong" means concluding that the population mean is different from $3.29 when it is not. Statisticians call this definition of "wrong" the *Type I error*. By choosing a significance level when you perform a hypothesis test, you control the probability of making a Type I error.

When you choose a significance level, you define the reference probability. For a significance level of 10%, the reference probability is 0.10, which is the significance level expressed in decimals. The reference probability is called the α-level (pronounced "alpha level") for the statistical test.

When you perform a statistical test, if the *p*-value is less than the reference probability (the α-level), you conclude that the result is statistically significant. Suppose you decide to perform a statistical test at the 10% significance level, which means that the α-level is 0.10. If the *p*-value is less than 0.10, you conclude that the result is statistically significant at the 10% significance level. If the *p*-value is more than 0.10, you conclude that the result is not statistically significant at the 10% significance level.

As a more concrete example, suppose you perform a statistical test at the 10% significance level, and the *p*-value is 0.002. You limited the risk of making a Type I error to 10%, giving you an α-level of 0.10. Your test statistic value could occur only about 2 times in 1000 by chance if the null hypothesis were true. Because the *p*-value of 0.002 is less than the α-level of 0.10, you reject the null hypothesis. The test is statistically significant at the 10% level.

Choosing a Significance Level

The choice of the significance level (or α-level) depends on the risk you are willing to take of making a Type I error. Three significance levels are commonly used: 0.10, 0.05, and 0.01. These are often referred to as moderately significant, significant, and highly significant, respectively.

Your situation should help you choose a level of significance. If you are challenging an established principle, then your null hypothesis is that this established principle is true. In this case, you want to be very careful not to reject the null hypothesis if it is true. You should choose a very small α-level (for example, 0.01). If, however, you have a situation where the consequences of rejecting the null hypothesis are not so severe, then an α-level of 0.05 or 0.10 might be more appropriate.

When choosing the significance level, think about the acceptable risks. You might risk making the wrong conclusion about 5 times out of 100, or about 1 time out of 100. Typically, you would not choose an extremely small significance level such as 0.0001, which limits your risk of making a wrong decision to about 1 time in 10,000.

More on *p*-values

As described in this book, hypothesis testing involves rejecting or failing to reject a null hypothesis at a predetermined α-level. You make a decision about the truth of the null hypothesis. But, in some cases, you might want to use the *p*-value only as a summary measure that describes the evidence against the null hypothesis. This situation occurs most often when conducting basic, exploratory research. The *p*-value describes the evidence against the null hypothesis. The smaller a *p*-value is, the more you doubt the null hypothesis. A *p*-value of 0.003 provides strong evidence against the null hypothesis, whereas a *p*-value of 0.36 provides less evidence.

Another Type of Error

In addition to the Type I error, there is another type of error called the *Type II error*. A Type II error occurs when you fail to reject the null hypothesis when it is false. You can control this type of error when you choose the sample size for your study. The Type II error is not discussed in this book because design of experiments is not discussed. If you are planning an experiment (where you choose the sample size), consult a statistician to ensure that the Type II error is controlled.

The following table shows the relationships between the true underlying situation (which you will never know unless you measure the entire population) and your conclusions. When you conclude that there is not enough evidence to reject the null hypothesis, and this matches the unknown underlying situation, you make a correct conclusion. When you reject the null hypothesis, and this matches the unknown underlying situation, you make a correct conclusion. When you reject the null hypothesis, and this does not match the unknown underlying situation, you make a Type I error. When you fail to reject the null hypothesis, and this does not match the unknown underlying situation, you make a Type II error.

	Unknown Underlying Situation	
Your Conclusion	**Null Hypothesis True**	**Alternative Hypothesis True**
Fail to Reject the Null Hypothesis	Correct	Type II error
Reject the Null Hypothesis	Type I error	Correct

Practical Significance

Practical significance is based on common sense. Sometimes, a *p*-value indicates statistical significance where the difference is not practically significant. This situation can occur with large data sets, or when there is a small amount of variation in the data.

Other times, a *p*-value indicates non-significant differences where the difference is important from a practical standpoint. This situation can occur with small data sets, or when there is a large amount of variation in the data.

You need to base your final conclusion on both statistical significance and common sense. Think about the size of difference that would lead you to spend time or money. Then, when looking at the statistical results, look at both the *p*-value and the size of the difference.

Example

Suppose you gave a proficiency test to employees who were divided into one of two groups: those who just completed a training program (trainees), and those who have been on the job for a while (experienced employees). Possible scores on the test range from 0 to 100. You want to find out if the mean scores are different, and you want to use the result to decide whether to change the training program. You perform a hypothesis test at the 5% significance level, and you obtain a *p*-value of 0.025. From a statistical standpoint, you can conclude that the two groups are significantly different.

What if the average scores for the two groups are very similar? If the average score for trainees is 83.5, and the average score for experienced employees is 85.2, is the difference practically significant? Should you spend time and money to change the training program based on this small difference? Common sense says no. Finding statistical significance without practical significance is likely to occur with large sample sizes or small variances. If your sample contains several hundred employees in each group, or if the variation within each group is small, you are likely to find statistical significance without practical significance.

Continuing with this example, suppose your proficiency test has a different outcome. You obtain a *p*-value of 0.125, which indicates that the two groups are not significantly different from a statistical standpoint (at the 5% significance level). However, the average score for the trainees is 63.8, and the average score for the experienced employees is 78.4. Common sense says that you need to think about making some changes in the training program. Finding practical significance without statistical significance can occur with small sample sizes or large variances in each group. Your sample could contain only a few employees in each group, or there could be a lot of variability in each group. In these cases, you are likely to find practical significance without statistical significance.

The important thing to remember about practical and statistical significance is that you should not take action based on the *p*-value alone. Sample sizes, variability in the data, and incorrect assumptions about your data can produce *p*-values that indicate one action, where common sense indicates another action. To help avoid this situation, consult a statistician when you design your study. A statistician can help you plan a study so that your sample size is large enough to detect a practically significant difference, but not so large that it would detect a practically insignificant difference. In this way, you control both Type I and Type II errors.

Summary

Key Ideas

- Populations and samples are both collections of values, but a population contains the entire group of interest, and a sample contains a subset of the population.

- A sample is a simple random sample if the process that is used to collect the sample ensures that any one sample is as likely to be selected as any other sample.

- Summary measures for a population are called parameters, and summary measures for a sample are called statistics. Different notation is used for each.

- The normal distribution is a theoretical distribution with important properties. A normal distribution has a bell-shaped curve that is smooth and symmetric. For a normal distribution, the mean=mode=median.

- The Empirical Rule gives a quick way to summarize data from a normal distribution. The Empirical Rule says the following:

  □ About 68% of the values are within one standard deviation of the mean.

  □ About 95% of the values are within two standard deviations of the mean.

  □ More than 99% of the values are within three standard deviations of the mean

- A formal statistical test is used to test for normality.

- Informal methods of checking for normality include the following:

  □ Viewing a histogram to check the shape of the data and to compare the shape with a normal curve.

  □ Viewing a box plot to compare the mean and median.

  □ Checking if skewness and kurtosis are close to 0.

  □ Comparing the data with a normal distribution in the normal quantile plot.

  □ Using a stem-and-leaf plot to check the shape of the data.

- If you find potential outlier points when checking for normality, investigate them. Do not assume that outlier points can be omitted.

- Hypothesis testing involves choosing a null hypothesis and alternative hypothesis, calculating a test statistic from the data, obtaining the *p*-value, and comparing the *p*-value to an α-level.

- By choosing the α-level for the hypothesis test, you control the probability of making a Type I error, which is the probability of rejecting the null hypothesis when it is true.

Syntax

To test for normality using PROC UNIVARIATE

PROC UNIVARIATE DATA=*data-set-name* **NORMAL PLOT;**
   VAR *variables*;
   HISTOGRAM *variables* **/ NORMAL(COLOR=***color* **NOPRINT);**
   PROBPLOT *variables* **/ NORMAL(MU=EST SIGMA=EST**
                              COLOR=*color***);**

data-set-name	is the name of a SAS data set.
variables	lists one or more variables. When testing for normality, use interval or ratio variables only. If you omit the *variables* from the VAR statement, the procedure uses all continuous variables. If you omit the *variables* from the HISTOGRAM and PROBPLOT statements, the procedure uses all variables listed in the VAR statement.
NORMAL	in the PROC UNIVARIATE statement, this option performs formal tests for normality.
PLOT	in the PROC UNIVARIATE statement, this option adds line printer stem-and-leaf plot, box plot, and normal quantile plot.
HISTOGRAM	creates a histogram of the data.
NORMAL	in the HISTOGRAM statement, this option adds a fitted normal curve to the histogram. This option also creates the Parameters for Normal Distribution table, Goodness of Fit Tests for Normal Distribution table, Histogram Bin Percents for Normal Distribution table, and Quantiles for Normal Distribution table.
COLOR=	in the HISTOGRAM statement, this option specifies the color of the fitted normal curve. This option is not required.

NOPRINT	in the HISTOGRAM statement, this option suppresses the tables that are created by the NORMAL option in this statement. This option is not required.
PROBPLOT	creates a normal probability plot of the data.
NORMAL	in the PROBPLOT statement, this option specifies a normal probability plot. This option is not required for a simple probability plot. If you want to add the reference line for the normal distribution, then this option is required.
MU=EST	in the PROBPLOT statement, this option specifies to use the estimated mean from the data for the fitted normal distribution.
SIGMA=EST	in the PROBPLOT statement, this option specifies to use the estimated standard deviation from the data for the fitted normal distribution.
COLOR=	in the PROBPLOT statement, this option specifies the color of the fitted normal distribution line. This option is not required.

To include only selected observations using a WHERE statement

WHERE *expression*;

expression	is a combination of variables, comparison operators, and values that identify the observations to use in the procedure. The list below shows sample expressions:

```
where gender='m' ;
where age >=65;
where country='US' and region='southeast' ;
where country='China' or country='India' ;
where state is missing ;
where name contains 'Smith' ;
where income between 5000 and 10000;
```

See the list of references in Appendix 1 for papers and books about the WHERE statement, and see the SAS documentation for complete details.

Example

The program below produces all of the output shown in this chapter:

```
options nodate nonumber ls=80 ps=60;
data falcons;
   input aerieht @@;
   label aerieht='Aerie Height in Meters';
   datalines;
15.00 3.50 3.50 7.00 1.00 7.00 5.75 27.00 15.00 8.00
4.75 7.50 4.25 6.25 5.75 5.00 8.50 9.00 6.25 5.50
4.00 7.50 8.75 6.50 4.00 5.25 3.00 12.00 3.75 4.75
6.25 3.25 2.50
;
run;

proc univariate data=falcons normal plot;
   var aerieht;
   histogram aerieht / normal(color=red);
   probplot aerieht / normal(mu=est sigma=est color=red);
title 'Normality Test for Prairie Falcon Data';
run;

proc univariate data=falcons normal plot;
   where aerieht<15;
   var aerieht;
   histogram aerieht / normal(color=red noprint);
   probplot aerieht / normal(mu=est sigma=est color=red);
title 'Normality Test for Subset of Prairie Falcon Data';
run;
```

C h a p t e r **6**

Estimating the Mean

Like Chapter 5, this chapter focuses more on statistical concepts than on using SAS. It shows how to use **PROC MEANS** to find a confidence interval for the mean. Confidence intervals are based on statistical concepts that are illustrated with simulated data. This chapter discusses the following major topics:

- estimating the mean with a single number

- exploring the effect of sample size when estimating the mean

- exploring the effect of population variance when estimating the mean

- understanding the distribution of sample averages

- building confidence intervals for the mean

SAS creates confidence intervals for numeric variables only. However, the general statistical concepts discussed in this chapter apply to all types of variables.

Using One Number to Estimate the Mean

To estimate the mean of a normally distributed population, use the arithmetic average of a random sample that is taken from the population. Non-normal distributions also often use the average of a random sample to estimate the mean. If your data is not normally distributed, you might want to consult a statistician because another statistic might be better.

The sample average gives a *point estimate* of the population mean. The sample average gives one number, or point, to estimate the mean. The sample size and the population variance both affect the precision of this point estimate. Generally, larger samples produce more precise estimates, or produce estimates closer to the true unknown population mean. In general, samples from a population with a small variance produce more precise estimates than samples from a population with a large variance. Sample size and population variance interact. Large samples from a population with a small variance are likely to produce the most precise estimates. The next two sections describe the effect of sample size and population variance.

Effect of Sample Size

Suppose you want to estimate the average monthly starting salary of recently graduated male software quality engineers in the United States. You cannot collect the salaries of every recent male graduate in the United States, so you take a random sample of salaries and use the sample values to calculate the average monthly starting salary. This average is an estimate of the population mean. How close do you think the sample average is to the population mean? One way to answer this question is to take many samples, and then compute the average for each sample. This collection of sample averages will cluster around the population mean. If you knew what the population mean actually was, you could examine the collection to see how closely the sample averages are clustered around the mean.

In this salary example, suppose you could take 200 samples. How would the averages for the different samples vary? The rest of this section explores the effect of different sample sizes on the sample averages. This section uses simulation, which is the process of using the computer to generate data, and then exploring the data to understand concepts.

Suppose you know that the true population mean is $4850 per month, and that the true population standard deviation is $625. Now, suppose you sample 100 values from the population, and you find the average for this sample. Then, you repeat this process 199 more times, for a total of 200 different samples. From each sample, you get a sample average that estimates the population mean.

To demonstrate what would happen, we simulated this example in SAS. The simulation created 200 sample averages. Figure 6.1 shows the results of using **PROC UNIVARIATE** to plot a histogram of the sample averages.

Figure 6.1 Distribution of 200 Samples with n=100, μ=4850, and σ=625

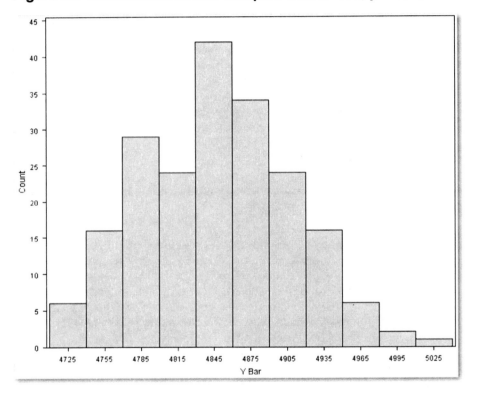

Figure 6.1 shows a histogram of the 200 sample averages, where each average is based on a sample that contains 100 values. Most of the sample averages are close to the true population mean of $4850. PROC UNIVARIATE calculates the 2.5% and 97.5% values to be 4737 and 4961, rounded to the nearest dollar. Thus, about 95% of the values are between $4737 and $4961.

Reducing the Sample Size

What happens if you reduce the sample size to 50 values? For each of the 200 samples, you calculate the average of 50 values. Then, you create a histogram that summarizes the averages from the 200 samples.

What happens if you reduce the sample size even further, so that each sample contains only 10 values? What happens if you reduce the sample size to 5 values?

Figure 6.2 shows the results of reducing the sample size. All of the histograms have the same scale to make comparing them easier.

The averages for the larger sample sizes are more closely grouped around the true population mean of $4850. The range of values for the larger sample sizes is smaller than it is for the smaller sample sizes. The following table lists the 2.5% and 97.5% values for these samples. This range encompasses about 95% of the averages.

Sample Size	Range That Includes 95% of Averages (2.5%–97.5%)
100	4737-4961
50	4704-5006
10	4556-5233
5	4304-5400

The population with the smallest sample size of 5 values has the widest range that includes 95% of the averages. This matches the results in Figure 6.2, where the histogram for the sample size of 5 has the widest range of values.

Figure 6.2 Effect of Reducing Sample Size

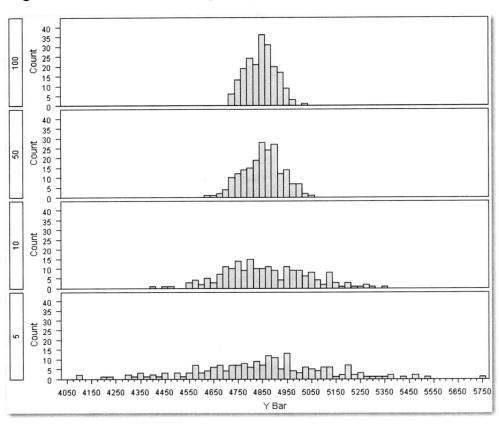

This example shows how larger sample sizes produce more precise estimates. An estimate from a larger sample size is more likely to be closer to the true population mean.

However, collecting sample data requires time and often costs money. At some point, increasing the sample size does not provide practical value. For example, increasing the sample size to 10,000 for the salary example is unlikely to provide an estimate where the increased precision is worth the time and effort to collect such a large sample.

Effect of Population Variability

This section continues with the salary example and explores how population variability affects the estimates from a sample. Recall that one measure of population variability is the standard deviation—σ. As the population standard deviation decreases, the sample average is a more precise estimate of the population mean.

Estimation with a Smaller Population Standard Deviation

Suppose that the true population standard deviation is $300, instead of $625. Note that you cannot change the population standard deviation the way you can change the sample size. The true population standard deviation is fixed, so you are now collecting samples from a different distribution of the population with a true population mean of $4850, and a true population standard deviation of $300. Suppose you collect 200 samples of size 10 from a population with a true standard deviation of $300. What happens to the histogram of sample averages?

What if the true population standard deviation is even smaller, say, $150? Now the samples are from another distribution of the population with a true population standard deviation of $150. What happens if the true population standard deviation is even smaller, say, $75?

Figure 6.3 shows the results for smaller population standard deviations. All of the histograms have the same scale to make comparing them easier.

The averages for the populations with smaller standard deviations are more closely grouped around the true population mean of $4850. The range of values for the smaller standard deviations is smaller than it is for the larger standard deviations.

Figure 6.3 Effect of Smaller Population Standard Deviations

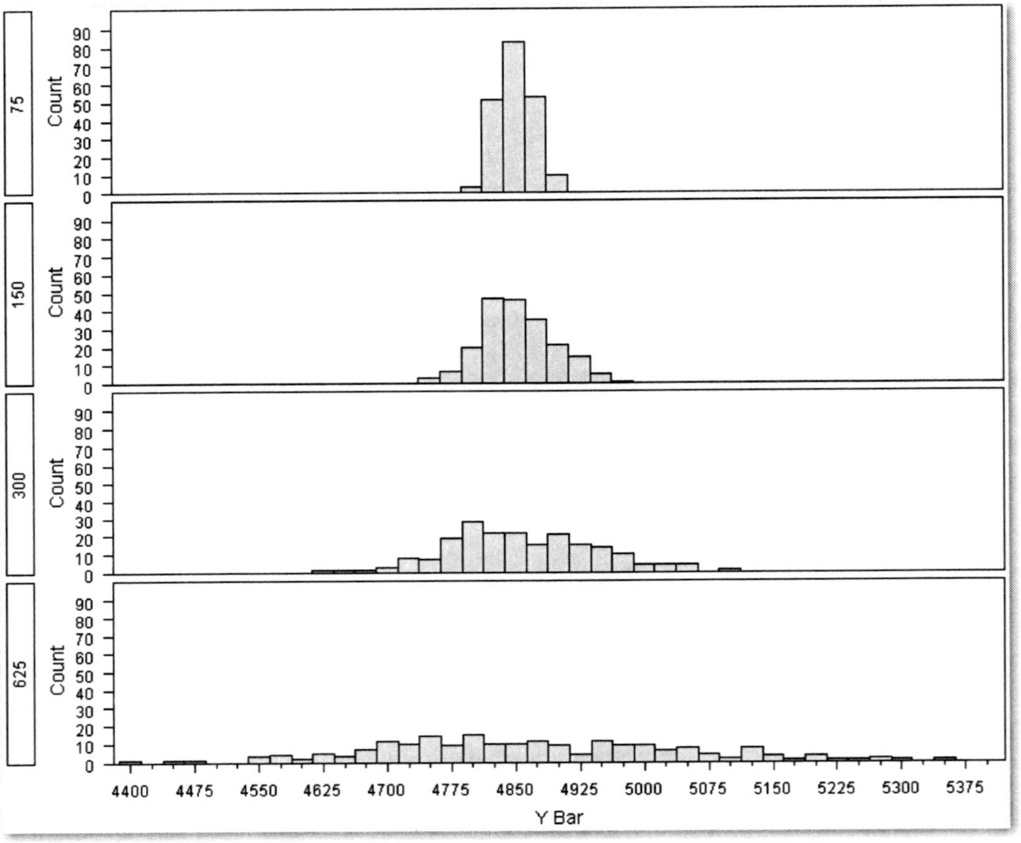

The following table lists the 2.5% and 97.5% values for these samples. This range encompasses about 95% of the averages.

Population Standard Deviation	Range That Includes 95% of Averages (2.5%–97.5%)
625	4556-5233
300	4709-5034
150	4779-4942
75	4814-4896

The population with the largest standard deviation of 625 has the widest range that includes 95% of the averages. This matches the results in Figure 6.3, where the histogram for the population standard deviation of 625 has the widest range of values.

This example shows how samples from populations with smaller standard deviations produce more precise estimates. An estimate from a population with a smaller standard deviation is more likely to be closer to the true population mean.

The Distribution of Sample Averages

Figures 6.2 and 6.3 show how the sample size and the population standard deviation affect the estimate of the population mean. Look again at all of the histograms in Figures 6.2 and 6.3. Just as individual values in a sample have a distribution, so do the sample averages. This section discusses the distribution of sample averages.

The Central Limit Theorem

The histogram of sample averages in Figure 6.1 is roughly bell-shaped. Normal distributions are bell-shaped. Combining these two facts, you conclude that sample averages are approximately normally distributed.

You might think that this initial conclusion is correct only because the samples are from a normal distribution (because salaries can conceivably be from a normal distribution). The fact is that your initial conclusion is correct even for populations that cannot conceivably be from a normal distribution. As an example, consider another type of distribution—the exponential distribution. One common use for this type of distribution is to estimate the time to failure. For example, think about laptops, which typically have a 3-hour battery. Suppose you measure the time to failure for the battery for 100 laptops. Figure 6.4 shows a histogram of this sample.

Figure 6.4 Simulated Laptop Time to Failure (Exponential Distribution)

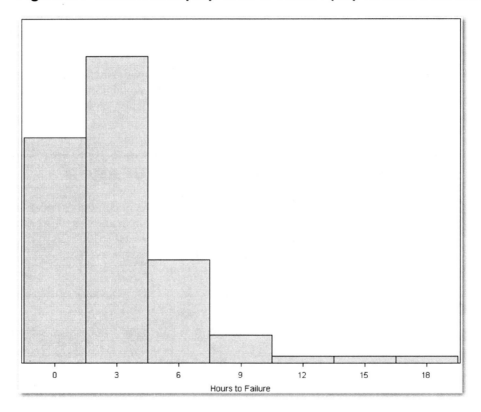

Now, suppose you measure the time to failure for the battery in 30 laptops, and repeat this process 200 times. (Note that this description simplifies the experiment. In a real experiment, you would need to control the types of laptops, the types of batteries, and so on. However, for this example, assume that the experiment has been run correctly.) Figure 6.5 shows a histogram of the averages.

Figure 6.5 Distribution of Averages for 200 Exponential Distributions

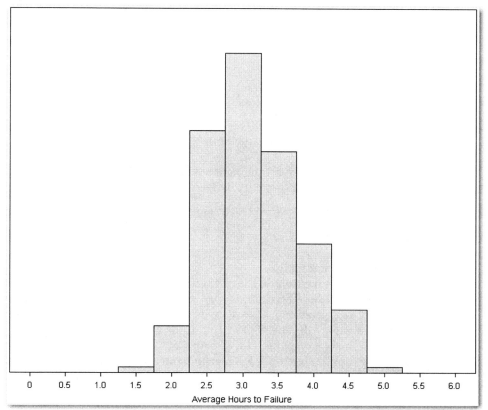

The distribution is roughly bell-shaped. This leads you to initially conclude that the sample averages are approximately normally distributed. The fact that sample averages from non-normal distributions are approximately normally distributed is one of the most important foundations of statistical theory.

The Central Limit Theorem is a formal mathematical theorem that supports your initial conclusion. Essentially, the Central Limit Theorem says the following:

If you have a simple random sample of n observations from a population with mean μ and standard deviation σ

and if n is large

then the sample average $\overline{X}$ is approximately normally distributed with mean μ and standard deviation $\sigma/\sqrt{n}$.

The list below shows two important practical implications of the Central Limit Theorem:

- Even if the sample values are not normally distributed, the sample average is approximately normally distributed.

- Because the sample average is approximately normally distributed, you can use the Empirical Rule to summarize the distribution of sample averages.

How Big Is Large?

The description of the Central Limit Theorem says "and if n is large." How big is large? The answer depends on how non-normal the population is. For moderately non-normal populations, sample averages based on as few as 5 observations ($n=5$) tend to be approximately normally distributed. But, if the population is highly skewed or has heavy tails (for example, it contains a lot of extremely large or extremely small values), then a much larger sample size is needed for the sample average to be approximately normally distributed. For practical purposes, samples with more than 5 observations are usually collected.

One reason for collecting larger samples is because larger sample sizes produce more precise estimates of the population mean. See Figure 6.2 for an illustration.

A second reason for collecting larger samples is to increase the chance that the sample is a representative sample. Small samples, even if they are random, are less likely to be representative of the entire population. For example, an opinion poll that is a random sample of 150 people might not be as representative of the entire population as an opinion poll that is a random sample of 1000 people. (Note that stratified random sampling, mentioned in Chapter 5, can enable you to use a smaller sample. Collecting larger random samples might not always be the best way to sample.)

The Standard Error of the Mean

The Central Limit Theorem says that sample averages are approximately normally distributed. The mean of the normal distribution of sample averages is μ, which is also the population mean. The standard deviation of the distribution of sample averages is $\sigma/\sqrt{n}$, where σ is the standard deviation of the population. The standard deviation of the distribution of sample averages is called the *standard error of the mean.*

The Empirical Rule and the Central Limit Theorem

One of the practical implications of the Central Limit Theorem is that you can use the Empirical Rule to summarize the distribution of sample averages.

- About 68% of the sample averages are between $\mu - \sigma/\sqrt{n}$ and $\mu + \sigma/\sqrt{n}$.

- About 95% of the sample averages are between $\mu - 2\sigma/\sqrt{n}$ and $\mu + 2\sigma/\sqrt{n}$.

- More than 99% of the sample averages are between $\mu - 3\sigma/\sqrt{n}$ and $\mu + 3\sigma/\sqrt{n}$.

From now on, this book uses "between $\mu \pm 2\sigma/\sqrt{n}$" to mean "between $\mu - 2\sigma/\sqrt{n}$ and $\mu + 2\sigma/\sqrt{n}$."

The following table uses the second statement of the Empirical Rule to calculate the range of values that includes 95% of the sample averages. The last column of the table shows the 2.5% and 97.5% percentiles for the monthly salary data (which includes 95% of the sample averages). With different data, the range for the data would change. Because the Empirical Rule is based on the population mean, the population standard deviation, and the sample size, the range for the Empirical Rule stays the same. For this data, the ranges for the actual data are close to the ranges for the Empirical Rule. Sometimes the ranges are higher, and sometimes lower, but generally the ranges are close.

Sample Size	Population Standard Deviation	Range for 95% of Values from Empirical Rule	Range That Includes 95% of values (2.5%–97.5%)
100	625	4725–4975	4737–4961
50	625	4673–5027	4704–5006
5	625	4291–5409	4304–5400
10	625	4455–5245	4556–5233
10	300	4660–5040	4709–5034
10	150	4755–4945	4779–4942
10	75	4803–4897	4814–4896

With many sample averages, you can get a good estimate of the population mean by looking at the distribution of sample averages. You can use the distribution of sample averages to determine how close your estimate should be to the true population mean. However, typically, you have only one sample average. This one sample average gives a point estimate of the population mean, but a point estimate might not be close to the true population mean (as you have seen in the previous examples in this chapter). What if you want to get an estimate with upper and lower limits around it? To do this, get a confidence interval for the mean. The next section describes how.

Getting Confidence Intervals for the Mean

A *confidence interval* for the mean gives an *interval estimate* for the population mean. This interval estimate places upper and lower limits around a point estimate for the mean. The examples in this chapter show how the sample average estimates the population mean, and how the sample size and population standard deviation affect the precision of the estimate. The chapter also explains how the Central Limit Theorem and the Empirical Rule can be used to summarize the distribution of sample averages. A confidence interval uses all of these ideas.

A 95% confidence interval is an interval that is very likely to contain the true population mean (μ). Specifically, if you collect a large number of samples, and then calculate a confidence interval for each sample, 95% of the confidence intervals would contain the true population mean (μ), and 5% would not. Unfortunately, with only one sample, you don't know whether the confidence interval you calculate is in the 95%, or in the 5%.

However, you can consider your interval to be one of many intervals. And, any one of these many intervals has a 95% chance of containing the true population mean. Thus, the term "confidence interval" indicates your degree of belief or confidence that the interval contains the true population mean. The term "confidence interval" is often abbreviated as CI.

Recall the 95% bound on estimates suggested by the Central Limit Theorem and the Empirical Rule. A conceptual bound to use in a 95% confidence interval can be determined with the formula $\pm 2\sigma/\sqrt{n}$ because this is the bound that contains 95% of the distribution of sample averages. However, in most real-life applications, the population standard deviation (σ) is unknown. A natural substitution is to use the sample standard deviation (s) to replace the population standard deviation (σ) in the formula. After you make this replacement, the conceptual bound formula is $\pm 2s/\sqrt{n}$. However, if you replace σ with s in the formula, the number 2 in the formula is not correct.

The normal distribution is completely defined by the population mean (μ) and the population standard deviation (σ). If you replace σ with s, the use of the normal distribution (and, thus, the Empirical Rule) is not exactly correct. A t distribution should be used instead of the normal distribution. A t distribution is very similar to the normal distribution, and allows you to adjust for different sample sizes. Now, the conceptual bound formula becomes $\pm t\, s/\sqrt{n}$. The value of t (or the t-value) is based on the sample size and level of confidence you choose. The formula for a confidence interval for the mean is as follows:

$$\overline{X} \pm t_{df,\,1-\alpha/2}\,\frac{s}{\sqrt{n}}$$

$\overline{X}$ is the sample average.

$t_{df,\,1-\alpha/2}$ is the t-value for a given df and α. The df is one less than the sample size ($df=n-1$), and is called the *degrees of freedom* for the t-value. As n gets large, the t-value approaches the value for a normal distribution. The confidence level for the interval is $1-\alpha$. For a 95% confidence level, $\alpha = 0.05$. Calculating a 95% confidence interval for a data set with 12 observations requires $t_{11,0.975}$ (which is 2.201). When finding the t-value, divide α by 2 before subtracting it from 1.

s is the sample standard deviation.

n is the sample size.

You can use **PROC MEANS** to get a confidence interval for the mean. As discussed in Chapter 4, you can use *statistics-keywords* to request statistics that the procedure does not automatically display in the output.

The data in Table 6.1 consists of interest rates for mortgages from several banks. Because mortgage rates differ based on discount points, the size of the loan, the length and type of loan, and whether an origination fee is charged, the data consists of interest rates for the same type of loan. Specifically, the interest rates in the data set are for a 30-year fixed-rate loan for $295,000, with no points and a 1% origination fee. Also, because interest rates change quickly (sometimes, even daily), all of the information was collected on the same day. The actual bank names are not used.

Table 6.1 Mortgage Data

Mortgage Rates				
5.750	5.750	5.875	5.750	5.750
5.750	5.750	5.625	5.875	5.500
5.500	5.625	5.750	5.625	5.625
5.750	5.750	5.750	5.750	5.750
5.500	5.875	5.750	5.750	5.500
5.750	5.625	5.875	5.500	5.625

This data is available in the **rates** data set in the sample data for the book. In SAS, the following code calculates confidence intervals:

```
data rates;
   input mortrate @@;
   label mortrate='Mortgage Rate';
   datalines;
5.750 5.750 5.500 5.750 5.500 5.750 5.750 5.750 5.625
5.750 5.875 5.625 5.875 5.625 5.750 5.750 5.750 5.875
5.750 5.875 5.625 5.750 5.750 5.500 5.750 5.500 5.625
5.750 5.500 5.625
;
run;

proc means data=rates n mean stddev clm maxdec=3;
   var mortrate;
title 'Summary of Mortgage Rate Data';
run;
```

The PROC MEANS statement uses *statistics-keywords* discussed in Chapter 4 (**N**, **MEAN**, and **STDDEV**), and uses a new keyword, **CLM**, for the confidence limits for the mean.

The **MAXDEC=** option controls the number of decimal places that are displayed. This option can be useful in reports or presentations.

Figure 6.6 shows the output. The confidence limits are displayed under the headings **Lower 95% CL for Mean** and **Upper 95% CL for Mean**.

Figure 6.6 Confidence Intervals for Mortgage Data

```
                    Summary of Mortgage Rate Data

                        The MEANS Procedure

            Analysis Variable : mortrate Mortgage Rate

                                          Lower 95%        Upper 95%
     N         Mean        Std Dev       CL for Mean      CL for Mean
     30        5.700        0.117          5.656            5.744
```

Confidence intervals are often shown enclosed in parentheses, with the lower and upper confidence limits separated by a comma—for example, (5.656, 5.744). For this data, you can conclude with 95% confidence that the true mean mortgage rate for this type of loan is between 5.656% and 5.744%.

Changing the Confidence Level

PROC MEANS automatically creates a 95% confidence interval. As discussed in Chapter 5 regarding significance levels, your situation determines the confidence level you choose.

Using the **rates** data, suppose you want a 90% confidence interval. In SAS, the following code calculates this confidence interval:

```
proc means data=rates n mean stddev clm alpha=0.10;
   var mortrate;
title 'Summary of Mortgage Rate Data with 90% CI';
run;
```

The **ALPHA=** option specifies the confidence level for the confidence interval. For a 90% confidence interval, α=0.10.

Figure 6.7 shows the output. The confidence limits are displayed under the headings **Lower 90% CL for Mean** and **Upper 90% CL for Mean**. Figure 6.7 shows how PROC MEANS automatically displays statistics when the MAXDEC= option is omitted.

Figure 6.7 90% Confidence Interval for Mortgage Data

```
          Summary of Mortgage Rate Data with 90% CI

                     The MEANS Procedure

          Analysis Variable : mortrate Mortgage Rate

                                       Lower 90%        Upper 90%
   N          Mean        Std Dev      CL for Mean      CL for Mean

   30      5.7000000     0.1165229      5.6638526        5.7361474
```

The 90% confidence interval for the mortgage rate is (5.66, 5.74). When you decrease the confidence level from 95% to 90%, the confidence interval is narrower. To clarify, compare the two confidence intervals in the table below:

Confidence Level	Lower Limit	Upper Limit
90%	5.664	5.736
95%	5.656	5.744

The 95% confidence interval is wider than the 90% confidence interval. When you decrease the **alpha** level from 0.10 to 0.05, the confidence interval with the lower alpha level is wider.

The general form of the statements to add confidence intervals is shown below:

PROC MEANS DATA=*data-set-name* **N MEAN STDDEV CLM**
MAXDEC=*number* **ALPHA**=*value*;
  VAR *variables*;

data-set-name is the name of a SAS data set, and *variables* lists one or more variables in the data set.

The **MAXDEC=** option controls the number of decimal places to display.

The **ALPHA=** option must be between 0 and 1. The reason is because confidence intervals must be between 0% and 100%. The automatic value is **ALPHA=0.05**, which produces 95% confidence intervals.

You can use the **MAXDEC=** option and **ALPHA=** option together. You can use other *statistics-keywords* in the PROC MEANS statement. See Chapter 4 for examples.

Identifying ODS Tables

PROC MEANS includes the confidence interval in the table of summary statistics. Table 6.2 identifies the ODS table name.

Table 6.2 ODS Table Name for PROC MEANS Confidence Intervals

ODS Table Name	Output Table
Summary	Summary Statistics

Summary

Key Ideas

- Larger sample sizes produce more precise estimates of the population mean than smaller sample sizes.

- Samples from populations with small standard deviations produce more precise estimates of the population mean than samples from populations with large standard deviations.

- The Central Limit Theorem says that for large samples, the sample average is approximately normally distributed, even if the sample values are not normally distributed.

- To calculate confidence intervals for the mean when the population standard deviation is unknown, use the following formula:

$$\overline{X} \pm t_{df,1-\alpha/2} \frac{s}{\sqrt{n}}$$

$\overline{X}$ is the sample average, n is the sample size, and s is the sample standard deviation. The value of $t_{df,1-\alpha/2}$ is a *t*-value that is based on the sample size and the confidence level. The degrees of freedom (*df*) is one less than the sample size (*df=n-1*), and 1- α is the confidence level for the confidence interval. For a 95% confidence interval for a data set with 12 observations, the *t* needed is $t_{11,0.975}$.

Syntax

To create confidence intervals using PROC MEANS, change the number of decimal places displayed, and change the confidence level

PROC MEANS DATA=*data-set-name* **N MEAN STD CLM**
              MAXDEC=*number* **ALPHA=***value***;**
    VAR *variables***;**

data-set-name	is the name of a SAS data set.
variables	lists one or more variables in the data set.
MAXDEC=	specifies the number of places to print after the decimal point.
ALPHA=	must be between 0 and 1. The reason is because confidence intervals must be between 0% and 100%. The automatic value is **ALPHA=0.05**, which produces 95% confidence intervals.

You can use the MAXDEC= option and ALPHA= options together. You can also use other *statistics-keywords* in the PROC MEANS statement. See Chapter 4 for examples.

Example

The program that produced output for the simulations is not shown. The program below produces the output shown in this chapter for the **rates** data set.

```
data rates;
   input mortrate @@;
   label mortrate='Mortgage Rate';
   datalines;
5.750 5.750 5.500 5.750 5.500 5.750 5.750 5.750 5.625
5.750 5.875 5.625 5.875 5.625 5.750 5.750 5.750 5.875
5.750 5.875 5.625 5.750 5.750 5.500 5.750 5.500 5.625
5.750 5.500 5.625
;
run;

proc means data=rates n mean stddev clm maxdec=3;
   var mortrate;
title 'Summary of Mortgage Rate Data';
run;

proc means data=rates n mean stddev clm alpha=0.10;
   var mortrate;
title 'Summary of Mortgage Rate Data with 90% CI';
run;
```

Part **3**

Comparing Groups

Chapter 7

Comparing Paired Groups

Are students' grades on an achievement test higher after they complete a special course on how to take tests? Do employees who follow a regular exercise program for a year have lower resting pulse rates than they had when they started the program?

These questions involve comparing paired groups of data. The variable that classifies the data into two groups should be nominal. The response variable can be ordinal or continuous, but must be numeric. This chapter discusses the following topics:

- deciding whether you have independent or paired groups

- summarizing data from paired groups

- building a statistical test of hypothesis to compare the two groups

- deciding which statistical test to use

- performing statistical tests for paired groups

For the various tests, this chapter shows how to use SAS to perform the test, and how to interpret the test results.

Deciding between Independent and Paired Groups

When comparing two groups, you want to know whether the means for the two groups are different. The first step is to decide whether you have independent or paired groups. This chapter discusses paired groups. Chapter 8 discusses independent groups.

Independent Groups

Independent groups of data contain measurements for two unrelated samples of items. Suppose you have random samples of the salaries for male and female accountants. The salary measurements for men and women form two distinct and separate groups. The goal of analysis is to compare the average salaries for men and women, and decide whether the difference between the average salaries is greater than what could happen by chance. As another example, suppose a researcher selects a random sample of children, some who use fluoride toothpaste, and some who do not. There is no relationship between the children who use fluoride toothpaste and the children who do not. A dentist counts the number of cavities for each child. The goal of analysis is to compare the average number of cavities for children who use fluoride toothpaste and for children who do not.

Paired Groups

Paired groups of data contain measurements for one sample of items, but there are two measurements for each item. A common example of paired groups is before-and-after measurements, where the goal of analysis is to decide whether the average change from before to after is greater than what could happen by chance. For example, a doctor weighs 30 people before they begin a program to quit smoking, and weighs them again six months after they have completed the program. The goal of analysis is to decide whether the average weight change is greater than what could happen by chance.

Summarizing Data from Paired Groups

With paired groups, the first step is to find the paired differences for each observation in the sample. The second step is to summarize the differences.

Finding the Differences between Paired Groups

Sometimes, the difference between the two measurements for an observation is already calculated. For example, you might be analyzing data from a cattle-feeding program, and the data gives the weight change of each steer. If you already have the difference in the weight of each steer, you can skip the rest of this section.

Other times, only the before-and-after data is available. Before you can summarize the data, you need to find the differences. Instead of finding the differences with a calculator, you can compute them in SAS. To do this, you add *program statements* to the **DATA** step. There are many types of program statements in SAS. The simplest program statement creates a new variable.

Table 7.1 shows data from an introductory statistics class, STA6207.[1] The table shows scores from two exams for the 20 students in the class. Both exams covered the same material. Each student has a pair of scores, one score for each exam. The professor wants to find out if the exams appear to be equally difficult. If they are, the average difference between the two exam scores should be small.

---

[1] Data is from Dr. Ramon Littell, University of Florida. Used with permission.

Table 7.1 Exam Data

Student	Exam 1 Score	Exam 2 Score
1	93	98
2	88	74
3	89	67
4	88	92
5	67	83
6	89	90
7	83	74
8	94	97
9	89	96
10	55	81
11	88	83
12	91	94
13	85	89
14	70	78
15	90	96
16	90	93
17	94	81
18	67	81
19	87	93
20	83	91

This data is available in the STA6207 data set in the sample data for the book.

The following SAS statements create a data set and use a program statement to create a new variable:

```
data STA6207;
   input student exam1 exam2 @@;
   scorediff = exam2 - exam1 ;
   label scorediff='Differences in Exam Scores';
   datalines;
1 93 98 2 88 74 3 89 67 4 88 92 5 67 83 6 89 90
7 83 74 8 94 97 9 89 96 10 55 81 11 88 83 12 91 94
13 85 89 14 70 78 15 90 96 16 90 93 17 94 81
18 67 81 19 87 93 20 83 91
;
run;
```

The spacing that was used for the program statement in the example was chosen for readability. None of the spaces are required. The following statement produces the same results as the previous statement:

```
scorediff=exam2-exam1;
```

Table 7.2 outlines the steps for creating new variables with program statements.

Table 7.2 Creating New Variables with Program Statements

To create a new variable in the DATA step, follow these steps:

1. Type the **DATA** and **INPUT** statements for the data set.

2. Choose the name of the new variable. (Use the rules for SAS names in Table 2.3.) Type the name of the variable, and follow the name with an equal sign. If the new variable is **DIFF**, you would type the following:

   ```
   diff =
   ```

3. Decide what combination of other variables will be used to create the new variable. You must have defined these other variables in the INPUT statement. Type the combination of variables after the equal sign. For paired groups, you are usually interested in the difference between two variables. If the variable DIFF is the difference between **AFTER** and **BEFORE**, you would type the following:

   ```
   diff = after - before
   ```

4. End the program statement with a semicolon. For this example, here is the complete program statement:

```
diff = after - before;
```

5. Follow the program statement with the **DATALINES** statement, and then with the lines of data to complete the DATA step. Or, if you are including data from an external source, use the **INFILE** statement.

You can create several variables in a single DATA step. See "Technical Details: Program Statements" for examples.

Technical Details: Program Statements

You can create several variables in the same DATA step by including a program statement for each new variable. In addition, many combinations of variables can be used to create variables. Here are some examples:

Total cavities is the sum of cavities at the beginning and new cavities:

```
totcav = begcav + newcav;
```

The new variable is the old variable plus 5:

```
newvar = oldvar + 5;
```

Dollars earned is the product of the number sold and the cost for each one:

```
dollars = numsold*unitcost;
```

Yield per acre is the total yield divided by the number of acres:

```
yldacre = totyield/numacres;
```

Credit limit is the square of the amount in the checking account:

```
credlimt = checkamt ** 2;
```

In general, use a plus sign (+) to indicate addition, a minus sign (–) for subtraction, an asterisk (*) for multiplication, a slash (/) for division, and two asterisks and a number (**n) to raise a variable to the n-th power.

Summarizing Differences with PROC UNIVARIATE

After creating the differences, you can summarize them using **PROC UNIVARIATE**. Because the difference is a single variable, you can use the statements and options discussed in Chapters 4 and 5. (This procedure provides an extensive summary, histograms, and box plots. Later on, the section "Summarizing Differences with Other Procedures" discusses how you can also use the **MEANS**, **FREQ**, **GCHART**, and **CHART** procedures.)

```
proc univariate data=STA6207;
   var scorediff;
   histogram;
title 'Summary of Exam Score Differences';
run;
```

Figure 7.1 shows the **Basic Statistical Measures** and the **Extreme Observations** tables from **PROC UNIVARIATE** output. The statements above also create other tables in the output, which are not shown in Figure 7.1. Figure 7.1 also shows the histogram.

Figure 7.1 Summary of Exam Score Differences

```
              Summary of Exam Score Differences

                  The UNIVARIATE Procedure
      Variable:  scorediff  (Differences in Exam Scores)

                  Basic Statistical Measures

        Location                        Variability

    Mean      2.550000      Std Deviation          10.97113
    Median    4.000000      Variance              120.36579
    Mode      3.000000      Range                  48.00000
                            Interquartile Range     9.50000


                    Extreme Observations

        ----Lowest----              ----Highest---

        Value      Obs              Value      Obs

          -22        3                  8       14
          -14        2                  8       20
          -13       17                 14       18
           -9        7                 16        5
           -5       11                 26       10
```

The average difference between exam scores is 2.55. The minimum difference is –22, and the maximum difference is 26.

Figure 7.1 Summary of Exam Score Differences (continued)

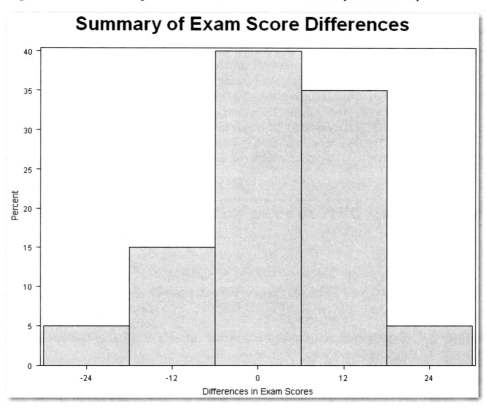

The histogram shows that many of the exam score differences are between −6 and 6 (the bar labeled **0** in the figure). The table below summarizes the midpoints and range of values for each bar in Figure 7.1.

Midpoint	Range of Values
−24	$-30 \leq \text{scorediff} < -18$
−12	$-18 \leq \text{scorediff} < -6$
0	$-6 \leq \text{scorediff} < 6$
12	$6 \leq \text{scorediff} < 18$
24	$18 \leq \text{scorediff} < 30$

However, some of the score differences are large, in both the positive and negative directions. You might think that the average difference in exam scores is not statistically significant. However, you cannot be certain from simply looking at the data.

Use the PROC UNIVARIATE output to check for errors in the new difference variable. In general, check the minimum and maximum values to see whether these values make sense for your data. For example, with the exam scores, the minimum cannot be less than –100, and cannot be more than 100 (in other words, no points on one exam, and a perfect score on the other exam). The count of differences should match the number of pairs in your data. If the count is different, then check for an error in the program statement that creates the difference variable. You might have an idea of the expected average difference—in that case, check the average value to see whether it makes sense.

Summarizing Differences with Other Procedures

Because the difference is a single variable, you can use the other procedures and options discussed in Chapter 4.

Add the **PLOT** option to PROC UNIVARIATE to create a box plot and a stem-and-leaf plot. Add the **FREQ** option to create a frequency table. Or, use PROC FREQ to create the frequency table. Use PROC MEANS for a concise summary. You can use the GCHART or CHART procedure to create bar charts, but the histogram is a better choice for the difference variable. Because the difference is a continuous variable, using a histogram is more appropriate.

Building Hypothesis Tests to Compare Paired Groups

So far, this chapter has discussed how to summarize differences between paired groups. Suppose you want to know how important the differences are, and if the differences are large enough to be significant. In statistical terms, you want to perform a hypothesis test. This section discusses hypothesis testing when comparing paired groups. (See Chapter 5 for the general idea of hypothesis testing.)

In building a test of hypothesis, you work with two hypotheses. When comparing paired groups, the null hypothesis is that the mean difference is 0, and the alternative hypothesis is that the mean difference is different from 0. In this case, the notation is the following:

$H_0: \mu_D = 0$

$H_a: \mu_D \neq 0$

μ_D indicates the mean difference.

In statistical tests that compare paired groups, the hypotheses are tested by calculating a test statistic from the data, and then comparing its value to a reference value that would be the result if the null hypothesis were true. The test statistic is compared to different reference values that are based on the sample size. In part, this is because smaller differences can be detected with larger sample sizes. As a result, if you use the reference value for a large sample size on a smaller sample, you are more likely to incorrectly conclude the mean difference is zero when it is not.

This concept is similar to the concept of confidence intervals, in which different *t*-values are used (discussed in Chapter 6). With confidence intervals, the *t*-value is based on the degrees of freedom, which are determined by the sample size.

Deciding Which Statistical Test to Use

Because you test different hypotheses for independent and paired groups, you use different tests. In addition, there are parametric and nonparametric tests for each type of group. When deciding which statistical test to use, first, decide whether you have independent groups or paired groups.

Then, decide whether you should use a parametric or a nonparametric test. This second decision is based on whether the assumptions for the parametric test seem reasonable. In general, if the assumptions for the parametric test seem reasonable, use the parametric test. Although nonparametric tests have fewer assumptions, these tests typically are less powerful in detecting differences between groups.

The rest of this chapter describes the test to use in the two situations for paired groups. See Chapter 8 for details on the tests for the two situations for independent groups. Table 7.3 summarizes the tests.

Table 7.3 Statistical Tests for Comparing Two Groups

Type of Test	Groups	
	Independent[2]	Paired
Parametric	Two-sample *t*-test	Paired-difference *t*-test
Nonparametric	Wilcoxon Rank Sum test	Wilcoxon Signed Rank test

The steps for performing the analyses to compare paired groups are the following:

1. Create a SAS data set.

2. Check the data set for errors.

3. Choose the significance level for the test.

4. Check the assumptions for the test.

5. Perform the test.

6. Make conclusions from the test results.

The rest of this chapter uses these steps for all analyses.

Understanding Significance

For each of the four tests in Table 7.3, there are two possible results: the *p*-value is less than the reference probability, or it is not. (See Chapter 5 for discussions on significance levels and on understanding practical significance and statistical significance.)

Groups Significantly Different

If the *p*-value is less than the reference probability, the result is statistically significant, and you reject the null hypothesis. For paired groups, you conclude that the mean difference is significantly different from 0.

---

[2] Chapter 8 discusses comparing independent groups. For completeness, this table shows the tests for comparing both types of groups.

Groups Not Significantly Different

If the *p*-value is greater than the reference probability, the result is not statistically significant, and you fail to reject the null hypothesis. For paired groups, you conclude that the mean difference is not significantly different from 0.

Do not conclude that the mean difference is 0. You do not have enough evidence to conclude that the mean difference is 0. The results of the test indicate only that the mean difference is not significantly different from 0. You never have enough evidence to conclude that the population is exactly as described by the null hypothesis. In statistical hypothesis testing, do not accept the null hypothesis. Instead, either reject or fail to reject the null hypothesis. For a more detailed discussion of hypothesis testing, see Chapter 5.

Performing the Paired-Difference *t*-test

The paired-difference *t*-test is a parametric test for comparing paired groups.

Assumptions

The two assumptions for the paired-difference *t*-test are the following:

- each pair of measurements is independent of other pairs of measurements
- differences are from a normal distribution

Because the test analyzes the differences between paired observations, it is appropriate for continuous variables.

To illustrate this test, consider an experiment in liquid chromatography.[3] A chemist is investigating synthetic fuels produced from coal, and wants to measure the naphthalene values by using two different liquid chromatography methods. Each of the 10 fuel samples is divided into two units. One unit is measured using standard liquid chromatography. The other unit is measured using high-pressure liquid chromatography. The goal of analysis is to test whether the mean difference between the two methods is different from 0 at the 5% significance level. The chemist is willing to accept a 1-in-20 chance of saying that the mean difference is significantly different from 0, when, in fact, it is not. Table 7.4 shows the data.

---

[3] Data is from C. K. Bayne and I. B. Rubin, *Practical Experimental Designs and Optimization Methods for Chemists* (New York: VCH Publishers, 1986). Used with permission.

Table 7.4 Liquid Chromatography Data

Sample	High-Pressure	Standard
1	12.1	14.7
2	10.9	14.0
3	13.1	12.9
4	14.5	16.2
5	9.6	10.2
6	11.2	12.4
7	9.8	12.0
8	13.7	14.8
9	12.0	11.8
10	9.1	9.7

This data is available in the **chromat** data set in the sample data for the book.

Applying the six steps of analysis.

1. Create a SAS data set.

2. Check the data set for errors. Use **PROC UNIVARIATE** to check for errors. Based on the results, assume that the data is free of errors.

3. Choose the significance level for the test. Choose a 5% significance level, which requires a p-value less than 0.05 to conclude that the groups are significantly different.

4. Check the assumptions for the test. The first assumption is independent pairs of measurements. The assumption seems reasonable because there are 10 different samples that are divided into two units. The second assumption is that the differences are from a normal distribution. Use program statements to create a new variable named **methdiff** (which is **std** subtracted from **hp**, or the standard method subtracted from the high-pressure method). Use **PROC UNIVARIATE** to check for normality. Although there are only a few data points, the second assumption seems reasonable.

What if the second assumption was not reasonable? With only 10 samples, you do not have a very complete picture of the distribution. Think of weighing 10 people, and then deciding whether the weights are normally distributed based only on the weights of these 10 people. Sometimes you have additional knowledge about the population that enables you to use tests that require a normally distributed population. For example, you might know that the weights are generally normally distributed, and you can use that information. However, this approach should be used with caution. Use it only when you have substantial additional knowledge about the population. Otherwise, if you are concerned about the assumption of normality, or if a test for normality indicates that your data is not a sample from a normally distributed population, consider performing a Wilcoxon Signed Rank test, which is discussed later in this chapter.

5. Perform the test. See "Building Hypothesis Tests to Compare Paired Groups" earlier in this chapter. For paired groups, the null hypothesis is that the mean difference (μ_D) is 0. The alternative hypothesis is that the mean difference (μ_D) is different from 0.

   SAS provides two ways to test these hypotheses. In some cases, you have only the difference between the pairs. See "Using PROC UNIVARIATE to Test Paired Differences" for this case. In other cases, you have the raw data with the values for each pair. You can still create the difference variable and use PROC UNIVARIATE, or you can use **PROC TTEST**. See "Using PROC TTEST to Test Paired Differences" for this case.

6. Make conclusions from the test results. See the section "Finding the *p*-value" for each test.

Using PROC UNIVARIATE to Test Paired Differences

For the **chromat** data, the difference variable is available. PROC UNIVARIATE automatically performs a test that the mean of this variable is 0. The SAS statements below create the data set and perform the statistical test. These statements use the **ODS** statement discussed at the end of Chapter 4. The ODS statement limits the output to the test results. (Assume that you have checked the difference variable for errors and tested it for normality.)

```
data chromat;
   input hp std @@;
   methdiff=hp-std;
   datalines;
12.1 14.7 10.9 14.0 13.1 12.9 14.5 16.2 9.6 10.2 11.2 12.4
9.8 12.0 13.7 14.8 12.0 11.8 9.1 9.7
;
run;

ods select TestsForLocation;
proc univariate data=chromat;
   var methdiff;
title 'Testing for Differences between Chromatography Methods';
run;
```

Figure 7.2 shows the output. This output shows the results from three statistical tests. The heading in the procedure identifies the variable being tested. The **Tests for Location** table identifies the hypothesis being tested. PROC UNIVARIATE automatically tests the hypothesis that the mean difference is 0, shown by the heading **Mu0=0**. The **Test** column identifies the statistical test. This particular section discusses the paired-difference t-test. The nonparametric Wilcoxon Signed Rank test is discussed later in the chapter.

Figure 7.2 Comparing Paired Groups with PROC UNIVARIATE

```
Testing for Differences between Chromatography Methods

                    The UNIVARIATE Procedure
                      Variable:  methdiff

                  Tests for Location: Mu0=0

      Test                 -Statistic-      -----p Value------

      Student's t      t   -3.56511    Pr > |t|      0.0061
      Sign             M         -3    Pr >= |M|     0.1094
      Signed Rank      S      -24.5    Pr >= |S|     0.0078
```

Finding the *p*-value

Look for the **Student's t** row in the Tests for Location table. The **Statistic** column identifies the *t*-statistic of **−3.565** (rounded). The **p Value** column identifies the *p*-value as **Pr > |t|** of **0.0061**, which is less than the significance level of 0.05. You conclude that the mean difference in naphthalene values from the two methods is significantly different from 0.

In general, to interpret SAS results, look at the *p*-value that appears to the right of the **Pr > |t|** heading. If the *p*-value is less than the significance level, then you can conclude that the mean difference between the paired groups is significantly different from 0. If the *p*-value is greater than the significance level, then you can conclude that the mean difference is not significantly different from 0. Do not conclude that the mean difference is 0. (See the discussion earlier in this chapter.)

Technical Details: Formula for Student's *t*

The test statistic for the paired-difference *t*-test is calculated as the following:

$$t = \frac{\overline{X} - \mu_0}{s / \sqrt{n}}$$

$\overline{X}$ is the average difference, *s* is the standard deviation of the difference, and *n* is the sample size. For testing paired groups, μ_0 is 0 (testing the null hypothesis of no difference between groups).

The general form of the statements to perform the paired-difference *t*-test using PROC UNIVARIATE is shown below:

> **PROC UNIVARIATE DATA=***data-set-name*;
>
> **VAR** *difference-variable*;

data-set-name is the name of a SAS data set, and *difference-variable* is the difference between the pairs of values. If you omit the *difference-variable* from the VAR statement, then the procedure uses all numeric variables. You can specify the TestsForLocation table in an ODS statement, which limits the output to the test results.

Using PROC TTEST to Test Paired Differences

So far, this section has discussed how to test paired differences using a difference variable. This approach is important because sometimes you receive data and you don't have the original pairs of information. When you do have the original pairs, you do not need to create a difference variable. PROC TTEST provides another way to test paired differences.

```
proc ttest data=chromat;
   paired hp*std;
title 'Paired Differences with PROC TTEST';
run;
```

The **PAIRED** statement identifies the two variables that contain the original pairs of information. In the PAIRED statement, an asterisk separates the two variables. This special use of the asterisk is not interpreted as multiply. Instead, the asterisk tells the procedure to subtract the second variable from the first variable. Figure 7.3 shows the output.

Figure 7.3 Comparing Paired Groups with PROC TTEST

```
             Paired Differences with PROC TTEST

                    The TTEST Procedure

                 Difference:  hp - std

   N       Mean      Std Dev      Std Err     Minimum      Maximum

  10     -1.2700      1.1265      0.3562     -3.1000       0.2000

     Mean          95% CL Mean          Std Dev      95% CL Std Dev

  -1.2700      -2.0758  -0.4642         1.1265       0.7748   2.0565

                    DF      t Value     Pr > |t|

                     9       -3.57       0.0061
```

This output shows how SAS creates the difference variable, gives descriptive statistics for the difference variable, and performs the paired-difference *t*-test.

Finding the *p*-value

Look below the heading **Pr > |t|** in Figure 7.3. The value is **0.0061**, which is less than the significance level of 0.05. This test has the same result as the analysis of the difference variable. You conclude that the mean difference in naphthalene values from the two methods is significantly different from 0. You interpret the *p*-value for this test the exact same way as described for PROC UNIVARIATE.

The subheading **Difference: hp - std** indicates how PROC TTEST calculates the paired difference. **Difference** is the definition of the paired difference. SAS subtracts the second variable in the PAIRED statement from the first variable in the statement.

The list below summarizes the **T tests** section of output for the paired-difference *t*-test.

DF	Degrees of freedom for the test. In general, this is the number of pairs minus 1. For the **chromat** data, DF is **9**, which is 10–1.

t Value

Test statistic for the *t*-test, calculated as the following:

$$t = \frac{\overline{X} - \mu_0}{s / \sqrt{n}}$$

$\overline{X}$ is the average difference, *s* is the standard deviation the difference, and *n* is the sample size. For testing paired groups, μ_0 is 0 (testing the null hypothesis of no difference between groups). PROC TTEST and PROC UNIVARIATE perform the same test.

Pr > |t|

Gives the *p*-value associated with the test. This *p*-value is for a two-sided test, which tests the alternative hypothesis that the mean difference between the paired groups is significantly different from 0.

To perform the test at the 5% significance level, you conclude that the mean difference between the paired groups is significantly different from zero if the *p*-value is less than 0.05. To perform the test at the 10% significance level, you conclude that the mean difference between the paired groups is significantly different from zero if the *p*-value is less than 0.10. Conclusions for tests at other significance levels are made in a similar manner.

Understanding Other Items in the Output

The list below summarizes the **Statistics** and **Confidence Limits** tables in the output in Figure 7.3.

N

Sample size.

Mean

Average of the difference variable. This statistic appears in both output tables.

Std Dev

Standard deviation of the difference variable. This statistic appears in both output tables.

Std Err	Standard error of the difference variable. It is calculated as $s/\sqrt{n}$, where s is the standard deviation of the difference variable and n is the sample size.
Minimum Maximum	Minimum and maximum values of the difference variable.
95% CL Mean[4]	Upper and lower 95% confidence limits for the mean. If the confidence interval includes 0, then the test will not be significant. If the confidence interval does not include 0, then the test will be significant.
	See Chapter 6 for more discussion of confidence limits.
95% CL Std Dev[5]	Upper and lower 95% confidence limits for the standard deviation of the difference variable.

Caution: Use the six steps of analysis regardless of the SAS procedure. Although both PROC UNIVARIATE and PROC TTEST perform the paired-difference *t*-test, the test is the last step in the analysis. Before performing the test, be sure to check the data set for errors, choose the significance level, and check the assumptions for the test.

The general form of the statements to perform the paired-difference *t*-test using PROC TTEST is shown below:

PROC TTEST DATA=*data-set-name*;

   PAIRED *first-variable* * *second-variable*;

data-set-name is the name of a SAS data set. SAS subtracts the *second-variable* from the *first-variable* to create the differences. The PAIRED statement is required, and the asterisk in the statement is required. The variables in the PAIRED statement must be numeric and should be continuous variables for a valid *t*-test.

Identifying ODS Tables

PROC UNIVARIATE and PROC TTEST automatically create the output tables that test for differences in paired groups. Table 7.5 identifies the ODS table names.

---

[4] If you are running a version of SAS earlier than 9.2, you will see headings of Upper CL Mean and Lower CL Mean. The descriptive statistics appear in a different order.

[5] If you are running a version of SAS earlier than 9.2, you will see headings of Lower CL Std Dev and Upper CL Std Dev. The descriptive statistics appear in a different order.

Table 7.5 ODS Table Names for Testing Paired Groups

SAS Procedure	ODS Name	Output Table
UNIVARIATE	TestsForLocation	Tests for Location table, which contains both the *t*-test and Wilcoxon Signed Rank test
TTEST	Statistics	Row with descriptive statistics
	ConfLimits	Rows showing mean, standard deviation, and associated confidence limits
	TTests	Row with test statistic, degrees of freedom, and *p*-value

Performing the Wilcoxon Signed Rank Test

The Wilcoxon Signed Rank test is a nonparametric analogue to the paired-difference *t*-test. The Wilcoxon Signed Rank test assumes that each pair of measurements is independent of other pairs of measurements. This test can be used with ordinal and continuous variables. This section uses the **chromat** data as an example.

Apply the six steps of analysis to the **chromat** data:

1. Create a SAS data set.

2. Check the data set for errors. Assume that you checked and found no errors.

3. Choose the significance level for the test. Choose a 5% significance level.

4. Check the assumptions for the test. Check the assumption of independent pairs of measurements. (See the earlier discussion for the paired-difference *t*-test.)

5. Perform the test. For this test, you need to create the difference variable.

6. Make conclusions from the test results.

PROC UNIVARIATE automatically performs the Wilcoxon Signed Rank test. The following statements were shown earlier in the chapter. Figure 7.2 shows the output.

```
ods select TestsForLocation;
proc univariate data=chromat;
   var methdiff;
title 'Testing for Differences between Chromatography Methods';
run;
```

The next two topics discuss the output.

Finding the *p*-value

In Figure 7.2, look for the **Signed Rank** row in the Tests for Location table. The **p Value** column identifies the *p*-value as **Pr >= |S|** of **0.0078**, which is less than the significance level of 0.05. You conclude that the mean difference in naphthalene values from the two methods is significantly different from 0. In general, you interpret the *p*-value for the Wilcoxon Signed Rank test the exact same way as described for the paired-difference *t*-test.

Understanding Other Items in the Output

The Signed Rank row in the Tests for Location table shows the value of the Wilcoxon Signed Rank test statistic, **S.** For the chromat data, this value is **−24.5**.

The **Sign** row in the Tests for Location table shows the results of the sign test. The sign test tests the null hypothesis that the median of the difference variable is 0 against the alternative hypothesis that it is not. Interpret the *p*-value the exact same way as described for the other two tests. The sign test is generally less likely to detect differences between paired groups than the Wilcoxon Signed Rank test.

Identifying ODS Tables

PROC UNIVARIATE creates the Tests for Location table, which contains both the *t*-test and the Wilcoxon Signed Rank test. Table 7.5 identifies the ODS name for this table as TestsForLocation.

Summary

Key Ideas

- Paired groups contain paired measurements for each item.

- Use program statements in the DATA step to create a difference variable. Use the same SAS procedures that summarize a single variable to summarize the new difference variable.

- To choose a statistical test, first, decide whether the data is from independent or paired groups. Then, decide whether to use a parametric or a nonparametric test. The second decision is based on whether the assumptions for a parametric test seem reasonable. Tests for the four cases are:

	Groups	
Type of Test	**Independent**[6]	**Paired**
Parametric	Two-sample *t*-test	Paired-difference *t*-test
Nonparametric	Wilcoxon Rank Sum test	Wilcoxon Signed Rank test

- Regardless of the statistical test you choose, the six steps of analysis are:
    1. Create a SAS data set.
    2. Check the data set for errors.
    3. Choose the significance level for the test.
    4. Check the assumptions for the test.
    5. Perform the test.
    6. Make conclusions from the test results.

---

[6] Chapter 8 discusses comparing independent groups. For completeness, this table shows the tests for comparing both types of groups.

- Regardless of the statistical test, to make conclusions, compare the *p*-value for the test with the significance level.

  □ If the *p*-value is less than the significance level, then you reject the null hypothesis and conclude that the mean difference between the paired groups is significantly different from 0.

  □ If the *p*-value is greater than the significance level, then you do not reject the null hypothesis. You conclude that the mean difference between the paired groups is not significantly different from 0. (Remember, do not conclude that the mean difference is 0.)

- Test for normality using PROC UNIVARIATE. For paired groups, check the difference variable.

- When comparing paired groups, PROC UNIVARIATE provides both parametric and nonparametric tests for the difference variable. PROC TTEST can provide the paired-difference *t*-test using the two original pairs.

- SAS creates a confidence interval for the mean difference with PROC TTEST. PROC MEANS can also display this confidence interval. (See Chapter 6 for details.) If the confidence interval includes 0, then the test will not be significant. If the confidence interval does not include 0, then the test will be significant.

Syntax

To summarize data from paired groups

- To summarize data from paired groups, use PROC UNIVARIATE with the new difference variable just as you would for a single variable. For a concise summary, use PROC MEANS. To create a frequency table, use PROC FREQ.

- To plot the difference variable, use a HISTOGRAM statement in PROC UNIVARIATE. Or, use PROC GCHART or PROC CHART for a bar chart.

To check the assumptions (step 4)

- To check the assumption of independent observations, you need to think about your data and whether this assumption seems reasonable. This is not a step where using SAS will answer this question.

- To check the assumption of normality, use PROC UNIVARIATE. For paired groups, check the difference variable.

To perform the test (step 5)

- To perform the paired-difference *t*-test or the Wilcoxon Signed Rank test on a difference variable, use PROC UNIVARIATE.

  PROC UNIVARIATE DATA=_data-set-name_;
  VAR _difference-variable_;

  data-set-name is the name of a SAS data set, and **_difference-variable_** is the difference between the pairs of values. If you omit the **_difference-variable_** from the VAR statement, then the procedure uses all numeric variables. You can specify the TestsForLocation table in an ODS statement, which limits the output to the test results.

- To perform the paired-difference *t*-test with the original pairs, use PROC TTEST and identify the pair of variables in the PAIRED statement.

  PROC TTEST DATA=_data-set-name_;
  PAIRED _first-variable_ * _second-variable_;

  data-set-name is the name of a SAS data set. SAS subtracts the **_second-variable_** from the **_first-variable_** to create the difference. The PAIRED statement is required, and the asterisk in the statement is required. The variables in the PAIRED statement must be numeric and should be continuous variables for a valid *t*-test.

Example

The program below produces the output shown in this chapter:

```
data STA6207;
   input student exam1 exam2 @@;
   scorediff = exam2 - exam1 ;
   label scorediff='Differences in Exam Scores';
   datalines;
1 93 98 2 88 74 3 89 67 4 88 92 5 67 83 6 89 90
7 83 74 8 94 97 9 89 96 10 55 81 11 88 83 12 91 94
13 85 89 14 70 78 15 90 96 16 90 93 17 94 81
18 67 81 19 87 93 20 83 91
;
run;
```

```
proc univariate data=STA6207;
   var scorediff;
   histogram;
title 'Summary of Exam Score Differences';
run;

data chromat;
   input hp std @@;
   methdiff=hp-std;
   datalines;
12.1 14.7 10.9 14.0 13.1 12.9 14.5 16.2 9.6 10.2 11.2 12.4
 9.8 12.0 13.7 14.8 12.0 11.8 9.1 9.7
;
run;

ods select TestsForLocation;
proc univariate data=chromat;
   var methdiff;
title 'Testing for Differences between Chromatography Methods';
run;

proc ttest data=chromat;
   paired hp*std;
title 'Paired Differences with PROC TTEST';
run;
```

Chapter 8

Comparing Two Independent Groups

Do male accountants earn different salaries than female accountants? Do people who are given a new shampoo use a different amount of shampoo than people who are given an old shampoo? Do cows that are fed a grain supplement and hay gain a different amount of weight than cows that are fed only hay?

These questions involve comparing two *independent* groups of data (as defined in the previous chapter). The variable that classifies the data into two groups should be nominal. The response variable can be ordinal or continuous, but must be numeric. This chapter discusses the following topics:

- summarizing data from two independent groups

- building a statistical test of hypothesis to compare the two groups

- deciding which statistical test to use

- performing statistical tests for independent groups

For the various tests, this chapter shows how to use SAS to perform the test, and how to interpret the test results.

Deciding between Independent and Paired Groups

When comparing two groups, you want to know whether the means for the two groups are different. The first step is to decide whether you have independent or paired groups. See Chapter 7 for more detail on this decision.

Summarizing Data

This section explains how to summarize data from two independent groups. These methods provide additional understanding of the statistical results.

Table 8.1 shows the percentages of body fat for several men and women. These people participated in unsupervised aerobic exercise or weight training (or both) about three times per week for a year. Then, they were measured once at the end of the year. Chapter 1 introduced this data, and Chapter 4 summarized the frequency counts for men and women in the fitness program.

Table 8.1 Body Fat Data

Group	Body Fat Percentage				
Male	13.3	8	20	12	12
	19	18	31	16	24
	20	22	21		
Female	22	16	21.7	21	30
	26	12	23.2	28	23

This data is available in the **bodyfat** data set in the sample data for the book.

The next five topics discuss methods for summarizing this data.

Using PROC MEANS for a Concise Summary

Just as you can use **PROC MEANS** for a concise summary of one variable, you can use the procedure for a concise summary of two independent groups. You add the **CLASS** statement to identify the variable that classifies the data into groups. The SAS statements below create the data set and provide a brief summary of the body fat percentages for men and women.

```
proc format;
value $gentext 'm' = 'Male'
               'f' = 'Female';
run;

data bodyfat;
    input gender $ fatpct @@;
    format gender $gentext.;
    label fatpct='Body Fat Percentage';
    datalines;
m 13.3 f 22 m 19 f 26 m 20 f 16 m 8 f 12 m 18 f 21.7
m 22 f 23.2 m 20 f 21 m 31 f 28 m 21 f 30 m 12 f 23
m 16 m 12 m 24
;
run;

proc means data=bodyfat;
    class gender;
    var fatpct;
title 'Brief Summary of Groups';
run;
```

Figure 8.1 shows the results.

Figure 8.1 Concise Summary of Two Groups

gender	N Obs	N	Mean	Std Dev	Minimum	Maximum
Female	10	10	22.2900000	5.3196596	12.0000000	30.0000000
Male	13	13	18.1769231	6.0324337	8.0000000	31.0000000

Brief Summary of Groups

The MEANS Procedure

Analysis Variable : fatpct Body Fat Percentage

Figure 8.1 shows descriptive statistics for each gender. The column labeled **gender** identifies the levels of the CLASS variable, and the other columns provide descriptive statistics. For example, the average percentage of body fat for women is 22.29. The average percentage of body fat for men is 18.18. The standard deviation for women is 5.32, and for men it is 6.03. The range of values for women is 12 to 30, and for men it is

8 to 31. If you plotted these values on a number line, you would see a wide range of overlapping values.

From this summary, you might initially conclude that the means for the two groups are different. You don't know whether this difference is greater than what could happen by chance. Finding out whether the difference is due to chance or is real requires a statistical test. This statistical test is discussed in "Building Hypothesis Tests to Compare Two Independent Groups" later in this chapter.

The general form of the statements to create a concise summary of two independent groups using PROC MEANS is shown below:

PROC MEANS DATA=*data-set-name*;

  CLASS *class-variable*;

  VAR *measurement-variables*;

data-set-name is the name of a SAS data set, *class-variable* is a variable that classifies the data into groups, and *measurement-variables* are the variables that you want to summarize.

To compare two independent groups, the CLASS statement is required. The *class-variable* can be character or numeric and should be nominal. If you omit the *measurement-variables* from the VAR statement, then the procedure uses all numeric variables.

You can use the *statistical-keywords* described in Chapter 4 with PROC MEANS and a CLASS statement.

Using PROC UNIVARIATE for a Detailed Summary

Just as you can use **PROC UNIVARIATE** for a detailed summary of one variable, you can use the procedure for a detailed summary of two independent groups. You add the CLASS statement to identify the variable that classifies the data into groups. The SAS statements below provide a detailed summary of the body fat percentages for men and women.

```
ods select moments basicmeasures extremeobs plots;
proc univariate data=bodyfat plot;
   class gender;
   var fatpct;
title 'Detailed Summary of Groups';
run;
```

The **ODS** statement controls the tables displayed in the output. Figure 8.2 shows the results.

Figure 8.2 shows descriptive statistics for each gender. For each of the tables that PROC UNIVARIATE creates, the output shows the statistics for each level of the CLASS variable. For example, look at the **Basic Statistical Measures** tables. The **gender=Female** table identifies statistics for women, and the **gender=Male** table identifies statistics for men.

Figure 8.2 shows a missing mode for the females. For data where no repetition of a value occurs, PROC UNIVARIATE does not print a mode. Table 8.1 shows the data, and you can see that each value of the body fat percentage is unique for the women. For the men, the values of 12 and 20 each occur twice. Figure 8.2 shows the mode for men as **12**, and prints a note stating that this is the smallest of two modes in the data.

The **PLOT** option creates box plots, stem-and-leaf plots, and normal quantile plots for each gender. Figure 8.2 does not show the normal quantile plots. Chapter 4 discusses using these plots to check for errors in the data. Chapter 5 discusses how these plots are useful in checking the data for normality. You can visually compare the plots for the two groups. However, you might prefer comparative plots instead. See the next three topics for comparative histograms, comparative bar charts, and side-by-side box plots.

This more extensive PROC UNIVARIATE summary supports your initial thoughts that the means for the two groups are different. Finding out whether the difference is due to chance or is real requires a statistical test. This statistical test is discussed in "Building Hypothesis Tests to Compare Two Independent Groups" later in this chapter.

Figure 8.2 Detailed Summary of Two Independent Groups

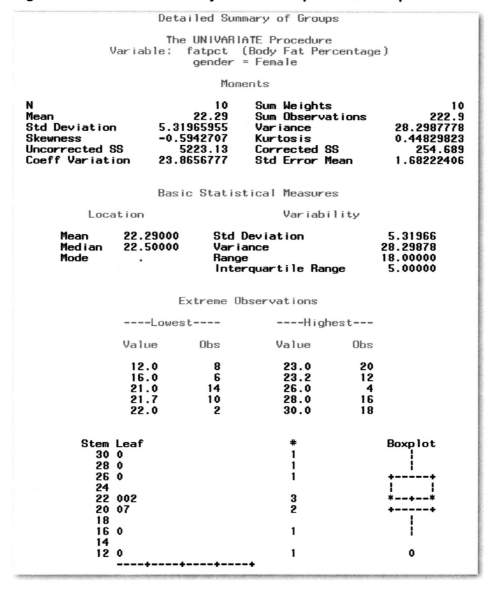

```
                    Detailed Summary of Groups

                     The UNIVARIATE Procedure
              Variable:   fatpct   (Body Fat Percentage)
                      gender = Female

                             Moments

N                          10      Sum Weights                10
Mean                    22.29      Sum Observations        222.9
Std Deviation      5.31965955      Variance           28.2987778
Skewness           -0.5942707      Kurtosis           0.44829823
Uncorrected SS        5223.13      Corrected SS          254.689
Coeff Variation    23.8656777      Std Error Mean     1.68222406


                  Basic Statistical Measures

         Location                        Variability

    Mean       22.29000     Std Deviation          5.31966
    Median     22.50000     Variance              28.29878
    Mode           .        Range                 18.00000
                            Interquartile Range    5.00000


                   Extreme Observations

       ----Lowest----           ----Highest---

       Value      Obs           Value      Obs

        12.0        8            23.0       20
        16.0        6            23.2       12
        21.0       14            26.0        4
        21.7       10            28.0       16
        22.0        2            30.0       18


   Stem Leaf                        #          Boxplot
     30 0                           1             |
     28 0                           1             |
     26 0                           1          +-----+
     24                                        |     |
     22 002                         3          *--+--*
     20 07                          2          +-----+
     18                                           |
     16 0                           1             |
     14
     12 0                           1             0
        ----+----+----+----+
```

Figure 8.2 Detailed Summary of Two Independent Groups (continued)

```
                    Detailed Summary of Groups

                      The UNIVARIATE Procedure
                 Variable:  fatpct  (Body Fat Percentage)
                         gender = Male

                             Moments

N                            13    Sum Weights                    13
Mean                  18.1769231    Sum Observations            236.3
Std Deviation          6.03243371   Variance               36.3902564
Skewness               0.33437873   Kurtosis                0.51998374
Uncorrected SS          4731.89     Corrected SS           436.683077
Coeff Variation        33.1873204   Std Error Mean          1.67309608


                    Basic Statistical Measures

           Location                      Variability

      Mean      18.17692     Std Deviation          6.03243
      Median    19.00000     Variance              36.39026
      Mode      12.00000     Range                 23.00000
                             Interquartile Range    7.70000

NOTE: The mode displayed is the smallest of 2 modes with a count of 2.


                      Extreme Observations

            ----Lowest----            ----Highest---

           Value      Obs          Value      Obs

             8.0        7            20        13
            12.0       22            21        17
            12.0       19            22        11
            13.3        1            24        23
            16.0       21            31        15


    Stem Leaf                         #          Boxplot
       3 1                            1             |
       2                                            |
       2 00124                        5          +-----+
       1 689                          3          *--+--*
       1 223                          3          +-----+
       0 8                            1             |
         ----+----+----+----+
    Multiply Stem.Leaf by 10**+1
```

The general form of the statements to create a detailed summary of two independent groups using PROC UNIVARIATE is shown below:

PROC UNIVARIATE DATA=*data-set-name*;

  CLASS *class-variable*;

  VAR *measurement-variables*;

Items in italic were defined earlier in the chapter.

To compare two independent groups, the CLASS statement is required. The *class-variable* can be character or numeric and should be nominal. If you omit the *measurement-variables* from the VAR statement, then the procedure uses all numeric variables.

You can use the options described in Chapters 4 and 5 with PROC UNIVARIATE and a CLASS statement.

Adding Comparative Histograms to PROC UNIVARIATE

Just as you can use PROC UNIVARIATE for a histogram of one variable, you can use the procedure for comparative histograms of two independent groups. You add the CLASS statement to identify the variable that classifies the data into groups. You also add the **HISTOGRAM** statement to request histograms. (To create high-resolution histograms, you must have SAS/GRAPH software licensed.) The SAS statements below provide comparative histograms. The **NOPRINT** option suppresses the printed output tables with the detailed summary of the body fat percentages for men and women.

```
proc univariate data=bodyfat noprint;
   class gender;
   var fatpct;
   histogram fatpct;
title 'Comparative Histograms of Groups';
run;
```

Figure 8.3 shows the results.

Figure 8.3 Comparative Histograms for Two Independent Groups

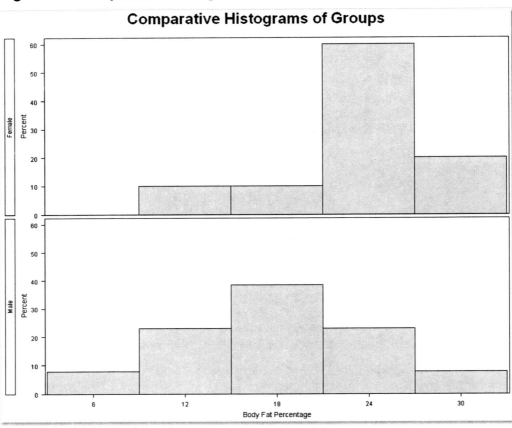

The HISTOGRAM statement creates the pair of histograms in Figure 8.3. These histograms are plotted on the same axis scales. This approach helps you visually compare the distribution of the two groups.

The distribution of values for the men is more symmetrical and more spread out. It has more values at the lower end of the range. These graphs support your initial thoughts that the means for the two groups are different.

The general form of the statements to create comparative histograms of two independent groups using PROC UNIVARIATE is shown below:

PROC UNIVARIATE DATA=*data-set-name* **NOPRINT;**

  CLASS *class-variable*;

  VAR *measurement-variables*;

  HISTOGRAM *measurement-variables*;

Items in italic were defined earlier in the chapter. You can omit the NOPRINT option and create the detailed summary and histograms at the same time. If you omit the *measurement-variables* from the HISTOGRAM statement, then the procedure uses all numeric variables in the VAR statement. If you omit the *measurement-variables* from the HISTOGRAM and VAR statements, then the procedure uses all numeric variables. You cannot specify variables in the HISTOGRAM statement unless the variables are also specified in the VAR statement.

Using PROC CHART for Side-by-Side Bar Charts

Figure 8.2 shows the line printer stem-and-leaf plots separately for the men and women in the fitness program. Figure 8.3 shows high-resolution histograms for these two independent groups. If you do not have SAS/GRAPH software licensed, you can use **PROC CHART** to create side-by-side bar charts that are low resolution (line printer quality).

```
proc chart data=bodyfat;
   vbar fatpct / group=gender;
   title 'Charts for Fitness Program';
run;
```

You can list only one variable in the **GROUP=** option. This variable can be either character or numeric, and it should have a limited number of levels. Variables such as height, weight, and temperature are not appropriate variables for the GROUP= option because they have many levels. Producing a series of side-by-side bar charts with one observation in each bar does not summarize the data well.

PROC CHART arranges the levels of the grouping variable in increasing numeric or alphabetical order. It uses the same bar midpoints for each group.

Figure 8.4 shows the results.

These bar charts are similar to the histograms in that they help you visually compare the distribution of the two groups. Again, the distribution of values for the men is more spread out than for the women, and has more values at the lower end of the range. These graphs support your initial thoughts that the means for the two groups are different.

Figure 8.4 Comparative Bar Charts for Two Independent Groups

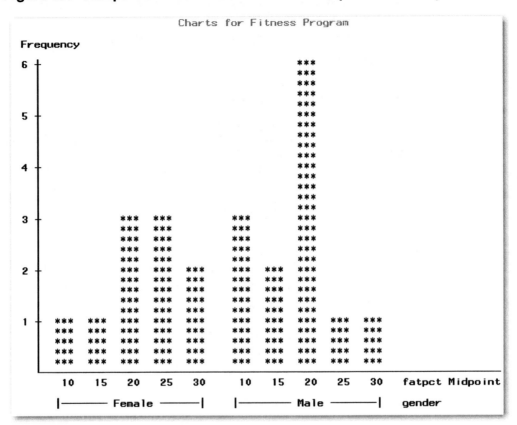

Compare Figures 8.3 and 8.4, which have different midpoints for the bars. In Figure 8.4, the symmetry of the distribution of values for the men is not as obvious as in Figure 8.3.

The general form of the statements to create comparative bar charts for two independent groups using PROC CHART is shown below:

PROC CHART DATA=*data-set-name***;**

    VBAR *measurement-variables* **/ GROUP=***class-variable***;**

Items in italic were defined earlier in the chapter. For comparative horizontal bar charts, replace **VBAR** with **HBAR**.

Like other options in PROC CHART, the GROUP= option appears after the slash. You can use the GROUP= option with other PROC CHART options discussed in Chapter 4.

Although you can list only one variable in the GROUP= option, you can specify several *measurement-variables* to be charted, and you can produce several side-by-side bar charts simultaneously. If you list several *measurement-variables* before the slash, SAS creates side-by-side bar charts for each variable. Suppose you measure the calories, calcium content, and vitamin D content of several cartons of regular and skim milk. The variable that classifies the data into groups is milktype. Suppose you want side-by-side bar charts for each of the three measurement variables. In SAS, you type the statements below:

```
proc chart data=dairy;
   vbar calories calcium vitamind / group=milktype;
 run;
```

The statements produce three side-by-side bar charts, one for each of the measurement variables listed.

Using PROC BOXPLOT for Side-by-Side Box Plots

Figure 8.2 shows separate line printer box plots for the men and women in the fitness program. You can use **PROC BOXPLOT** to create high-resolution plots and display side-by-side box plots. (To create high-resolution box plots, you must have SAS/GRAPH software licensed.) The SAS statements below provide comparative box plots:

```
proc sort data=bodyfat;
   by gender;
run;

proc boxplot data=bodyfat;
   plot fatpct*gender;
title 'Comparative Box Plots of Groups';
run;
```

PROC BOXPLOT expects the data to be sorted by the variable that classifies the data into groups. If your data is already sorted, you can omit the **PROC SORT** step. The first variable in the **PLOT** statement identifies the measurement variable, and the second variable identifies the variable that classifies the data into groups. Figure 8.5 shows the results.

Figure 8.5 Comparative Box Plots for Two Independent Groups

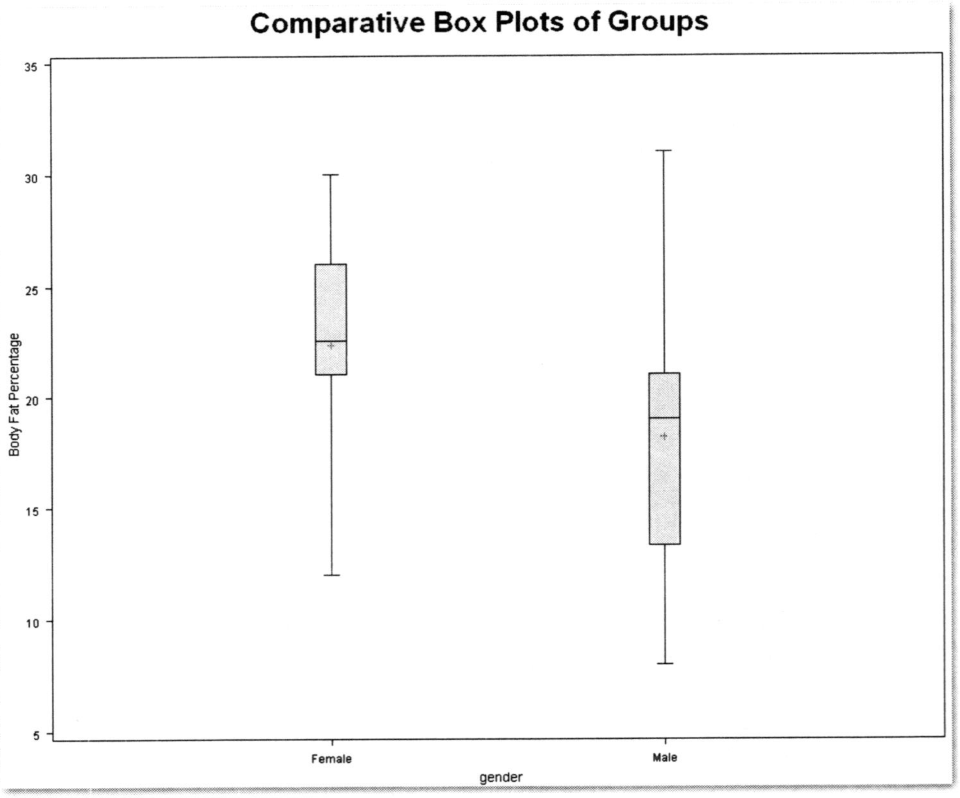

From the box plots, the distribution of values for males is wider than it is for females. You can also see this in the comparative histograms in Figure 8.3 and the bar charts in Figure 8.4. The box plots add the ability to display the mean, median, and interquartile range for each group. The lines at the ends of the whiskers are the minimum and maximum values for each group. Figure 8.5 shows the mean and median for males are lower than the mean and median for females. Figure 8.5 also shows the ends of the box for males (the 25th and 75th percentiles) are lower than the ends of the box for females.

The box plots support your initial thoughts that the means for the two groups are different.

The general form of the statements to create side-by-side box plots for two independent groups using PROC BOXPLOT is shown below:

    PROC BOXPLOT DATA=data-set-name;

        PLOT measurement-variable*class-variable;

Items in italic were defined earlier in the chapter. Depending on your data, you might need to first sort by the **class-variable** before creating the plots. The asterisk in the PLOT statement is required.

Building Hypothesis Tests to Compare Two Independent Groups

So far, this chapter has discussed how to summarize differences between two independent groups. Suppose you want to know how important the differences are, and if the differences are large enough to be significant. In statistical terms, you want to perform a hypothesis test. This section discusses hypothesis testing when comparing two independent groups. (See Chapter 5 for the general idea of hypothesis testing.)

In building a test of hypothesis, you work with two hypotheses. When comparing two independent groups, the null hypothesis is that the two means for the independent groups are the same, and the alternative hypothesis is that the two means are different. In this case, the notation is the following:

    $H_o: \mu_A = \mu_B$

    $H_a: \mu_A \neq \mu_B$

H_o indicates the null hypothesis, H_a indicates the alternative hypothesis, and μ_A and μ_B are the population means for independent groups A and B.

In statistical tests that compare two independent groups, the hypotheses are tested by calculating a test statistic from the data, and then comparing its value to a reference value that would be the result if the null hypothesis were true. The test statistic is compared to different reference values that are based on the sample size. In part, this is because smaller differences can be detected with larger sample sizes. As a result, if you use the reference value for a large sample size on a smaller sample, you are more likely to incorrectly conclude that the means for the groups are different when they are not.

This concept is similar to the concept of confidence intervals, in which different t-values are used (discussed in Chapter 6). With confidence intervals, the t-value is based on the degrees of freedom, which were determined by the sample size. Similarly, when comparing two independent groups, the degrees of freedom for a reference value are based on the sample sizes of the two groups. Specifically, the degrees of freedom are equal to $N-2$, where N is the total sample size for the two groups.

Deciding Which Statistical Test to Use

Because you test different hypotheses for independent and paired groups, you use different tests. In addition, there are parametric and nonparametric tests for each type of group. When deciding which statistical test to use, first, decide whether you have independent groups or paired groups.

Then, decide whether you should use a parametric or a nonparametric test. This second decision is based on whether the assumptions for the parametric test seem reasonable. In general, if the assumptions for the parametric test seem reasonable, use the parametric test. Although nonparametric tests have fewer assumptions, these tests typically are less powerful in detecting differences between groups.

The rest of this chapter describes the test to use for two independent groups. See Chapter 7 for details on the tests for the two situations for paired groups. Table 8.3 summarizes the tests.

Table 8.3 Statistical Tests for Comparing Two Groups

Type of Test	Groups	
	Independent	**Paired**[1]
Parametric	Two-sample t-test	Paired-difference t-test
Nonparametric	Wilcoxon Rank Sum test	Wilcoxon Signed Rank test

The steps for performing the analyses to compare independent groups are the following:

1. Create a SAS data set.

2. Check the data set for errors.

---

[1] Chapter 7 discusses comparing paired groups. For completeness, this table shows the tests for comparing both types of groups.

3. Choose the significance level for the test.

4. Check the assumptions for the test.

5. Perform the test.

6. Make conclusions from the test results.

The rest of this chapter uses these steps for all analyses.

Understanding Significance

For each of the four tests in Table 8.3, there are two possible results: the *p*-value is less than the reference probability, or it is not. (See Chapter 5 for discussions on significance levels and on understanding practical significance and statistical significance.)

Groups Significantly Different

If the *p*-value is less than the reference probability, the result is statistically significant, and you reject the null hypothesis. For independent groups, you conclude that the means for the two groups are significantly different.

Groups Not Significantly Different

If the *p*-value is greater than the reference probability, the result is not statistically significant, and you fail to reject the null hypothesis. For independent groups, you conclude that the means for the two groups are not significantly different.

Do not conclude that the means for the two groups are the same. You do not have enough evidence to conclude that the means are the same. The results of the test indicate only that the means are not significantly different. You never have enough evidence to conclude that the population is exactly as described by the null hypothesis. In statistical hypothesis testing, do not accept the null hypothesis. Instead, either reject or fail to reject the null hypothesis. For a more detailed discussion of hypothesis testing, see Chapter 5.

Performing the Two-Sample *t*-test

The two-sample *t*-test is a parametric test for comparing two independent groups.

Assumptions

The three assumptions for the two-sample *t*-test are the following:

- observations are independent
- observations in each group are a random sample from a population with a normal distribution
- variances for the two independent groups are equal

Because of the second assumption, the two-sample *t*-test applies to continuous variables only.

Apply the six steps of analysis:

1. Create a SAS data set.

2. Check the data set for errors. Use **PROC UNIVARIATE** to check for errors. Based on the results, assume that the data is free of errors.

3. Choose the significance level for the test. Choose a 5% significance level, which requires a *p*-value less than 0.05 to conclude that the groups are significantly different.

4. Check the assumptions for the test.

    The first assumption is independent observations. The assumption seems reasonable because each person's body fat measurement is unrelated to every other person's body fat measurement.

    The second assumption is that the observations in each group are from a normal distribution. You can test this assumption for each group to determine whether this assumption seems reasonable. (See Chapter 6.)

    The third assumption of equal variances seems reasonable based on the similarity of the standard deviations of the two groups. From Figure 8.2, the standard deviation is 5.32 for females and 6.03 for males. See "Testing for Equal Variances" for a statistical test to check this assumption.

5. Perform the test. **PROC TTEST** provides the exact two-sample *t*-test, which assumes equal variances, and an approximate test, which can be used if the assumption of equal variances isn't met. See "Testing to Compare Two Means."

6. Make conclusions from the test results. See "Finding the *p*-value" for each test.

Testing for Equal Variances

PROC TTEST automatically tests the assumption of equal variances, and provides a test to use when the assumption is met, and a test to use when it is not met. The SAS statements below perform these tests:

```
proc ttest data=bodyfat;
   class gender;
   var fatpct;
title 'Comparing Groups in Fitness Program';
run;
```

The CLASS statement identifies the variable that classifies the data set into groups. The procedure automatically uses the formatted values of the variable to compare groups. For numeric values, the lower value is subtracted from the higher value (for example, 2–1). For alphabetic values, the value later in the alphabet is subtracted from the value earlier in the alphabet (for example, F–M).

The VAR statement identifies the measurement variable that you want to analyze. Figure 8.6 shows the **Equality of Variances** table. Figure 8.7 shows the other tables created by the SAS statements above.

Figure 8.6 Testing for Equal Variances

	Equality of Variances			
Method	Num DF	Den DF	F Value	Pr > F
Folded F	**12**	**9**	**1.29**	**0.7182**

The Folded F test has a null hypothesis that the variances are equal, and an alternative hypothesis that they are not equal. Because the *p*-value of **0.7182** is much greater than the significance level of 0.05, do not reject the null hypothesis of equal variances for the two groups. In fact, there is almost a 72% chance that this result occurred by random chance. You proceed based on the assumption that the variances for the two groups are equal.

The list below describes other items in the **Equality of Variances** table:

Method — identifies the test as the Folded *F* test.

Num DF
Den DF — specifies the degrees of freedom. **Num DF** is the degrees of freedom for the numerator, which is n_1-1, where n_1 is the size of the group with the larger variance (males in the **bodyfat** data). **Den DF** is the degrees of freedom for the denominator, which is n_2-1, where n_2 is the size of the group with the smaller variance (females in the **bodyfat** data).

F Value — is the ratio of the larger variance to the smaller variance. For the **bodyfat** data, this is the ratio of the variance for males to the variance for females.

Pr > F — is the significance probability for the test of unequal variances. You compare this value to the significance level you selected in advance. Typically, values less than 0.05 lead you to conclude that the variances for the two groups are not equal. Statisticians have different opinions on the appropriate significance level for this test. Instead of 0.05, many statisticians use higher values such as 0.10 or 0.20.

Testing to Compare Two Means

You have checked the data set for errors, chosen a significance level, and checked the assumptions for the test. You are ready to compare the means.

PROC TTEST provides a test to use when the assumption of equal variances is met, and a test to use when it is not met. Figure 8.7 shows the **Statistics** and **Confidence Limits** tables, and the **T tests** section of output. The T tests section is at the bottom of the figure and shows the results of the statistical tests. The SAS statements in "Testing for Equal Variances" create these results.

Figure 8.7 Comparing Two Means with PROC TTEST

```
                    Comparing Groups in Fitness Program

                           The TTEST Procedure

                    Variable:  fatpct  (Body Fat Percentage)

gender          N          Mean      Std Dev      Std Err     Minimum     Maximum

Female          10      22.2900       5.3197       1.6822     12.0000     30.0000
Male            13      18.1769       6.0324       1.6731      8.0000     31.0000
Diff (1-2)               4.1131       5.7378       2.4135

   gender         Method              Mean        95% CL Mean         Std Dev

   Female                          22.2900     18.4845   26.0955      5.3197
   Male                           18.1769     14.5316   21.8223      6.0324
   Diff (1-2)     Pooled           4.1131     -0.9060    9.1321      5.7378
   Diff (1-2)     Satterthwaite    4.1131     -0.8277    9.0539

      gender         Method            95% CL Std Dev

      Female                         3.6590    9.7116
      Male                           4.3258    9.9580
      Diff (1-2)     Pooled          4.4144    8.1997
      Diff (1-2)     Satterthwaite

   Method          Variances        DF      t Value     Pr > |t|

   Pooled          Equal             21        1.70      0.1031
   Satterthwaite   Unequal       20.539        1.73      0.0980
```

Finding the *p*-value

To compare two independent groups, use the T tests section in Figure 8.7. Look at the line with **Variances** of **Equal**, and find the value in the column labeled **Pr > |t|**. This is the *p*-value for testing that the means for the two groups are significantly different under the assumption that the variances are equal. Figure 8.7 shows the value of **0.1031**, which is greater than 0.05. This value indicates that the body fat averages for men and women are not significantly different at the 5% significance level.

Suppose the assumption of equal variances is not reasonable. In this case, you look at the line with **Variances** of **Unequal**, and find the value in the column labeled **Pr > |t|**. This is the *p*-value for testing that the means for the two groups are significantly different without assuming that the variances are equal. For the unequal variances case, the value is **0.0980**. This value also indicates that the body fat averages for men and women are not significantly different at the 5% significance level. The reason for the difference in *p*-values is the calculation for the standard deviations. See "Technical Details: Pooled and Unpooled" for formulas.

Other studies show that the body fat averages for men and women are significantly different. Why don't you find a significant difference in this set of data? One possible reason is the small sample size. Only a few people were measured. Perhaps a larger data set would have given a more accurate picture of the population. Another possible reason is uncontrolled factors affecting the data. Although these people were exercising regularly during the year, their activities and diet were not monitored. These factors, as well as other factors, could have had enough influence on the data to obscure differences that do indeed exist.

The two groups would be significantly different at the 15% significance level. Now you see why you always choose the significance level first. Your criterion for statistical significance should not be based on the results of the test.

In general, to find the *p*-value for comparing two means, use the T **tests** section in the output. Look at the Equal row if the assumption of equal variances seems reasonable, or the Unequal row if it does not.

Understanding Information in the Output

The list below describes the T **tests** section in Figure 8.7.

Method	identifies the test. The **Pooled** row gives the test for equal variances, and the **Satterthwaite** row gives the test for unequal variances.
Variances	displays the assumption of variances as **Equal** or **Unequal**.
DF	specifies the degrees of freedom for the test. This is the second key distinction between the equal and unequal variances tests. For the equal variances test (**Pooled**), the degrees of freedom are calculated by adding the sample size for each group, and then subtracting 2 (21=10+13−2). For the Satterthwaite test, the degrees of freedom use a different calculation.
t Value	is the value of the *t*-statistic. This value is the difference between the averages of the two groups, divided by the standard error. Because the standard deviation is involved in calculating the standard error, and the two tests calculate the standard deviation as pooled and unpooled, the values of the *t*-statistic differ.
	Generally, a value of the *t*-statistic greater than 2 (or less than −2) indicates significant difference between the two groups at the 95% confidence level.

Pr > |t| is the *p*-value associated with the test. This *p*-value is for a two-sided test, which tests the alternative hypothesis that the means for the two independent groups are different.

To perform the test at the 5% significance level, you conclude that the means for the two independent groups are significantly different if the *p*-value is less than 0.05. To perform the test at the 10% significance level, you conclude that the means for the two independent groups are significantly different if the *p*-value is less than 0.10. Conclusions for tests at other significance levels are made in a similar manner.

For both the **Statistics** and **Confidence Limits** tables in the output, the first column identifies the classification variable. For Figure 8.7, this variable is **gender**.

The **Statistics** table contains the sample size (**N**), average (**Mean**), standard deviation (**Std Dev**), standard error (**Std Err**), minimum (**Minimum**), and maximum (**Maximum**) for each group identified by the classification variable. The third row of the **Statistics** table contains statistics for the difference between the two groups, and identifies how this difference is calculated. For the **bodyfat** data, SAS calculates the difference by subtracting the average for males from the average for females. SAS labels this row as **Diff (1–2)**.

The list below describes the **Confidence Limits** table in Figure 8.7.[2]

Variable   identifies the name of the classification variable. For Figure 8.7, the variable is **gender**.

Method   contains a value that depends on the given row in the table. For the first two rows that show data values, the **Method** column is blank. The **Pooled** row shows statistics based on the pooled estimate for standard deviation, and the **Satterthwaite** row shows statistics based on the alternative method for unequal variances.

Mean   shows the average for each group, or the difference of the averages between the two groups, depending on the row in the table.

---

[2] If you are running a version of SAS earlier than 9.2, you will see headings of **Upper CL Mean**, **Lower CL Mean**, **Upper CL Std Dev**, and **Lower CL Std Dev**. The descriptive statistics appear in a different order. The output for earlier versions of SAS consists of the **Statistics** table and the **T tests** section.

95% CL Mean[3]   shows the 95% confidence limits for the mean of each group for the first two rows. For the second two rows, it shows the 95% confidence limits for the mean difference. The confidence limits differ because of the different estimates for standard deviation that SAS uses for equal and unequal variances.

Std Dev   shows the standard deviation for each group and the pooled estimate for standard deviation. PROC TTEST does not automatically show the estimate of standard deviation for the situation of unequal variances.

95% CL Std Dev[4]   shows the 95% confidence limits for the mean of each group for the first two rows. For the third row, it shows a 95% confidence interval for the pooled estimate of the pooled standard deviation. PROC TTEST does not automatically show the confidence interval for standard deviation for the situation of unequal variances.

As discussed in Chapter 4, you can use the ODS statement to choose the output tables to display. Table 8.4 identifies the output tables for PROC TTEST.

Table 8.4 ODS Table Names for PROC TTEST

ODS Name	Output Table
Statistics	Statistics
ConfLimits[5]	Confidence Limits
Equality	Equality of Variances
Ttests	T Tests

---

[3] If you are running a version of SAS earlier than 9.2, you will see headings of Upper CL Mean and Lower CL Mean. The descriptive statistics appear in a different order.

[4] If you are running a version of SAS earlier than 9.2, you will see headings of Lower CL Std Dev and Upper CL Std Dev. The descriptive statistics appear in a different order.

[5] This output table is not available in versions of SAS earlier than 9.2. Instead, the confidence limits appear in the Statistics table.

For example, the statements below create output that contains only the **Equality of Variances** table.

```
ods select equality;
proc ttest data=bodyfat;
   class gender;
   var fatpct;
run;
```

The general form of the statements to perform tests for equal variances and tests to compare two means using PROC TTEST is shown below:

PROC TTEST DATA=*data-set-name*;

  CLASS *class-variable*;

  VAR *measurement-variables*;

Items in italic were defined earlier in the chapter.

The CLASS statement is required. The *class-variable* can have only two levels. While the VAR statement is not required, you should use it. If you omit the VAR statement, then PROC TTEST performs analyses for every numeric variable in the data set.

You can use the **BY** statement to perform several *t*-tests corresponding to the levels of another variable. However, another statistical analysis might be more appropriate. Consult a statistical text or a statistician before you do this analysis.

Technical Details: Pooled and Unpooled

Here are the formulas for how SAS calculates the *t*-statistic for the pooled and unpooled tests. For the equal variances (pooled) test, SAS uses the following calculation:

$$t = \frac{\overline{X}_1 - \overline{X}_2}{s_p \sqrt{1/n_1 + 1/n_2}}$$

$\overline{X}_1$ and $\overline{X}_2$ are the averages for the two groups.

$$s_p^2 = \frac{(n_1 - 1)s_1^2 + (n_2 - 1)s_2^2}{n_1 + n_2 - 2}$$

n_1 and n_2 are sample sizes for the two groups, and s_1 and s_2 are the standard deviations for the two groups.

For the unequal variances (unpooled) test, SAS uses the following calculation:

$$t' = \frac{\overline{X}_1 - \overline{X}_2}{\sqrt{\dfrac{s_1^2}{n_1} + \dfrac{s_2^2}{n_2}}}$$

n_1, n_2, s_1, and s_2 are defined above.

Changing the Alpha Level for Confidence Intervals

SAS automatically creates 95% confidence intervals. You can change the alpha level for the confidence intervals with the **ALPHA=** option in the PROC TTEST statement. The statements below create 90% confidence intervals and produce Figure 8.8.[6]

```
ods select conflimits;
proc ttest data=bodyfat alpha=0.10;
   class gender;
   var fatpct;
run;
```

---

[6] The **Confidence Limits** output table is not created for versions of SAS earlier than 9.2, so the ODS statement will not work for versions of SAS earlier than 9.2. Instead, use an ODS statement to request only the **Statistics** table, which contains confidence limits for versions of SAS earlier than 9.2. Also, see the footnote for Figure 8.7—the same differences in headings occur in Figure 8.8 for versions of SAS earlier than 9.2.

Figure 8.8 Adjusting Alpha Levels in PROC TTEST

```
                    Comparing Groups in Fitness Program

                           The TTEST Procedure

                    Variable:  fatpct  (Body Fat Percentage)

gender          Method              Mean        90% CL Mean       Std Dev

Female                             22.2900   19.2063   25.3737    5.3197
Male                               18.1769   15.1950   21.1589    6.0324
Diff (1-2)     Pooled               4.1131   -0.0399    8.2660    5.7378
Diff (1-2)     Satterthwaite        4.1131    0.0262    8.1999

               gender       Method          90% CL Std Dev

               Female                       3.8799    8.7519
               Male                         4.5573    9.1411
               Diff (1-2)   Pooled          4.6002    7.7231
               Diff (1-2)   Satterthwaite
```

Compare Figures 8.7 and 8.8. The column headings in each figure identify the confidence levels. Figure 8.7 contains 95% confidence limits, and Figure 8.8 contains 90% confidence limits. All the other statistics are the same.

The general form of the statements to specify the alpha level using PROC TTEST is shown below:

PROC TTEST DATA=*data-set-name* **ALPHA=***level*;

   CLASS *class-variable*;

   VAR *measurement-variables*;

level gives the confidence level. The value of *level* must be between 0 and 1. A *level* value of 0.05 gives the automatic 95% confidence intervals. Items in italic were defined earlier in the chapter.

Performing the Wilcoxon Rank Sum Test

The Wilcoxon Rank Sum test is a nonparametric test for comparing two independent groups. It is a nonparametric analogue to the two-sample *t*-test, and is sometimes referred to as the Mann-Whitney *U* test. The null hypothesis is that the two means for the independent groups are the same. The only assumption for this test is that the observations are independent.

To illustrate the Wilcoxon Rank Sum test, consider an experiment to analyze the content of the gastric juices of two groups of patients.[7] The patients are divided into two groups: patients with peptic ulcers, and normal or control patients without peptic ulcers. The goal of the analysis is to determine whether the mean lysozyme levels for the two groups are significantly different at the 5% significance level. (Lysozyme is an enzyme that can destroy the cell walls of some types of bacteria.) Table 8.5 shows the data.

Table 8.5 Gastric Data

Group	Lysozyme Levels									
Ulcers	0.2	10.9	1.1	16.2	3.3	20.7	4.8	40.0	5.3	60.0
	10.4	0.4	12.4	2.1	18.9	4.5	25.4	5.0	50.0	9.8
	0.3	11.3	2.0	17.6	3.8	24.0	4.9	42.2	7.5	
Normal	0.2	5.7	0.7	8.7	1.5	10.3	2.4	16.5	3.6	20.7
	5.4	0.4	7.5	1.5	9.1	2.0	16.1	2.8	20.0	4.8
	0.3	5.8	1.2	8.8	1.9	15.6	2.5	16.7	4.8	33.0

This data is available in the **gastric** data set in the sample data for the book. Although not shown here, you could use PROC UNIVARIATE to investigate the normality of each group. The data is clearly not normal. As a result, the parametric two-sample *t*-test should not be used to compare the two groups.

The first four steps of analysis are complete.

1. Create a SAS data set.

2. Check the data set for errors. PROC UNIVARIATE was used to check for errors. Although outlier points appear in the box plots, the authors of the original data did not find any underlying reasons for these outlier points. Assume that the data is free of errors.

3. Choose the significance level for the test. Choose a 5% significance level, which requires a *p*-value less than 0.05 to conclude that the groups are significantly different.

4. Check the assumptions for the test. The only assumption for this test is that the observations are independent. This seems reasonable, because each observation is a different person. The lysozyme level in one person is not dependent on the

[7] Data is from K. Myer et al., "Lysozyme activity in ulcerative alimentary disease," in *American Journal of Medicine 5* (1948): 482-495. Used with permission.

lysozyme level in another person. (This fact ignores the possibility that two people could be causing stress—or ulcers—in each other!)

5. Perform the test. See "Using PROC NPAR1WAY for the Wilcoxon Rank Sum Test."

6. Make conclusions from the test results. See "Finding the *p*-value."

Using PROC NPAR1WAY for the Wilcoxon Rank Sum Test

SAS performs nonparametric tests to compare two independent groups with the **NPAR1WAY** procedure. To create the data set and perform the Wilcoxon Rank Sum test, submit the following code:

```
data gastric;
   input group $ lysolevl @@;
   datalines;
U 0.2 U 10.4 U 0.3 U 10.9 U 0.4 U 11.3 U 1.1 U 12.4 U 2.0
U 16.2 U 2.1 U 17.6 U 3.3 U 18.9 U 3.8 U 20.7 U 4.5
U 24.0 U 4.8 U 25.4 U 4.9 U 40.0 U 5.0 U 42.2 U 5.3
U 50.0 U 7.5 U 60.0 U 9.8
N 0.2 N 5.4 N 0.3 N 5.7 N 0.4 N 5.8 N 0.7 N 7.5 N 1.2 N 8.7
N 1.5 N 8.8 N 1.5 N 9.1 N 1.9 N 10.3 N 2.0 N 15.6 N 2.4
N 16.1 N 2.5 N 16.5 N 2.8 N 16.7 N 3.6 N 20.0
N 4.8 N 20.7 N 4.8 N 33.0
;
run;

proc npar1way data=gastric wilcoxon;
   class group;
   var lysolevl;
   title 'Comparison of Ulcer and Control Patients';
run;
```

The **WILCOXON** option requests the Wilcoxon Rank Sum test. The CLASS statement identifies the variable that classifies the data into two groups. The VAR statement identifies the variable to use to compare the two groups. Figure 8.9 shows the results.

Figure 8.9 shows the **Scores, Wilcoxon Two-Sample Test**, and **Kruskal-Wallis Test** tables. Chapter 9 discusses the Kruskal-Wallis Test table, which is appropriate for more than two groups. The next two topics discuss the other two tables.

Figure 8.9 Wilcoxon Rank Sum Test for Gastric Data

```
              Comparison of Ulcer and Control Patients

                      The NPAR1WAY Procedure

         Wilcoxon Scores (Rank Sums) for Variable lysolevl
                  Classified by Variable group

                    Sum of      Expected      Std Dev         Mean
  group    N        Scores      Under H0      Under H0        Score

  U        29        976.0        870.0       65.943928    33.655172
  N        30        794.0        900.0       65.943928    26.466667

              Average scores were used for ties.


                    Wilcoxon Two-Sample Test

        Statistic              976.0000

        Normal Approximation
        Z                        1.5998
        One-Sided Pr >  Z        0.0548
        Two-Sided Pr > |Z|       0.1096

        t Approximation
        One-Sided Pr >  Z        0.0575
        Two-Sided Pr > |Z|       0.1151

        Z includes a continuity correction of 0.5.


                      Kruskal-Wallis Test

        Chi-Square               2.5838
        DF                            1
        Pr > Chi-Square          0.1080
```

Finding the *p*-value

Figure 8.9 shows the results of the Wilcoxon Rank Sum test. First, answer the research question, "Are the mean lysozyme levels significantly different for patients without ulcers and patients with ulcers?"

In Figure 8.9, look in the Wilcoxon Two-Sample Test table and find the first row with the heading **Two-Sided Pr > |Z|**. Look at the number to the right, which gives the *p*-value for the Wilcoxon Rank Sum test. For the gastric data, this value is **0.1096**, which is greater than the significance level of 0.05. You conclude that the mean lysozyme levels for the patients without ulcers and for the patients with ulcers are not significantly different at the 5% significance level.

In general, to interpret results, look at the number to the right of the first row labeled as **Two-Sided Pr > |Z|**. If the *p*-value is less than the significance level, you conclude that the means for the two groups are significantly different. If the *p*-value is greater than the significance level, you conclude that the means are not significantly different. Do not conclude that the means for the two groups are the same. (See the discussion earlier in this chapter.)

Understanding Tables in the Output

The list below describes the other items in the Wilcoxon Two-Sample Test table in Figure 8.9.

Statistic	is the Wilcoxon statistic, which is the sum of scores for the group with the smaller sample size.
Normal Approximation Z	is the normal approximation for the standardized test statistic.
Normal Approximation One-Sided Pr > Z	is the one-sided *p*-value, which displays when Z is greater than 0. When Z is less than 0, SAS displays One-Sided Pr < Z instead.
Normal Approximation Two-Sided Pr > |Z|	is the two-sided *p*-value, which was described earlier.
t Approximation One-Sided Pr > Z	Approximate *p*-values for the two-sample *t*-test. If the *t*-test is appropriate for your data, use the exact results from PROC TTEST instead.
t Approximation Two-Sided Pr > |Z|	Approximate *p*-values for the two-sample *t*-test. If the *t*-test is appropriate for your data, use the exact results from PROC TTEST instead.
Z includes a continuity correction of 0.5	Explains that PROC NPAR1WAY has automatically applied a continuity correction to the test statistic. Although you can suppress this continuity correction with the **CORRECT=NO** option in the PROC NPAR1WAY statement, most statisticians accept the automatic application of the continuity correction.

This list below describes the **Scores** table in Figure 8.9. The title for this table identifies the measurement variable and the classification variable.

Variable
identifies the name of the classification variable and the levels of this variable. The title and content of this column depend on the variable listed in the **CLASS** statement. In this example, the group variable defines the groups and has the levels **U** (ulcer) and **N** (normal).

N
is the number of observations in each group.

Sum of Scores
is the sum of the Wilcoxon scores for each group. To get the scores, SAS ranks the variable from lowest to highest, assigning the lowest value 1, and the highest value n, where n is the sample size. SAS sums the scores for each group to get the sum of scores.

Expected Under H0
lists the Wilcoxon scores expected under the null hypothesis of no difference between the groups. If the sample sizes for the two groups are the same, this value for both groups will also be the same.

Std Dev Under H0
gives the standard deviation of the sum of scores under the null hypothesis.

Mean Score
is the average score for each group, calculated as (Sum of Scores)/N.

Average scores were used for ties
describes how ties were handled by the procedure. Ties occur when you arrange the data from highest to lowest, and two values are the same. Suppose the observations that would be ranked 7 and 8 have the same value for the measurement variable. Then, these observations are both assigned a rank of 7.5 (=15/2). This message is for information only. There isn't an option to control how ties are handled.

As discussed in Chapter 4, you can use the ODS statement to choose the output tables to display. Table 8.6 identifies the output tables for **PROC NPAR1WAY** and the Wilcoxon Rank Sum test.

Table 8.6 ODS Table Names for PROC NPAR1WAY

ODS Name	Output Table
WilcoxonScores	Wilcoxon Scores
WilcoxonTest	Wilcoxon Two-Sample Test

For example, the statements below create output that contains only the Wilcoxon Two-Sample Test table:

```
ods select wilcoxontest;
proc npar1way data=gastric wilcoxon;
    class group;
    var lysolevl;
run;
```

The general form of the statements to perform tests for equal variances and tests to compare two means using PROC NPAR1WAY is shown below:

PROC NPAR1WAY DATA=*data-set-name*;

   CLASS *class-variable*;

   VAR *measurement-variables*;

Other items were defined earlier.

The CLASS statement is required. The *class-variable* can have only two levels. While the VAR statement is not required, you should use it. If you omit the VAR statement, then PROC NPAR1WAY performs analyses for every numeric variable in the data set.

You can use the BY statement to perform several *t*-tests corresponding to the levels of another variable. However, another statistical analysis might be more appropriate. Consult a statistical text or a statistician before you do this analysis.

Summary

Key Ideas

- Independent groups of data contain measurements for two unrelated samples of items.

- To summarize data for independent groups, summarize each group separately.

- To choose a statistical test, first, decide whether the data is from independent or paired groups. Then, decide whether to use a parametric or a nonparametric test. The second decision is based on whether the assumptions for the parametric test seem reasonable. Tests for the four cases are:

	Groups	
Type of Test	**Independent**	**Paired**[8]
Parametric	Two-sample *t*-test	Paired-difference *t*-test
Nonparametric	Wilcoxon Rank Sum test	Wilcoxon Signed Rank test

- Regardless of the statistical test you choose, the steps of analysis are:
    1. Create a SAS data set.

    2. Check the data set for errors.

    3. Choose the significance level for the test.

    4. Check the assumptions for the test.

    5. Perform the test.

    6. Make conclusions from the test results.

- Regardless of the statistical test, to make conclusions, compare the *p*-value for the test with the significance level.

    □ If the *p*-value is less than the significance level, then you reject the null hypothesis, and you conclude that the means for the groups are significantly different.

---

[8] Chapter 7 discusses comparing paired groups. For completeness, this table shows the tests for comparing both types of groups.

□   If the *p*-value is greater than the significance level, then you fail to reject the null hypothesis. You conclude that the means for the groups are not significantly different. (Remember, do not conclude that the means for the two groups are the same!)

- Test for normality using PROC UNIVARIATE as described in Chapter 5. For independent groups, test each group separately.

- Test for equal variances using PROC TTEST when comparing two independent groups.

- When comparing two independent groups, what type of SAS analysis to perform depends on the outcome of the unequal variances test. In either case, use PROC TTEST and choose either the **Equal** or **Unequal** test.

- You can change the confidence levels in PROC TTEST with the ALPHA= option.

- Use PROC NPAR1WAY to perform the Wilcoxon Rank Sum test when the assumptions for the *t*-test are not met.

Syntax

To summarize data from two independent groups

- To create a concise summary of two independent groups:

  PROC MEANS DATA=_data-set-name_**;**
     CLASS _class-variable_**;**
     VAR _measurement-variables_**;**

  data-set-name   is the name of a SAS data set.

  class-variable   is the variable that classifies the data into groups.

  measurement-variables
               are the variables you want to summarize.

  The CLASS statement is required. You should use the VAR statement. You can use any of the options described in Chapter 4.

- To create a detailed summary of two independent groups with comparative histograms:

  PROC UNIVARIATE DATA=*data-set-name* **NOPRINT;**
    CLASS *class-variable*;
    VAR *measurement-variables*;
    HISTOGRAM *measurement-variables*;

  Items in italic were defined earlier. The CLASS statement is required. You should use the VAR statement. You can use any of the options described in Chapters 4 and 5. You might want to use the NOPRINT option to suppress output and view only the comparative histograms.

- To create comparative vertical bar charts as an alternative to high-resolution histograms:

  PROC CHART DATA=*data-set-name*;
    VBAR *measurement-variables* **/ GROUP=***class-variable*;

  Items in italic were defined earlier. For comparative horizontal bar charts, replace VBAR with HBAR. The GROUP= option appears after the slash. You can use any of the options described in Chapter 4.

- To create side-by-side box plots, you might need to first use PROC SORT to sort the data by the class variable.

  PROC SORT DATA=*data-set-name*;
    BY *class-variable*;
  PROC BOXPLOT DATA=*data-set-name*;
    PLOT *measurement-variable*class-variable*;

  Items in italic were defined earlier. The asterisk in the PLOT statement is required.

To check the assumptions (step 4)

- To check the assumption of independent observations, you need to think about your data and whether this assumption seems reasonable. This is not a step where using SAS will answer this question.

- To check the assumption of normality, use PROC UNIVARIATE. For independent groups, check each group separately.

- To test for equal variances for the two-sample *t*-test, use PROC TTEST and look at the *p*-value for the Folded *F* test. See the general form in the next step.

To perform the test (step 5)

▪ To perform the two-sample *t*-test to compare two independent groups, use PROC TTEST. Look at the appropriate *p*-value, depending on the results of testing the assumption for equal variances.

PROC TTEST DATA=*data-set-name***;**
  CLASS *class-variable***;**
  VAR *measurement-variables***;**

Items in italic were defined earlier. The CLASS statement is required. You should use the VAR statement.

▪ To perform the nonparametric Wilcoxon Rank Sum test to compare two independent groups, use PROC NPAR1WAY. Look at the two-sided *p*-value for the normal approximation.

PROC NPAR1WAY DATA=*data-set-name***;**
  CLASS *class-variable***;**
  VAR *measurement-variables***;**

Items in italic were defined earlier. The CLASS statement is required. You should use the VAR statement.

Enhancements

▪ To change the alpha level for confidence intervals for independent groups, use the ALPHA= option in the PROC TTEST statement.

Example

The program below produces the output shown in this chapter:

```
proc format;
value $gentext 'm' = 'Male'
               'f' = 'Female';
run;

data bodyfat;
   input gender $ fatpct @@;
   format gender $gentext.;
   label fatpct='Body Fat Percentage';
   datalines;
m 13.3 f 22 m 19 f 26 m 20 f 16 m 8 f 12 m 18 f 21.7
m 22 f 23.2 m 20 f 21 m 31 f 28 m 21 f 30 m 12 f 23
```

```
m 16 m 12 m 24
;
run;

proc means data=bodyfat;
   class gender;
   var fatpct;
title 'Brief Summary of Groups';
run;

ods select moments basicmeasures extremeobs plots;
proc univariate data=bodyfat plot;
   class gender;
   var fatpct;
title 'Detailed Summary of Groups';
run;

proc univariate data=bodyfat noprint;
   class gender;
   var fatpct;
   histogram fatpct;
title 'Comparative Histograms of Groups';
run;

proc chart data=bodyfat;
   vbar fatpct / group=gender;
   title 'Charts for Fitness Program';
run;

proc sort data=bodyfat;
   by gender;
run;

proc boxplot data=bodyfat;
   plot fatpct*gender;
title 'Comparative Box Plots of Groups';
run;

proc ttest data=bodyfat;
   class gender;
   var fatpct;
title 'Comparing Groups in Fitness Program';
run;
```

```
ods select conflimits;
proc ttest data=bodyfat alpha=0.10;
   class gender;
   var fatpct;
run;

data gastric;
   input group $ lysolevl @@;
   datalines;
U 0.2 U 10.4 U 0.3 U 10.9 U 0.4 U 11.3 U 1.1 U 12.4 U 2.0
U 16.2 U 2.1 U 17.6 U 3.3 U 18.9 U 3.8 U 20.7 U 4.5
U 24.0 U 4.8 U 25.4 U 4.9 U 40.0 U 5.0 U 42.2 U 5.3
U 50.0 U 7.5 U 60.0 U 9.8
N 0.2 N 5.4 N 0.3 N 5.7 N 0.4 N 5.8 N 0.7 N 7.5 N 1.2 N 8.7
N 1.5 N 8.8 N 1.5 N 9.1 N 1.9 N 10.3 N 2.0 N 15.6 N 2.4
N 16.1 N 2.5 N 16.5 N 2.8 N 16.7 N 3.6 N 20.0
N 4.8 N 20.7 N 4.8 N 33.0
;
run;

proc npar1way data=gastric wilcoxon;
   class group;
   var lysolevl;
   title 'Comparison of Ulcer and Control Patients';
run;
```

Chapter 9

Comparing More Than Two Groups

Suppose you are a greenhouse manager and you want to compare the effects of six different fertilizers on the growth of geraniums. You have 30 plants, and you randomly assigned five plants to each of the six fertilizer groups. You carefully controlled other factors: you used the same type of soil for all 30 plants, you made sure that they all received the same number of hours of light each day, and you applied the fertilizers on the same day and in the same amounts to all 30 plants. At the end of this designed experiment, you measured the height of each plant. Now, you want to compare the six fertilizer groups in terms of plant height. This chapter discusses the analysis of this type of situation. This chapter discusses the following topics:

- summarizing data from more than two groups

- building a statistical test of hypothesis to compare data in several groups

- performing a one-way analysis of variance

- performing a Kruskal-Wallis test

- exploring group differences with multiple comparison procedures

The methods that are discussed are appropriate when each group consists of measurements of different items. The methods are not appropriate when "groups" consist of measurements of the same item at different times (for example, if you measured the same 15 plants at two, four, and six weeks after applying fertilizer). (This is a repeated measures design. See Appendix 1, "Further Reading," for information.)

This chapter discusses analysis methods for an ordinal or continuous variable as the measurement variable. The variable that classifies the data into groups should be nominal and can be either character or numeric.

Summarizing Data from Multiple Groups

This section explains how to summarize data from multiple groups. These methods do not replace formal statistical tests, but they do provide a better understanding of the data and can help you understand the results from formal statistical tests.

Before you can summarize data, you need data to be in a SAS data set. Then, you check the data set for errors. For a small data set, you can print the data and compare the printout with the data that you collected. For a larger data set, you can use the methods discussed in Chapter 4. You can use the same SAS procedures to check for errors and to summarize the data.

Suppose you have data from an investigation of teachers' salaries for different subject areas. You want to know whether the teachers' salaries are significantly different for different subject areas. Table 9.1 shows the data.[1]

---

[1] Data is from Dr. Ramon Littell, University of Florida. Used with permission.

Table 9.1 Teacher Salaries

Subject Area	Annual Salary					
Special Ed	35584	41400	42360	31319	57880	39697
	43128	42222	39676	41899	45773	53813
	51096	44625	35762	35083	52616	31114
	36844					
Mathematics	27814	25470	34432	45480	25358	29360
	29610	25000	29363	25091	55600	31624
	47770	31908	33000	26355	39201	45733
	32000	37120	62655	36733	28521	28709
	26674	48836	25096	27038	27197	51655
	25125	27829	28935	31124	37323	32960
	26428	39908	34692	23663	32188	45268
	33957	34055	53282	43890	26000	27107
	31615	24032	56070	24530	40174	24305
	35560	35955				
Language	26162	23963	27403	30180	32134	29594
	33472	28535	34609	49100	44207	47902
	26705	44888	29969	27599	28662	38948
	27664	29612	47316	27556	35465	33042
	38250	30171	32022	33884	36980	30230
	33618	29485	31006	48411	33058	49881
	42485	26966	27878	27607	30665	34001
Music	21827	35787	30043	27847	24150	29954
	21635	46691	24895	22515	27827	24712
	46001	25666	27178	41161	23092	24720
	44444	28004	32040	26417	41220	22148
	38914	28770				
Science	44324	43075	30000	24532	34930	25276
	39784	32576	40330	59910	47475	
Social Science	42210	36133	49683	31993	38728	46969
	59704					

This data is available in the **salary** data set in the sample data for the book. The statements in the "Example" section at the end of the chapter create a SAS data set for the teacher salary data in Table 9.1.

To summarize the data in SAS, you can use the same methods that you use for two independent groups.

- **PROC MEANS** with a **CLASS** statement for a concise summary.

- **PROC UNIVARIATE** with a CLASS statement for a detailed summary. This summary contains output tables for each value of the *class-variable*. You can use the **PLOT** option to create line printer plots for each value of the *class-variable*.

- PROC UNIVARIATE with a CLASS statement and **HISTOGRAM** statement for comparative histograms.

- **PROC CHART** for side-by-side bar charts. These bar charts are a low-resolution alternative to comparative histograms.

- **PROC BOXPLOT** for side-by-side box plots.

Following the same approach as you do for two groups, use PROC MEANS. The SAS statements below produce a concise summary of the data:

```
proc means data=salary;
   class subjarea;
   var annsal;
title 'Brief Summary of Teacher Salaries';
run;
```

Figure 9.1 shows the results.

Figure 9.1 Concise Summary for Multiple Groups

```
                    Brief Summary of Teacher Salaries

                          The MEANS Procedure

                   Analysis Variable : annsal Annual Salary

                   N
Subject Area      Obs      N        Mean       Std Dev      Minimum       Maximum
Language           42      42    33840.12      7194.69     23963.00      49881.00
Mathematics        56      56    34221.04      9517.30     23663.00      62655.00
Music              26      26    30294.54      7953.11     21635.00      46691.00
Science            11      11    38382.91     10434.09     24532.00      59910.00
Social Science      7       7    43631.43      9343.73     31993.00      59704.00
Special Ed         19      19    42204.79      7480.01     31114.00      57880.00
```

Figure 9.1 shows descriptive statistics for each group. **Social Science** teachers have the highest average annual salary, and **Music** teachers have the lowest average annual salary. The sizes of the groups vary widely, from only seven **Social Science** teachers to 56 **Mathematics** teachers.

Looking at the averages and standard deviations, you might think that there are salary differences for the different subject areas. To find out whether these differences are greater than what could have happened by chance requires a statistical test.

Creating Comparative Histograms for Multiple Groups

Chapter 8 discussed comparative histograms for two independent groups. You can use the same approach for multiple groups. PROC UNIVARIATE automatically creates two rows of histograms on a single page. This is ideal for two independent groups, but PROC UNIVARIATE prints on multiple pages when there are more than two groups. For multiple groups, you can use the **NROWS=** option to display all of the groups on a single page. For the teacher salary data, type the following:

```
proc univariate data=salary noprint;
   class subjarea;
   var annsal;
   histogram annsal / nrows=6;
title 'Comparative Histograms by Subject Area';
run;
```

The statements use the **NOPRINT** option to suppress the display of the usual output tables. The NROWS=6 option specifies six rows of histograms, one row for each of the six subject areas. Figure 9.2 shows the results.

The histograms show the different distributions of salaries for the subject areas. Although there might be salary differences in the subject areas, you don't know whether these differences are greater than what could have happened by chance.

The histograms also highlight potential outliers to investigate. For this data, the **Mathematics** group shows a bar separated from the other bars. If you investigate, you find that this is a valid data value for the teacher earning $62,655.

Figure 9.2 Comparative Histograms for Multiple Groups

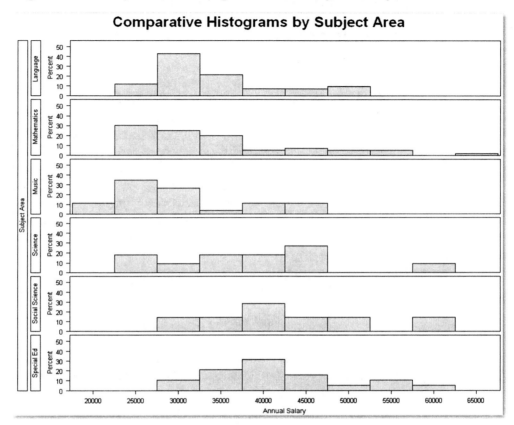

The general form of the statements to create comparative histograms for multiple groups using PROC UNIVARIATE is shown below:

PROC UNIVARIATE DATA=*data-set-name* **NOPRINT;**

  CLASS *class-variable*;

  VAR *measurement-variables*;

   HISTOGRAM *measurement-variables* **/ NROWS=***value*;

data-set-name is the name of a SAS data set, *class-variable* is a variable that classifies the data into groups, and *measurement-variables* are the variables that you want to summarize.

To compare groups, the CLASS statement is required. The *class-variable* can be character or numeric and should be nominal. If you omit the *measurement-variables* from the VAR statement, then the procedure uses all numeric variables.

You can use the options described in Chapters 4 and 5 with PROC UNIVARIATE and a CLASS statement. You can omit the NOPRINT option and create the detailed summary and histograms at the same time.

The NROWS= option controls the number of histograms that appear on a single page. SAS automatically uses NROWS=2. For your data, choose a *value* equal to the number of groups.

If any option is used in the HISTOGRAM statement, then the slash is required.

Creating Side-by-Side Box Plots for Multiple Groups

Chapter 8 discussed side-by-side box plots for two independent groups. You can use the same approach for multiple groups. When group sizes vary widely, the **BOXWIDTHSCALE=1** and **BWSLEGEND** options can be helpful. For the teacher salary data, type the following:

```
proc sort data=salary;
   by subjarea;
run;

proc boxplot data=salary;
   plot annsal*subjarea / boxwidthscale=1 bwslegend;
title 'Comparative Box Plots by Subject Area';
run;
```

PROC BOXPLOT expects the data to be sorted by the variable that classifies the data into groups. If your data is already sorted, you can omit the **PROC SORT** step.

The BOXWIDTHSCALE=1 option scales the width of the boxes based on the size of each group. The BWSLEGEND option adds a legend to the bottom of the comparative box plot. The legend explains the varying box widths. Figure 9.3 shows the results.

Figure 9.3 Side-by-Side Box Plots for Multiple Groups

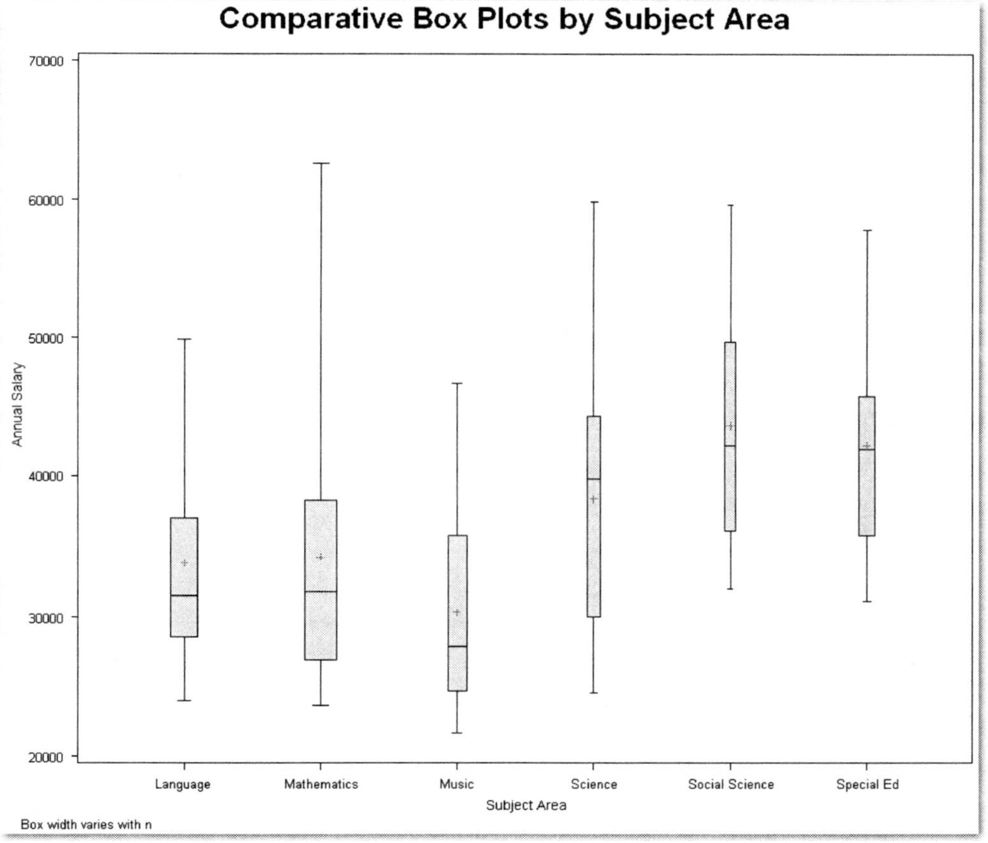

The box plots help you visualize the differences between the averages for the six subject areas. Looking at the median (the center line of the box), you can observe differences in the subject areas. But, you don't know whether these differences are greater than what could have happened by chance. To find out requires a statistical test.

Using BOXWIDTHSCALE=1 scales the widths of the boxes based on the sample size for each group. You can see the small sizes for the Science, Social Science, and Special Ed subject areas.

Looking at the box plots highlights the wide range of points for several subject areas. When checking the data for errors, you investigate the extreme values (using PROC UNIVARIATE) and compare them with the original data. You find that all of the data values are valid.

The general form of the statements to create side-by-side box plots for multiple groups using PROC BOXPLOT is shown below:

PROC BOXPLOT DATA=*data-set-name***;**

    PLOT *measurement-variable****class-variable**
        / BOXWIDTHSCALE=1 BWSLEGEND;

Items in italic were defined earlier in the chapter. Depending on your data, you might need to first sort by the *class-variable* before creating the plots. The asterisk in the PLOT statement is required.

The **BOXWIDTHSCALE=1** option scales the width of the boxes based on the sample size of each group. The **BWSLEGEND** option adds a legend to the bottom of the comparative box plot. The legend explains the varying box widths. If either option is used, then the slash is required. You can use the options alone or together, as shown above.

Building Hypothesis Tests to Compare More Than Two Groups

So far, this chapter has discussed how to summarize differences between groups. When you start a study, you want to know how important the differences between groups are, and whether the differences are large enough to be statistically significant. In statistical terms, you want to perform a hypothesis test. This section discusses hypothesis testing when comparing multiple groups. (See Chapter 5 for the general idea of hypothesis testing.)

When comparing multiple groups, the null hypothesis is that the means for the groups are the same, and the alternative hypothesis is that the means are different. The **salary** data has six groups. In this case, the notation is the following:

$$H_o: \quad \mu_A = \mu_B = \mu_C = \mu_D = \mu_E = \mu_F$$

$H_a:$ at least two means are different

H_o indicates the null hypothesis, H_a indicates the alternative hypothesis, and μ_A, μ_B, μ_C, μ_D, μ_E, and μ_F are the population means for the six groups. The alternative hypothesis does not specify which means are different from one another. It specifies simply that there are some differences among the means. The preceding notation shows the hypotheses for comparing six groups. For more groups or fewer groups, add or delete the appropriate number of means in the notation.

In statistical tests that compare multiple groups, the hypotheses are tested by partitioning the total variation in the data into variation due to differences between groups, and variation due to error. The error variation does not refer to mistakes in the data. It refers to the natural variation within a group (and possibly to the variation due to other factors that were not considered). The error variation is sometimes called the *within group* variation. The error variation represents the natural variation that would be expected by chance. If the variation between groups is large relative to the error variation, the means are likely to be different.

Because the hypothesis test analyzes the variation in the data, it is called an *analysis of variance* and is abbreviated as *ANOVA*. The specific case of analyzing only the variation between groups and the variation due to error is called a *one-way ANOVA*. Figure 9.4 shows how the total variation is partitioned into variation between groups and variation due to error.

Figure 9.4 How a One-Way ANOVA Partitions Variation

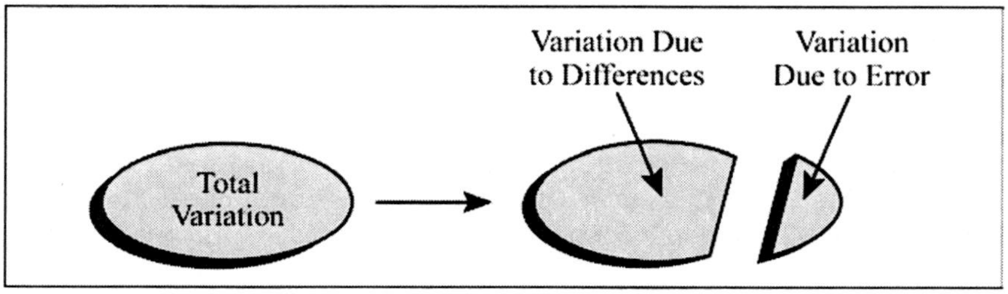

Another way to think about this idea is to use the following model:

variation in the *measurement-variable=*

(variation between levels of the *class-variable*)+

(variation within levels of the *class-variable*)

Abbreviated for the salary example, this model is the following:

annsal = subjarea + error

For a one-way ANOVA, the only term in the model is the term for the *class-variable*, which identifies the groups.

Using Parametric and Nonparametric Tests

In this book, the term "ANOVA" refers specifically to a parametric analysis of variance, which can be used if the assumptions seem reasonable. If the assumptions do not seem reasonable, use the nonparametric Kruskal-Wallis test, which has fewer assumptions. In general, if the assumptions for the parametric ANOVA seem reasonable, use the ANOVA. Although nonparametric tests like the Kruskal-Wallis test have fewer assumptions, these tests typically are less powerful in detecting differences between groups.

Balanced and Unbalanced Data

If all of the groups have the same sample size, the data is *balanced*. If the sample sizes are different, the data is *unbalanced*.

This distinction can be important for follow-up tests to an ANOVA, or for other types of experimental designs. However, for a one-way ANOVA or a Kruskal-Wallis test, you don't need to worry about whether the data is balanced. SAS automatically handles balanced and unbalanced data correctly in these analyses.

Understanding Significance

In either an ANOVA or a Kruskal-Wallis test, there are two possible results: the p-value is less than the reference probability, or it is not. (See Chapter 5 for discussions on significance levels and on understanding practical significance and statistical significance.)

Groups Significantly Different

If the p-value is less than the reference probability, the result is statistically significant, and you reject the null hypothesis. You conclude that the means for the groups are significantly different. You don't know which means are different from one another; you know simply that some differences exist in the means.

Your next step might be to use a *multiple comparison procedure* to further analyze the differences between groups. An alternative to using a multiple comparison procedure is performing comparisons that you planned before you collected data. If you have certain comparisons that you know you want to make, consult a statistician for help in designing your study.

Groups Not Significantly Different

If the *p*-value is greater than the reference probability, the result is not statistically significant, and you fail to reject the null hypothesis. You conclude that the means for the groups are not significantly different.

Do not "accept the null hypothesis" and conclude that the means for the groups are the same.

This distinction is very important. You do not have enough evidence to conclude that the means are the same. The results of the test indicate only that the means are not significantly different. You never have enough evidence to conclude that the population is exactly as described by the null hypothesis. In statistical hypothesis testing, do not accept the null hypothesis. Instead, either reject or fail to reject the null hypothesis.

A larger sample might have led you to conclude that the groups are different. Perhaps there are other factors that were not controlled, and these factors influenced the experiment enough to obscure the differences between the groups. For a more detailed discussion of hypothesis testing, see Chapter 5.

Performing a One-Way ANOVA

This section discusses the assumptions for a one-way ANOVA, how to test the assumptions, how to use SAS, and how to interpret the results. The steps for performing the analyses to compare multiple groups are the same as the steps for comparing two groups:

1. Create a SAS data set.

2. Check the data set for errors.

3. Choose the significance level for the test.

4. Check the assumptions for the test.

5. Perform the test.

6. Make conclusions from the test results.

Steps 1 and 2 have been done. For step 3, compare the groups at the 10% significance level. For significance, the *p*-value needs to be less than the reference probability value of 0.10.

Understanding Assumptions

The assumptions for an analysis of variance are the following:

- Observations are independent. The measurement of one observation cannot affect the measurement of another observation.

- Observations are a random sample from a population with a normal distribution. If there are differences between groups, there might be a different normal distribution for each group. (The assumption of normality requires that the measurement variable is continuous. If your measurement variable is ordinal, go to the section on the Kruskal-Wallis test later in this chapter.)

- Groups have equal variances.

The assumption of independent observations seems reasonable because the annual salary for one teacher is unrelated to and unaffected by the annual salary for another teacher.

The assumption of normality is difficult to verify. Testing this assumption requires performing a test of normality on each group. For the subject areas that have a larger number of observations, this test could be meaningful. But, for the subject areas that have only a few observations, you don't have enough data to perform an adequate test. And, because you need to verify the assumption of normality in each group, this is difficult. For now, assume normality, and perform an analysis of variance.

In general, if you have additional information (such as other data) that indicates that the population is normally distributed, and if your data is a representative sample of the population, you can usually proceed based on the idea that the assumption of normality seems reasonable. In addition, the analysis of variance works well even for data from non-normal populations. However, if you are uncomfortable with proceeding, or if you do not have other data or additional information about the distribution of measurements, or if you have large groups that don't seem to be samples from normally distributed populations, you can use the Kruskal-Wallis test.

The assumption of equal variances seems reasonable. Look at the similarity of the standard deviations in Figure 9.1, and the relatively similar heights of the box plots in

Figure 9.3. The standard deviations vary, but it is difficult to determine whether the variability is because of true changes in the variances or because of sample sizes. Look at the differences in sample sizes for the different subject areas. There are 7 observations for **Social Science**, and 56 observations for **Mathematics**. The standard deviations for these subject areas are relatively similar. However, compare the standard deviations for **Social Science** and **Special Ed**. It's more difficult to determine whether the difference in standard deviations (which is larger) is because of true changes in the variances or because of less precise estimates of variances as a result of smaller sample sizes. For now, assume equal variances, and perform an analysis of variance. See "Analysis of Variance with Unequal Variances" later in the chapter for an alternative approach.

Performing the Analysis of Variance

SAS includes many procedures that perform an analysis of variance in different situations. **PROC ANOVA** is the simplest of these procedures and assumes that the data is balanced. For balanced data, PROC ANOVA is more efficient (runs faster) than the other SAS procedures for analysis of variance.

For the special case of a one-way ANOVA, you can use PROC ANOVA for both balanced and unbalanced data. In general, however, you should use **PROC GLM** for unbalanced data.

To perform a one-way ANOVA for the **salary** data, type the following:

```
proc anova data=salary;
   class subjarea;
   model annsal=subjarea;
title 'ANOVA for Teacher Salaries';
run;
```

The **CLASS** statement identifies the variable that classifies the data into groups. The **MODEL** statement describes the relationship that you want to investigate. In this case, you want to find out how much of the variation in the salaries (**annsal**) is due to differences between the subject areas (**subjarea**). Figure 9.5 shows the results.

Figure 9.5 Analysis of Variance for Salary Data

```
                       ANOVA for Teacher Salaries

                        The ANOVA Procedure

                        Class Level Information

Class       Levels  Values

subjarea       6    Language Mathematics Music Science Social Science Special Ed


              Number of Observations Read        161
              Number of Observations Used        161
```

Figure 9.5 Analysis of Variance for Salary Data (continued)

```
                        ANOVA for Teacher Salaries

                          The ANOVA Procedure

Dependent Variable: annsal     Annual Salary

                                    Sum of
Source                    DF       Squares     Mean Square    F Value    Pr > F

Model                      5     2297379854      459475971       6.30    <.0001

Error                    155    11305093093       72936084

Corrected Total          160    13602472947


            R-Square      Coeff Var       Root MSE      annsal Mean

            0.168894      24.31512       8540.263         35123.25


Source                    DF      Anova SS     Mean Square    F Value    Pr > F

subjarea                   5     2297379854      459475971       6.30    <.0001
```

Understanding Results

Figure 9.5 shows the details of the one-way ANOVA. First, answer the research question, "Are the mean salaries different for subject areas?"

Finding the *p*-value

To find the *p*-value in the salary data, look in Figure 9.5 at the line listed in the **Source** column labeled **subjarea**. Then look at the column labeled **Pr > F**. The number in this column is the *p*-value for comparing groups. Figure 9.5 shows a value of **<.0001**. Because 0.0001 is less than the reference probability of 0.10, you conclude that the mean salaries for subject areas are significantly different.

To find the *p*-value in your data, find the **Source** line labeled with the name of the variable in the CLASS statement. The *p*-value is listed in the **Pr > F** column. It gives you the results of the hypothesis test for comparing the means of multiple groups.

In general, to perform the test at the 5% significance level, you conclude that the means for the groups are significantly different if the *p*-value is less than 0.05. To perform the test at the 10% significance level, you conclude that the means for the groups are significantly different if the *p*-value is less than 0.10. Conclusions for tests at other significance levels are made in a similar manner.

Understanding the First Page of Output

The first page of output contains two tables. The **Class Level** table identifies the *class-variable*, number of levels for this variable, and values for this variable. The **Number of Observations** table identifies the number of observations read from the data set, and the number of observations that PROC ANOVA used in analysis. For data for a one-way ANOVA with no missing values, these two numbers will be the same.

Understanding the Second Page of Output

The first line of the second page of output identifies the *measurement-variable* with the label **Dependent Variable**. For the **salary** data, the variable is **annsal**. Because the variable has a label in the data set, PROC ANOVA also displays the label. For your particular data, the procedure will display the *measurement-variable* name and might display the label.

After this first heading, the second page of output contains three tables. The **Overall ANOVA** table shows the analysis with terms for **Model** and **Error**. The **Fit Statistics** table contains several statistics. The **Model ANOVA** table contains the one-way ANOVA and the *p*-value.

For a one-way ANOVA, the only term in the model is *class-variable*, which identifies the variable that classifies the data into groups. As a result, for a one-way ANOVA, the *p*-values in the Overall ANOVA table and in the Model ANOVA table are the same.

The list below describes items in the Overall ANOVA table:

Source	lists the sources of variation in the data. For a one-way ANOVA, this table summarizes all sources of variation. Because the model has only one source of variation (the groups), the values in the line for **Model** in this table match the corresponding values in the Model ANOVA table.
DF	is degrees of freedom. For a one-way ANOVA, the DF for **Corrected Total** is the number of observations minus 1 (160=161−1). The DF for **Model** is the number of groups minus 1. For the **salary** data, 5=6−1. For your data, the DF for **Model** will be one less than the number of groups. The DF for **Error** is the difference between these two values (155=160−5).

Sum of Squares   Measures the amount of variation that is attributed to a given source. In this table, the **Sum of Squares** measures the amount of variation for the entire model. **Sum of Squares** is often abbreviated as **SS**. For the one-way ANOVA, the **Model SS** and the **subjarea SS** are the same. This is not true for more complicated models.

The **Sum of Squares** for the **Model** and **Error** sum to be equal to the **Corrected Total Sum of Squares**.

Mean Square   is the **Sum of Squares** divided by DF. SAS uses **Mean Square** to construct the statistical test for the model.

F Value   is the value of the test statistic. The **F Value** is the **Mean Square** for **Model** divided by the **Mean Square** for **Error**.

Pr > F   gives the *p*-value for the test for the overall model. Because the model for an ANOVA contains only one term, this *p*-value is the same as the *p*-value to compare groups.

The list below describes items in the **Fit Statistics** table:

R-Square   for a one-way analysis of variance, describes how much of the variation in the data is due to differences between groups. This number ranges from 0 to 1; the closer it is to 1, the more variation in the data is due to differences between groups.

Coeff Var   gives the coefficient of variation, which is calculated by multiplying the standard deviation by 100, and then dividing this value by the average.

Root MSE   gives an estimate of the standard deviation after accounting for differences between groups. This is the square root of the **Mean Square** for **Error**.

annsal Mean   gives the overall average for the *measurement-variable*. For your data, the value will list the name of your *measurement-variable*.

The list below describes items in the **Model ANOVA** table:

Source   lists the sources of variation in the data. For a one-way ANOVA, this table identifies the variable that classifies the data into groups (**subjarea** in this example). Because the model has only one source of variation (the groups), the values for **subjarea** in this table match the corresponding values in the **Overall ANOVA** table.

DF   is degrees of freedom for the variable that classifies the data into groups (**subjarea** in this example). The value should be the same as DF for **Model** because the model has only one source of variation (the groups).

Anova SS   in this table, the **Anova SS** measures the amount of variation that is attributed to the variable that classifies the data into groups (**subjarea** in this example).

Mean Square   is the **Anova SS** divided by DF. SAS uses **Mean Square** to construct the statistical test to compare groups.

F Value   is the value of the test statistic. This is the **Mean Square** for the variable that classifies the data into groups (**subjarea** in this example) divided by the **Mean Square** for Error. An analysis of variance involves partitioning the variation and testing to find out how much variation is because of differences between groups. Partitioning is done by the *F* test, which involves calculating an *F*-value from the mean squares.

Pr > F   gives the *p*-value for the test to compare groups. Compare this value with the reference probability value that you selected before you ran the test.

Analysis of Variance with Unequal Variances

Earlier, this chapter discussed the assumption of equal variances, and an ANOVA was performed after looking at standard deviations and box plots. Some statisticians perform an ANOVA without testing for equal variances. Unless the variances in the groups are very different, or unless there are many groups, an ANOVA detects appropriate differences when all of the groups are about the same size.

SAS provides tests for unequal variances, and provides an alternative to the usual ANOVA when the assumption of equal variances does not seem reasonable.

Testing for Equal Variances

For the **salary** data, suppose you want to test for equal variances at the 10% significance level. This decision gives a reference probability of 0.10. To test the assumption of equal variances in SAS, type the following:

```
proc anova data=salary;
   class subjarea;
   model annsal=subjarea;
   means subjarea / hovtest;
title 'Testing for Equal Variances';
run;
```

These statements produce the results shown in Figure 9.5. The **MEANS** statement is new and produces the results shown in Figures 9.6 and 9.7.

The **HOVTEST** option tests for homogeneity of variances, which is another way of saying equal variances. SAS automatically performs Levene's test for equal variances. Levene's test has a null hypothesis that the variances are equal, and an alternative hypothesis that they are not equal. This test is most widely used by statisticians and is considered to be the standard test. SAS provides options for other tests, which are discussed in the SAS documentation.

Figure 9.6 shows descriptive statistics for each of the groups.

Figure 9.6 Descriptive Statistics from the MEANS Statement

```
                    Testing for Equal Variances

                      The ANOVA Procedure

Level of                      ------------annsal-----------
subjarea              N              Mean              Std Dev

Language             42         33840.1190           7194.6884
Mathematics          56         34221.0357           9517.2981
Music                26         30294.5385           7953.1143
Science              11         38382.9091          10434.0929
Social Science        7         43631.4286           9343.7321
Special Ed           19         42204.7895           7480.0078
```

Figure 9.7 Tests for Equal Variances for Salary Data

```
                       Testing for Equal Variances

                          The ANOVA Procedure

           Levene's Test for Homogeneity of annsal Variance
               ANOVA of Squared Deviations from Group Means

                              Sum of        Mean
      Source          DF      Squares       Square     F Value    Pr > F

      subjarea         5      5.312E16      1.062E16      0.93     0.4647
      Error          155      1.775E18      1.145E16
```

To find the *p*-value, find the line labeled **subjarea** and look at the column labeled
Pr > F. The number in this column is the *p*-value for testing the assumption of equal
variances. Because the *p*-value of **0.4647** is much greater than the significance level of
0.10, do not reject the null hypothesis of equal variances for the groups. In fact, there is
almost a 47% chance that this result occurred by random chance. You conclude that there
is not enough evidence that the variances in annual salaries for the six subject areas are
significantly different. You proceed, based on the assumption that the variances for the
groups are equal.

In general, compare the *p*-value to the reference probability. If the *p*-value is greater than
the reference probability, proceed with the ANOVA based on the assumption that the
variances for the groups are equal. If the *p*-value is less than the reference probability, use
the Welch ANOVA (discussed in the next topic) to compare the means of the groups.

This table also contains other details for Levene's test, including the degrees of freedom,
sum of squares, mean square, and *F*-value. See the SAS documentation for more
information about Levene's test.

Performing the Welch ANOVA

Welch's ANOVA tests for differences in the means of the groups, while allowing for
unequal variances across groups. To perform Welch's ANOVA in SAS, type the following:

```
proc anova data=salary;
   class subjarea;
   model annsal=subjarea;
   means subjarea / welch;
title 'Welch ANOVA for Salary Data';
run;
```

These statements produce the results shown in Figures 9.5 and 9.6. The **WELCH** option in the **MEANS** statement performs the Welch's ANOVA and produces the output in Figure 9.8.

Figure 9.8 Welch ANOVA for Salary Data

```
            Welch ANOVA for Salary Data

                The ANOVA Procedure

            Welch's ANOVA for annsal

   Source              DF     F Value    Pr > F

   subjarea         5.0000       6.56     0.0002
   Error           34.7639
```

Find the column labeled **Pr > F** under the **Welch's ANOVA** heading. The number in this column is the *p*-value for testing for differences in the means of the groups. For the **salary** data, this value is **0.0002**. You conclude that the mean salaries are significantly different across subject areas. For the **salary** data, this is for example only. You already know that the assumption of equal variances seems reasonable and that you can apply the usual ANOVA test.

The **F Value** gives the value of the test statistic. **DF** gives the degrees of freedom for the **Source**. For the *class-variable* (**subjarea** in Figure 9.8), the degrees of freedom are calculated as the number of groups minus 1. The degrees of freedom for **Error** has a more complicated formula. See the SAS documentation for formulas.

If Welch's ANOVA test is used, then multiple comparison procedures (discussed later in the chapter) are not appropriate.

Summarizing PROC ANOVA

The general form of the statements to perform a one-way ANOVA, test the assumption of equal variances, and perform Welch's ANOVA using PROC ANOVA is shown below:

PROC ANOVA DATA=*data-set-name*;

  CLASS *class-variable*;

  MODEL *measurement-variable=class-variable*;

  MEANS *class-variable* **/ HOVTEST WELCH**;

Items in italic were defined earlier in the chapter.

The **PROC ANOVA, CLASS**, and **MODEL** statements are required.

The **HOVTEST** option performs a test for equal variances across the groups. The **WELCH** option performs the Welch ANOVA when there are unequal variances. If either option is used, then the slash is required. You can use the options alone or together, as shown above.

You can use the **BY** statement in PROC ANOVA to perform several one-way ANOVAs for levels of another variable. However, another statistical analysis might be more appropriate. Consult a statistical text or a statistician before you do this analysis.

As discussed in Chapter 4, you can use the ODS statement to choose the output tables to display. Table 9.2 identifies the output tables for PROC ANOVA.

Table 9.2 ODS Table Names for PROC ANOVA

ODS Name	Output Table
ClassLevels	Classification levels and values, on first page
Nobs	Number of observations read and used, on first page
OverallANOVA	Shows the analysis for **Model** and **Error**, first table on the second page
FitStatistics	Shows R-Square and other statistics, second table on the second page
ModelANOVA	Shows the *p*-value for the one-way ANOVA, third table on the second page
Means	Descriptive statistics created by the **MEANS** statement
HOVFTest	Results of Levene's Test for homogeneity of variances created by the **HOVTEST** option in the **MEANS** statement
WELCH	Welch's ANOVA created by the **WELCH** option in the **MEANS** statement

Performing a Kruskal-Wallis Test

The Kruskal-Wallis test is a nonparametric analogue to the one-way ANOVA. Use this test when the normality assumption does not seem reasonable for your data. If you do have normally distributed data, this test can be conservative and can fail to detect differences between groups. The steps of the analysis are the same as for an ANOVA.

In practice, the Kruskal-Wallis test is used for ordinal or continuous variables. The null hypothesis is that the populations for the groups are the same.

You have already performed the first three steps of the analysis. As for the ANOVA, use a 10% significance level to test for differences between groups.

Assumptions

To perform the fourth step for analysis, check the assumptions for the test.

The only assumption for this test is that the observations are independent.

For the **salary** data, this assumption seems reasonable because the salary for one teacher is independent of the salary for another teacher.

Using PROC NPAR1WAY

To perform a Kruskal-Wallis test with SAS, use **PROC NPAR1WAY**:

```
proc npar1way data=salary wilcoxon;
   class subjarea;
   var annsal;
title 'Nonparametric Tests for Teacher Salary Data';
run;
```

The **CLASS** statement identifies the variable that classifies the data into groups. The **VAR** statement identifies the variable that you want to analyze. Figure 9.9 shows the results.

Figure 9.9 shows two output tables. The first table contains Wilcoxon scores, and has the same information that Chapter 8 discussed for the Wilcoxon Rank Sum test. (In Chapter 8, see "Using PROC NPAR1WAY for the Wilcoxon Rank Sum Test" for details.) You can think of the Kruskal-Wallis test as an extension of the Wilcoxon Rank Sum test for more than two groups. The second table contains the results of the Kruskal-Wallis test.

Figure 9.9 Kruskal-Wallis Test for Salary Data

```
                  Nonparametric Tests for Teacher Salaries

                        The NPAR1WAY Procedure

             Wilcoxon Scores (Rank Sums) for Variable annsal
                     Classified by Variable subjarea

                          Sum of      Expected      Std Dev        Mean
   subjarea          N    Scores      Under H0      Under H0       Score

   Language         42    3280.0       3402.0     259.755654    78.095238
   Mathematics      56    4183.0       4536.0     281.744565    74.696429
   Music            26    1391.0       2106.0     217.680959    53.500000
   Science          11    1061.0        891.0     149.248116    96.454545
   Social Science    7     857.0        567.0     120.635816   122.428571
   Special Ed       19    2269.0       1539.0     190.848107   119.421053


                        Kruskal-Wallis Test

                   Chi-Square         29.8739
                   DF                       5
                   Pr > Chi-Square     <.0001
```

Understanding Results

First, answer the research question, "Are the mean salaries different for subject areas?"

Finding the *p*-value

Find the line labeled **Pr > Chi-Square** in Figure 9.9. The number in this line is the *p*-value for comparing groups. Figure 9.9 shows a value of **<.0001**. Because 0.0001 is less than the reference probability of 0.10, you can conclude that the mean salaries are significantly different across subject areas.

In general, to interpret results, look at the *p*-value in **Pr > Chi-Square**. If the *p*-value is less than the significance level, you conclude that the means for the groups are significantly different. If the *p*-value is greater, you conclude that the means for the groups are not significantly different. Do not conclude that the means for the groups are the same.

Understanding Other Items in the Tables

See "Using PROC NPAR1WAY for the Wilcoxon Rank Sum Test" in Chapter 8 for details on the items in the **Wilcoxon Scores** table.

The **Kruskal-Wallis Test** table displays the statistic for the test (**Chi-Square**) and the degrees of freedom for the test (**DF**). The degrees of freedom are calculated by subtracting 1 from the number of groups. For the **salary** data, this is 5=6–1.

Summarizing PROC NPAR1WAY

The general form of the statements to perform tests to compare multiple groups using PROC NPAR1WAY is shown below:

PROC NPAR1WAY DATA=*data-set-name* **WILCOXON;**

   CLASS *class-variable*;

   VAR *measurement-variables*;

Items in italic were defined earlier in the chapter.

The **CLASS** statement is required.

While the VAR statement is not required, you should use it. If you omit the VAR statement, then PROC NPAR1WAY performs analyses for every numeric variable in the data set.

You can use the BY statement in PROC NPAR1WAY to perform several analyses for levels of another variable. However, another statistical analysis might be more appropriate. Consult a statistical text or a statistician before you do this analysis.

As discussed in Chapter 4, you can use the ODS statement to choose the output tables to display. Table 9.3 identifies the output tables for PROC NPAR1WAY and the Kruskal-Wallis test.

Table 9.3 ODS Table Names for PROC NPAR1WAY

ODS Name	Output Table
WilcoxonScores	Wilcoxon Scores
KruskalWallisTest	Kruskal-Wallis Test

Understanding Multiple Comparison Procedures

When you conclude that the means are significantly different after performing an ANOVA, you still don't know enough detail to take action. Which means differ from other means? At this point, you know only that some means differ from other means. Your next step is to use a *multiple comparison procedure*, or a test that makes two or more comparisons of the group means.

Many statisticians agree that multiple comparison procedures should be used only after a significant ANOVA test. Technically, this significant ANOVA test is called a "prior significant *F* test" because the ANOVA performs an *F* test.

To make things more complicated, statisticians disagree about which tests to use and which tests are best. This chapter discusses some of the tests available in SAS. If you need a specific test, then check SAS documentation because the test might be available in SAS.

Chapter 5 compared choosing an alpha level to choosing the level of risk of making a wrong decision. To use multiple comparison procedures, you consider risk again. Specifically, you decide whether you want to control the risk of making a wrong decision (deciding that means are different when they are not) for all comparisons overall, or if you want to control the risk of making a wrong decision for each individual comparison.

To make this decision process a little clearer, think about the **salary** data. There are 6 subject areas, so there are 15 combinations of 2 means to compare. Do you want to control the risk of making a wrong decision (deciding that the 2 means are different when they are not) for all 15 comparisons at once? If you do, you are controlling the risk for the experiment overall, or the *experimentwise error rate*. Some statisticians call this the overall error rate. On the other hand, do you want to control the risk of making a wrong decision for each of the 15 comparisons? If you do, you are controlling the risk for each comparison, or the *comparisonwise error rate*. SAS documentation uses the abbreviation CER for the comparisonwise error rate, and the abbreviation MEER for the maximum experimentwise error rate.

Performing Pairwise Comparisons with Multiple *t*-Tests

Assume that you have performed an ANOVA in SAS, and you have concluded that there are significant differences between the means. Now, you want to compare all pairs of means. You already know one way to control the comparisonwise error rate when comparing pairs of means. You simply use the two-sample *t*-test at the appropriate α-level. The test does not adjust for the fact that you might be making many comparisons. In other words, the test controls the comparisonwise error rate, but not the experimentwise error rate. See "Technical Details: Overall Risk" for more discussion.

To perform multiple comparison procedures in an analysis of variance, use the MEANS statement in PROC ANOVA, and add an option to specify the test you want performed. To perform multiple pairwise *t*-tests, use the **T** option:

```
proc anova data=salary;
   class subjarea;
   model annsal=subjarea;
   means subjarea / t;
title 'Multiple Comparisons with t Tests';
run;
```

These statements produce the results shown in Figures 9.5 and 9.6. Figure 9.10 shows the results from the **T** option for the multiple pairwise *t*-tests.

Deciding Which Means Differ

Figure 9.10 shows a 95% confidence interval for the difference between each pair of means. When the confidence interval for the difference encloses 0, the difference between the two means is not significant. A confidence limit that encloses 0 says that the difference in means for the two groups might be 0 so there cannot be a significant difference between the two groups. A confidence limit that doesn't enclose 0 says that the difference in means for the two groups is not 0 so there is a significant difference between the two groups.

SAS highlights comparisons that are significant with three asterisks (***) in the right-most column. From these results, you can conclude the following:

- The mean annual salaries for **Social Science** and **Special Ed** are significantly different from **Mathematics**, **Language**, and **Music**.

- The mean annual salary for **Science** is significantly different from **Music**.

- The mean annual salaries for other pairs of subject areas are not significantly different.

Books sometimes use connecting lines to summarize multiple comparisons. For the **salary** data, a connecting lines report for multiple pairwise *t*-tests would look like the following:

Social-Science	Special-Ed	Science	Math	Language	Music

Subject areas without connecting lines are significantly different. For example, **Social-Science** does not have any line connecting it to **Math**, so the two groups have significantly different mean annual salaries. In contrast, **Science** and **Math** do have a connecting line, so the two groups do not have significantly different mean annual salaries.

Figure 9.10 Results for T Option in MEANS Statement

```
                 Multiple Comparisons with t tests

                       The ANOVA Procedure

                     t Tests (LSD) for annsal

NOTE: This test controls the Type I comparisonwise error rate, not the
                    experimentwise error rate.


               Alpha                        0.05
               Error Degrees of Freedom      155
               Error Mean Square        72936084
               Critical Value of t       1.97539


       Comparisons significant at the 0.05 level are indicated by ***.


                                  Difference
                  subjarea        Between       95% Confidence
                  Comparison       Means           Limits

   Social Science - Special Ed        1427     -6032      8886
   Social Science - Science           5249     -2908     13405
   Social Science - Mathematics       9410      2647     16174   ***
   Social Science - Language          9791      2904     16679   ***
   Social Science - Music            13337      6153     20521   ***
   Special Ed     - Social Science   -1427     -8886      6032
   Special Ed     - Science           3822     -2570     10214
   Special Ed     - Mathematics       7984      3505     12463   ***
   Special Ed     - Language          8365      3700     13029   ***
   Special Ed     - Music            11910      6819     17002   ***
   Science        - Social Science   -5249    -13405      2908
   Science        - Special Ed       -3822    -10214      2570
   Science        - Mathematics       4162     -1402      9726
   Science        - Language          4543     -1171     10257
   Science        - Music             8088      2020     14156   ***
   Mathematics    - Social Science   -9410    -16174     -2647   ***
   Mathematics    - Special Ed       -7984    -12463     -3505   ***
   Mathematics    - Science          -4162     -9726      1402
   Mathematics    - Language           381     -3063      3825
   Mathematics    - Music             3926       -77      7930
   Language       - Social Science   -9791    -16679     -2904   ***
   Language       - Special Ed       -8365    -13029     -3700   ***
   Language       - Science          -4543    -10257      1171
   Language       - Mathematics       -381     -3825      3063
   Language       - Music             3546      -664      7755
   Music          - Social Science  -13337    -20521     -6153   ***
   Music          - Special Ed      -11910    -17002     -6819   ***
   Music          - Science          -8088    -14156     -2020   ***
   Music          - Mathematics      -3926     -7930        77
   Music          - Language         -3546     -7755       664
```

Figure 9.10 shows the **subjarea** comparisons twice. As an example, the comparison between science and special ed appears both as **Science - Special Ed** and as **Special Ed - Science**. The statistical significance is the same for both comparisons.

Figure 9.10 contains a note reminding you that this test controls the comparisonwise error rate. The output gives you additional information about the tests that were performed.

Understanding Other Items in the Report

Figure 9.10 contains two reports. The first report is the **Information** report, which contains the statistical details for the tests. The second report is the **Pairs** report, which contains the results of the pairwise comparisons.

The heading for both reports is **t Tests (LSD) for annsal**. For your data, the heading will contain the *measurement-variable* that you specify in the MODEL statement.

The LSD abbreviation in the heading appears because performing multiple *t*-tests when all group sizes are equal is the same as using a test known as Fisher's Least Significant Difference test. This test is usually referred to as Fisher's LSD test. You don't need to worry about whether your sample sizes are equal or not. If you request multiple pairwise *t*-tests, SAS performs Fisher's LSD test if it can be used.

The list below summarizes items in the **Information** report:

Alpha	is the α-level for the test. SAS automatically uses 0.05.
Error Degrees of Freedom	is the same as the degrees of freedom for **Mean Square** for **Error** in the ANOVA table.
Error Mean Square	is the same as the **Mean Square** for **Error** in the ANOVA table.
Critical Value of t	is the critical value for the tests. If the test statistic is greater than the critical value, the difference between the means of two groups is significant at the level given by **Alpha**.

The **Pairs** report also contains a column that identifies the comparison. For your data, the column heading will contain the *measurement-variable*. The report contains a column for **Difference Between Means**, which shows the difference between the two group averages in the row.

Technical Details: Overall Risk

When you perform multiple comparison procedures that control the comparisonwise error rate, you control the risk of making a wrong decision about each comparison. Here, the word "wrong" means concluding that groups are significantly different when they are not.

Controlling the comparisonwise error rate at 0.05 might give you a false sense of security. The more comparison-wise tests you perform, the higher your overall risk of making a wrong decision.

The **salary** data has 15 two-way comparisons for the 6 groups. Here is the overall risk of making a wrong decision:

$$1 - (0.95)^{15} = 0.537$$

This means that there is about a 54% chance of incorrectly concluding that two means are different in the **salary** data when the multiple pairwise *t*-tests are used. Most statisticians agree that this overall risk is too high to be acceptable.

In general, here is the overall risk:

$$1 - (0.95)^{m}$$

m is the number of pairwise comparisons to be performed. These formulas assume that you perform each test at the 95% confidence level, which controls the comparisonwise error rate at 0.05. You can apply this formula to your own data to help you understand the overall risk of making a wrong decision.

The Bonferroni approach and the Tukey-Kramer test are two solutions for controlling the experimentwise error rate.

Using the Bonferroni Approach

To control the experimentwise error rate with multiple pairwise *t*-tests, you can use the Bonferroni approach. With the Bonferroni approach, you decrease the alpha level for each test so that the overall risk of incorrectly concluding that the means are different is 5%. Basically, each test is performed using a very low alpha level. If you have many groups, the Bonferroni approach can be very conservative and can fail to detect significant differences between groups.

For the **salary** data, you want to control the experimentwise error rate at 0.05. There are 15 two-way comparisons. Essentially, performing each test to compare two means results in the following:

$$\alpha = 0.05 / 15 = 0.0033$$

To use the Bonferroni approach, use the **BON** option in the MEANS statement in PROC ANOVA:

```
proc anova data=salary;
   class subjarea;
   model annsal=subjarea;
   means subjarea / bon;
title 'Multiple Comparisons with Bonferroni Approach';
run;
```

These statements produce the results shown in Figures 9.5 and 9.6. Figure 9.11 shows the results for the multiple comparisons using the Bonferroni approach (produced by the BON option).

Deciding Which Means Differ

Figure 9.11 is very similar to Figure 9.10. It shows a 95% confidence interval for the difference between each pair of means.

SAS highlights comparisons that are significant with three asterisks (***) in the right-most column. From these results, you can conclude the following:

- The mean annual salary for **Social Science** is significantly different from **Music**.

- The mean annual salary for **Special Ed** is significantly different from **Mathematics**, **Language**, and **Music**.

- The mean annual salaries for other pairs of subject areas are not significantly different.

Compare the Bonferroni conclusions with the conclusions from the multiple pairwise *t*-tests. When controlling the experimentwise error rate, you find fewer significant differences between groups. However, you decrease your risk of incorrectly concluding that the means are different from each other.

In contrast, the results from the multiple pairwise *t*-tests detect more significant differences between groups. You have an increased overall risk of making incorrect decisions. Although you know that each comparison has an alpha level of 0.05, you don't know the alpha level for the overall experiment.

Figure 9.11 Results for BON Option in MEANS Statement

```
                    Multiple Comparisons with Bonferroni Approach

                             The ANOVA Procedure

                      Bonferroni (Dunn) t Tests for annsal

NOTE: This test controls the Type I experimentwise error rate, but it generally
      has a higher Type II error rate than Tukey's for all pairwise comparisons.


                       Alpha                          0.05
                       Error Degrees of Freedom        155
                       Error Mean Square          72936084
                       Critical Value of t         2.98138


          Comparisons significant at the 0.05 level are indicated by ***.


                                      Difference      Simultaneous
                        subjarea       Between       95% Confidence
                       Comparison       Means            Limits

        Social Science - Special Ed      1427       -9831    12684
        Social Science - Science         5249       -7062    17559
        Social Science - Mathematics     9410        -797    19618
        Social Science - Language        9791        -603    20186
        Social Science - Music          13337        2495    24179    ***
        Special Ed     - Social Science  -1427     -12684     9831
        Special Ed     - Science          3822      -5825    13469
        Special Ed     - Mathematics      7984       1224    14744    ***
        Special Ed     - Language         8365       1325    15404    ***
        Special Ed     - Music           11910       4225    19595    ***
        Science        - Social Science  -5249     -17559     7062
        Science        - Special Ed      -3822     -13469     5825
        Science        - Mathematics      4162      -4235    12559
        Science        - Language         4543      -4081    13167
        Science        - Music            8088      -1070    17246
        Mathematics    - Social Science  -9410     -19618      797
        Mathematics    - Special Ed      -7984     -14744    -1224    ***
        Mathematics    - Science         -4162     -12559     4235
        Mathematics    - Language          381      -4816     5578
        Mathematics    - Music            3926      -2116     9969
        Language       - Social Science  -9791     -20186      603
        Language       - Special Ed      -8365     -15404    -1325    ***
        Language       - Science         -4543     -13167     4081
        Language       - Mathematics      -381      -5578     4816
        Language       - Music            3546      -2808     9899
        Music          - Social Science -13337     -24179    -2495    ***
        Music          - Special Ed     -11910     -19595    -4225    ***
        Music          - Science         -8088     -17246     1070
        Music          - Mathematics     -3926      -9969     2116
        Music          - Language        -3546      -9899     2808
```

Understanding Other Items in the Report

Figure 9.11 contains two reports. Both are very similar to the reports in Figure 9.10. The first report is the Information report, which contains the statistical details for the tests. The second report is the Pairs report, which contains the results of the pairwise comparisons. See the information for Figure 9.10 for details on items in Figure 9.11.

The **Critical Value of t** in Figure 9.11 is higher than it is in Figure 9.10. This is a result of the lower per-comparison alpha level for the tests.

The heading for the reports in Figure 9.11 identifies the alternate name "**Dunn**" for the Bonferroni tests. The note below this heading refers to the Tukey-Kramer test, which is discussed in the next section. This note explains that the Bonferroni approach controls the experimentwise error rate.

Performing the Tukey-Kramer Test

The Tukey-Kramer test controls the experimentwise error rate. Many statisticians prefer this test to using the Bonferroni approach. This test is sometimes called the HSD test, or the honestly significant difference test in the case of equal group sizes.

To perform the Tukey-Kramer test, use the **TUKEY** option in the MEANS statement in PROC ANOVA:

```
proc anova data=salary;
   class subjarea;
   model annsal=subjarea;
   means subjarea / tukey;
title 'Multiple Comparisons with Tukey-Kramer Test';
run;
```

These statements produce the results shown in Figures 9.5 and 9.6. Figure 9.12 shows the results for the multiple comparisons using the Tukey-Kramer test (produced by the TUKEY option).

The heading for the reports in Figure 9.12 identifies the test. The note below this heading explains that the Tukey-Kramer test controls the experimentwise error rate.

Figure 9.12 Results for TUKEY Option in MEANS Statement

```
                 Means Comparisons with Tukey-Kramer Test

                          The ANOVA Procedure

              Tukey's Studentized Range (HSD) Test for annsal

       NOTE: This test controls the Type I experimentwise error rate.


              Alpha                                     0.05
              Error Degrees of Freedom                   155
              Error Mean Square                      72936084
              Critical Value of Studentized Range    4.08100


        Comparisons significant at the 0.05 level are indicated by ***.


                                       Difference        Simultaneous
                       subjarea         Between          95% Confidence
                      Comparison         Means               Limits

     Social Science - Special Ed          1427      -9470    12323
     Social Science - Science             5249      -6667    17164
     Social Science - Mathematics         9410       -469    19290
     Social Science - Language            9791       -270    19852
     Social Science - Music              13337       2843    23831   ***
     Special Ed     - Social Science     -1427     -12323     9470
     Special Ed     - Science             3822      -5515    13159
     Special Ed     - Mathematics         7984       1441    14527   ***
     Special Ed     - Language            8365       1551    15178   ***
     Special Ed     - Music              11910       4472    19348   ***
     Science        - Social Science     -5249     -17164     6667
     Science        - Special Ed         -3822     -13159     5515
     Science        - Mathematics         4162      -3966    12290
     Science        - Language            4543      -3804    12890
     Science        - Music               8088       -776    16953
     Mathematics    - Social Science     -9410     -19290      469
     Mathematics    - Special Ed         -7984     -14527    -1441   ***
     Mathematics    - Science            -4162     -12290     3966
     Mathematics    - Language             381      -4650     5411
     Mathematics    - Music               3926      -1922     9775
     Language       - Social Science     -9791     -19852      270
     Language       - Special Ed         -8365     -15178    -1551   ***
     Language       - Science            -4543     -12890     3804
     Language       - Mathematics         -381      -5411     4650
     Language       - Music               3546      -2604     9695
     Music          - Social Science    -13337     -23831    -2843   ***
     Music          - Special Ed        -11910     -19348    -4472   ***
     Music          - Science            -8088     -16953      776
     Music          - Mathematics        -3926      -9775     1922
     Music          - Language           -3546      -9695     2604
```

Deciding Which Means Differ

Figure 9.12 is very similar to Figure 9.10 and Figure 9.11. It shows a 95% confidence interval for the difference between each pair of means.

SAS highlights comparisons that are significant with three asterisks (***) in the right-most column. From these results, you can conclude the following:

- The mean annual salary for **Social Science** is significantly different from **Music**.

- The mean annual salary for **Special Ed** is significantly different from **Mathematics**, **Language**, and **Music**.

- The mean annual salaries for other pairs of subject areas are not significantly different.

These results are the same as the Bonferroni results, which makes sense because both tests control the experimentwise error rate. For your data, however, you might not get the exact same results from these two tests.

Understanding Other Items in the Report

Figure 9.12 contains two reports. Both are very similar to the reports in Figure 9.10 and Figure 9.11. The first report is the Information report, which contains the statistical details for the tests. The second report is the Pairs report, which contains the results of the pairwise comparisons. See the information for Figure 9.10 for details on items in Figure 9.12.

The **Critical Value of Studentized Range** in Figure 9.12 identifies the test statistic for the Tukey-Kramer test.

Changing the Alpha Level

SAS automatically performs multiple comparison procedures and creates confidence intervals at the 95% confidence level. You can change the confidence level for any of the multiple comparison procedures with the **ALPHA=** option in the MEANS statement:

```
proc anova data=salary;
   class subjarea;
   model annsal=subjarea;
   means subjarea / tukey alpha=0.10;
title 'Tukey-Kramer Test at 90%';
run;
```

These statements produce the results shown in Figures 9.5 and 9.6. Figure 9.13 shows the results for the multiple comparisons using the Tukey-Kramer test with a 0.10 alpha level (produced by the TUKEY and ALPHA= options).

Figure 9.13 Results for TUKEY and ALPHA= Options

```
                         Tukey-Kramer Test at 90%

                           The ANOVA Procedure

                Tukey's Studentized Range (HSD) Test for annsal

        NOTE: This test controls the Type I experimentwise error rate.


              Alpha                                    0.1
              Error Degrees of Freedom                 155
              Error Mean Square                   72936084
              Critical Value of Studentized Range  3.69679


     Comparisons significant at the 0.1 level are indicated by ***.


                                   Difference      Simultaneous
                   subjarea         Between        90% Confidence
                  Comparison         Means            Limits

    Social Science - Special Ed        1427     -8444    11297
    Social Science - Science           5249     -5545    16042
    Social Science - Mathematics       9410       461    18360   ***
    Social Science - Language          9791       677    18905   ***
    Social Science - Music            13337      3831    22843   ***
    Special Ed     - Social Science   -1427    -11297     8444
    Special Ed     - Science           3822     -4636    12280
    Special Ed     - Mathematics       7984      2057    13911   ***
    Special Ed     - Language          8365      2192    14537   ***
    Special Ed     - Music            11910      5172    18648   ***
    Science        - Social Science   -5249    -16042     5545
    Science        - Special Ed       -3822    -12280     4636
    Science        - Mathematics       4162     -3201    11524
    Science        - Language          4543     -3019    12104
    Science        - Music             8088        59    16118   ***
    Mathematics    - Social Science   -9410    -18360     -461   ***
    Mathematics    - Special Ed       -7984    -13911    -2057   ***
    Mathematics    - Science          -4162    -11524     3201
    Mathematics    - Language           381     -4176     4938
    Mathematics    - Music             3926     -1371     9224
    Language       - Social Science   -9791    -18905     -677   ***
    Language       - Special Ed       -8365    -14537    -2192   ***
    Language       - Science          -4543    -12104     3019
    Language       - Mathematics       -381     -4938     4176
    Language       - Music             3546     -2025     9116
    Music          - Social Science  -13337    -22843    -3831   ***
    Music          - Special Ed      -11910    -18648    -5172   ***
    Music          - Science          -8088    -16118      -59   ***
    Music          - Mathematics      -3926     -9224     1371
    Music          - Language         -3546     -9116     2025
```

Figure 9.13 is very similar to Figure 9.12. The key differences in appearance are the **Alpha** value in the **Information** report, and the column heading that indicates **Simultaneous 90% Confidence Limits** in the **Pairs** report.

Deciding Which Means Differ

SAS highlights comparisons that are significant with three asterisks (***) in the right-most column. From these results, you can conclude the following:

- The mean annual salaries for **Social Science** and **Special Ed** are significantly different from **Mathematics**, **Language**, and **Music**.

- The mean annual salary for **Science** is significantly different from **Music**.

- The mean annual salaries for other pairs of subject areas are not significantly different.

Because you used a lower confidence level (90% instead of 95%), it makes sense that you find more significant differences between groups.

Using Dunnett's Test When Appropriate

Some experiments have a control group. For example, clinical trials to investigate the effectiveness of a new drug often include a placebo group, where the patients receive a look-alike pill. The placebo group is the control group. Dunnett's test is designed for this situation. For example purposes only, assume that **Mathematics** is the control group for the salary data:

```
proc anova data=salary;
   class subjarea;
   model annsal=subjarea;
   means subjarea / dunnett('Mathematics');
title 'Means Comparisons with Mathematics as Control';
run;
```

The **DUNNETT** option performs the test. For your data, provide the value of the control group in parentheses. These statements produce the results shown in Figures 9.5 and 9.6. Figure 9.14 shows the results for the multiple comparisons using Dunnett's test.

Figure 9.14 contains the **Information** report and the **Pairs** report. The heading for the reports identifies the test. The note below this heading explains that Dunnett's test controls the experimentwise error rate in the situation of comparing all groups against a control group.

Figure 9.14 Results for DUNNETT Option (Mathematics as Control)

```
                 Means Comparisons with Mathematics as Control

                          The ANOVA Procedure

                     Dunnett's t Tests for annsal

NOTE: This test controls the Type I experimentwise error for comparisons of all
                   treatments against a control.


                  Alpha                              0.05
                  Error Degrees of Freedom           155
                  Error Mean Square             72936084
                  Critical Value of Dunnett's t  2.58551


       Comparisons significant at the 0.05 level are indicated by ***.

                                        Difference      Simultaneous
                        subjarea        Between        95% Confidence
                       Comparison        Means            Limits

       Social Science - Mathematics       9410        558    18262   ***
       Special Ed     - Mathematics       7984       2121    13846   ***
       Science        - Mathematics       4162      -3120    11444
       Language       - Mathematics       -381      -4888     4126
       Music          - Mathematics      -3926      -9167     1314
```

Deciding Which Means Differ

SAS highlights comparisons that are significant with three asterisks (***) in the right-most column. From these results, you can conclude the following:

- The mean annual salaries for **Social Science** and **Special Ed** are significantly different from **Mathematics**.

- The mean annual salaries for other subject areas are not significantly different from the control group of **Mathematics**.

Understanding Other Items in the Report

Figure 9.14 contains two reports. Both are very similar to the reports in Figure 9.10. The first report is the Information report, which contains the statistical details for the tests. The second report is the Pairs report, which contains the results of the pairwise comparisons. See the information for Figure 9.10 for details on items in Figure 9.14.

The **Critical Value of Dunnett's t** in Figure 9.14 identifies the test statistic for Dunnett's test.

Recommendations

First, use a multiple comparison test only after an analysis of variance is significant. If the ANOVA indicates that the means are not significantly different, then multiple comparison procedures are not appropriate. All of the multiple comparison procedures are available as options in the MEANS statement.

If you are interested in comparing all pairwise comparisons, and you want to control only the comparisonwise error rate, you can use the multiple pairwise *t*-tests with the T option.

If you have a few pre-planned comparisons between groups, and you want to control the experimentwise error rate, you can use the Bonferroni approach with the BON option.

If you want to compare all of the means between groups, and you want to control the experimentwise error rate, use either the Tukey-Kramer test or the Bonferroni approach. The Bonferroni approach is more conservative, and many statisticians prefer the Tukey-Kramer test. Use the TUKEY option for the Tukey-Kramer test.

If you have the special case of a known control group, you can use Dunnett's test with the DUNNETT option.

Summarizing Multiple Comparison Procedures

SAS automatically displays results from multiple comparison procedures using confidence intervals when the data is unbalanced. The confidence intervals help you understand how much the group means differ. Confidence intervals also help you to assess the practical significance of the differences between groups. You can specify confidence intervals with the CLDIFF option.

SAS automatically uses connecting letters when the data is balanced, as shown in the next section. You can specify connecting letters with the LINES option. Consult with a statistician before using the LINES option with unbalanced data because differences in group sizes can mean that the approach is not appropriate.

The general form of the statements to perform multiple comparison procedures using PROC ANOVA is shown below:

PROC ANOVA DATA=*data-set-name*;

  CLASS *class-variable*;

  MODEL *measurement-variable=class-variable*;

    MEANS *class-variable / options*;

The **PROC ANOVA**, **CLASS**, and **MODEL** statements are required.

The **MEANS** statement *options* can be one or more of the following:

T	multiple pairwise *t*-tests for pairwise comparisons.
BON	Bonferroni approach for multiple pairwise *t*-tests.
TUKEY	Tukey-Kramer test.
DUNNETT('*control***')**	Dunnett's test, where **'***control***'** identifies the control group. If you use formats in the data set, then use the formatted value in this option.
ALPHA=*level*	controls the significance level. Choose a *level* between 0.0001 and 0.9999. SAS automatically uses 0.05.
CLDIFF	shows differences with confidence intervals. SAS automatically uses confidence intervals for unequal group sizes.
LINES	shows differences with connecting letters. SAS automatically uses connecting letters for equal group sizes.

If any option is used, then the slash is required. You can use the options alone or together, as shown above.

PROC ANOVA has options for many other multiple comparison procedures. See the SAS documentation or the references in Appendix 1, "Further Reading," for information.

Other items were defined earlier.

As discussed in Chapter 4, you can use the ODS statement to choose the output tables to display. Table 9.4 identifies the output tables for PROC ANOVA and multiple comparison procedures.

Table 9.4 ODS Table Names for Multiple Comparisons

ODS Name	Output Table
Information	Contains the statistical details for the test.
Pairs	Contains the results of the pairwise comparisons, using confidence intervals. This report appears automatically for unbalanced data.
MCLINES	Contains information about which pairs differ, using lines of characters. This report appears automatically for balanced data.

Using PROC ANOVA Interactively

PROC ANOVA is an interactive procedure. If you use line mode or the windowing environment mode, you can perform an analysis of variance, and then perform multiple comparison procedures without rerunning the analysis of variance. Type the PROC ANOVA, CLASS, and MODEL statements, and then add a **RUN** statement to see the analysis of variance. Then, add a MEANS statement, and a second RUN statement to perform multiple comparison procedures. These statements produce the results shown in Figures 9.5, 9.6, and 9.10:

```
proc anova data=salary;
   class subjarea;
   model annsal=subjarea;
   title 'ANOVA for Teacher Salary Data';
run;
   means subjarea / t;
   title 'Multiple Comparisons with t Tests';
run;
quit;
```

When you use the statements above, the second RUN statement does not end PROC ANOVA because the procedure is waiting to receive additional statements. The **QUIT** statement ends the procedure. The form of the QUIT statement is simply the word QUIT followed by a semicolon.

Although you can end the procedure by starting another DATA or PROC step, you might receive an error. If you use an ODS statement before the next PROC step, and the table you specify in the ODS statement is not available in PROC ANOVA (which is running interactively), SAS prints an error in the log and does not create output. To avoid this situation, use the QUIT statement.

Summarizing with an Example

This section summarizes the steps for performing the analyses to compare multiple groups. This data has equal group sizes, so it is balanced.

Step 1: Create a SAS data set.

The data is from an experiment that compares muzzle velocities for different types of gunpowder. The muzzle velocity is measured for eight cartridges from each of the three types of gunpowder. Table 9.5 shows the data.

Table 9.5 Muzzle Velocities for Three Types of Gunpowder

Gunpowder	Muzzle Velocity			
Blasto	27.3	28.1	27.4	27.7
	28.0	28.1	27.4	27.1
Zoom	28.3	27.9	28.1	28.3
	27.9	27.6	28.5	27.9
Kingpow	28.4	28.9	28.3	27.9
	28.2	28.9	28.8	27.7

This data is available in the **bullets** data set in the sample data for the book. The following SAS statements create the data set:

```
data bullets;
   input powder $ velocity @@;
   datalines;
BLASTO 27.3 BLASTO 28.1 BLASTO 27.4 BLASTO 27.7
BLASTO 28.0 BLASTO 28.1 BLASTO 27.4 BLASTO 27.1
ZOOM 28.3 ZOOM 27.9 ZOOM 28.1 ZOOM 28.3 ZOOM 27.9
ZOOM 27.6 ZOOM 28.5 ZOOM 27.9 KINGPOW 28.4 KINGPOW 28.9 KINGPOW
28.3 KINGPOW 27.9 KINGPOW 28.2 KINGPOW 28.9
KINGPOW 28.8 KINGPOW 27.7
;
run;
```

Step 2: Check the data set for errors.

Following similar steps as you did for the **salary** data, use PROC MEANS and PROC
BOXPLOT to summarize the data and check it for errors:

```
proc means data=bullets;
    class powder;
    var velocity;
title 'Brief Summary of Bullets Data';
run;

proc boxplot data=bullets;
    plot velocity*powder / boxwidthscale=1 bwslegend;
title 'Comparative Box Plots by Gunpowder';
run;
```

Figures 9.15 and 9.16 show the results.

From the statistics and the box plots, you initially conclude that **KINGPOW** has a higher
mean velocity than the other two types of gunpowder. But, you don't know whether this
observed difference is statistically significant. Because the three groups are the same size,
the box widths are all the same. Because PROC BOXPLOT does not scale the box
widths, the graph does not contain a legend.

The box plots don't reveal any outlier points or potential errors in the data. Proceed with
the analysis.

Figure 9.15 Summarizing the Bullets Data

```
                    Brief Summary of Bullets Data

                         The MEANS Procedure

                      Analysis Variable : velocity

              N
powder       Obs    N        Mean        Std Dev      Minimum      Maximum

BLASTO        8     8    27.6375000     0.3925648   27.1000000   28.1000000

KINGPOW       8     8    28.3875000     0.4549333   27.7000000   28.9000000

ZOOM          8     8    28.0625000     0.2924649   27.6000000   28.5000000
```

Step 3: Choose the significance level for the test.

Test for differences between the three groups of gunpowder using the 5% significance
level. The reference value is 0.05.

Figure 9.16 Box Plots for the Bullets Data

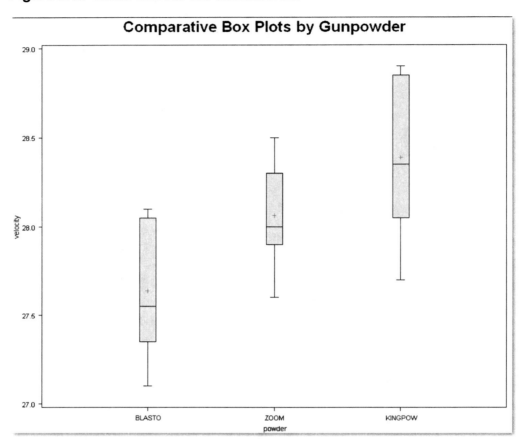

Comparative Box Plots by Gunpowder

Step 4: Check the assumptions for the test.

In checking the assumptions for an analysis of variance, note that the observations are independent. The standard deviations for the three groups are similar. With only eight observations per group, it is difficult to test for normality. For completeness, to check these assumptions, type the following:

```
ods select HOVFTest;
proc anova data=bullets;
   class powder;
   model velocity=powder;
   means powder / hovtest;
```

```
title 'Testing Equal Variances for Bullets Data';
run;
quit;

ods select TestsForNormality;
proc univariate data=bullets normal;
   class powder;
   var velocity;
title 'Testing Normality for Bullets Data';
run;
```

Figure 9.17 shows the results from testing for equal variances. Proceed with the analysis based on the assumption that the variances for the three groups are the same.

Figure 9.17 Testing for Equal Variances in Bullets Data

```
            Testing Equal Variances for Bullets Data

                     The ANOVA Procedure

        Levene's Test for Homogeneity of velocity Variance
            ANOVA of Squared Deviations from Group Means

                         Sum of        Mean
   Source        DF      Squares       Square     F Value    Pr > F

   powder         2      0.0454        0.0227        1.60    0.2255
   Error         21      0.2979        0.0142
```

Figure 9.18 shows the resulting text reports from testing for normality in each group. Even though the sample sizes are small, these results indicate that you can assume normality and proceed with the analysis. For completeness, you might want to perform informal checks using histograms, normal probability plots, skewness, and kurtosis. These informal checks can be difficult with such small groups, so this chapter does not show them.

Figure 9.18 Testing for Normality in Bullets Data

```
                 Testing Normality for Bullets Data

                      The UNIVARIATE Procedure
                      Variable:  velocity
                      powder = BLASTO

                     Tests for Normality

Test                      --Statistic---        -----p Value------

Shapiro-Wilk         W      0.890497     Pr < W         0.2365
Kolmogorov-Smirnov   D      0.227409     Pr > D        >0.1500
Cramer-von Mises     W-Sq   0.065783     Pr > W-Sq     >0.2500
Anderson-Darling     A-Sq   0.412575     Pr > A-Sq     >0.2500
```

```
                 Testing Normality for Bullets Data

                      The UNIVARIATE Procedure
                      Variable:  velocity
                      powder = KINGPOW

                     Tests for Normality

Test                      --Statistic---        -----p Value------

Shapiro-Wilk         W      0.917768     Pr < W         0.4120
Kolmogorov-Smirnov   D      0.192724     Pr > D        >0.1500
Cramer-von Mises     W-Sq   0.040928     Pr > W-Sq     >0.2500
Anderson-Darling     A-Sq   0.295023     Pr > A-Sq     >0.2500
```

```
                 Testing Normality for Bullets Data

                      The UNIVARIATE Procedure
                      Variable:  velocity
                      powder = ZOOM

                     Tests for Normality

Test                      --Statistic---        -----p Value------

Shapiro-Wilk         W      0.945521     Pr < W         0.6661
Kolmogorov-Smirnov   D      0.210765     Pr > D        >0.1500
Cramer-von Mises     W-Sq   0.054755     Pr > W-Sq     >0.2500
Anderson-Darling     A-Sq   0.310993     Pr > A-Sq     >0.2500
```

Step 5: Perform the test.

Use PROC ANOVA to perform an analysis of variance. For example purposes only, this step also shows the results of the Kruskal-Wallis test.

```
proc anova data=bullets;
   class powder;
   model velocity=powder;
   means powder / hovtest;
title 'ANOVA for Bullets Data';
run;
quit;

proc npar1way data=bullets wilcoxon;
   class powder;
   var velocity;
title 'Nonparametric Tests for Bullets Data';
run;
```

Figure 9.19 shows the results from the analysis of variance. Figure 9.20 shows the results from the Kruskal-Wallis test. PROC ANOVA prints a first page that shows the CLASS variable, the levels of the CLASS variable, and the values. It also shows the number of observations for the analysis. This first page is similar to the first page shown for the salary data in Figure 9.5.

Figure 9.19 ANOVA Results for Bullets Data

```
                    ANOVA for Bullets Data

                    The ANOVA Procedure

Dependent Variable: velocity

                                 Sum of
Source                 DF        Squares    Mean Square   F Value   Pr > F

Model                   2     2.26333333     1.13166667      7.60   0.0033

Error                  21     3.12625000     0.14886905

Corrected Total        23     5.38958333


            R-Square    Coeff Var    Root MSE    velocity Mean

            0.419946     1.376550    0.385836         28.02917


Source                 DF       Anova SS    Mean Square   F Value   Pr > F

powder                  2     2.26333333     1.13166667      7.60   0.0033
```

Figure 9.20 Kruskal-Wallis Results for Bullets Data

```
                Nonparametric Tests for Bullets Data

                      The NPARIWAY Procedure

            Wilcoxon Scores (Rank Sums) for Variable velocity
                      Classified by Variable powder

                         Sum of      Expected        Std Dev         Mean
      powder      N      Scores      Under H0        Under H0        Score
      BLASTO      8       56.50       100.0         16.255211       7.06250
      ZOOM        8      104.50       100.0         16.255211      13.06250
      KINGPOW     8      139.00       100.0         16.255211      17.37500

                  Average scores were used for ties.


                          Kruskal-Wallis Test

                  Chi-Square          8.6628
                  DF                       2
                  Pr > Chi-Square     0.0131
```

Step 6: Make conclusions from the test results.

The *p*-value for comparing groups is **0.0033**, which is less than the reference probability value of 0.05. You conclude that the mean velocity is significantly different between the three types of gunpowder. If the analysis of variance assumptions had not seemed reasonable, then the Kruskal-Wallis test would have led to the same conclusion, with *p*=**0.0131**.

Next, perform the Tukey-Kramer test. Use the following MEANS statement:

```
means powder / tukey;
title 'Tukey Kramer Test for Bullets Data';
run;
```

Figure 9.21 shows the results. The note for the test refers to another multiple comparison test, **REGWQ**, which is not discussed in this book. Compare Figures 9.21 and 9.12. When PROC ANOVA shows the results of the Tukey-Kramer test using confidence intervals, the heading for the test is slightly different from when the procedure shows the results using connecting lines.

Figure 9.21 Tukey-Kramer Results for Bullets Data

```
                    Tukey Kramer Test for Bullets Data

                          The ANOVA Procedure

              Tukey's Studentized Range (HSD) Test for velocity

NOTE: This test controls the Type I experimentwise error rate, but it generally
              has a higher Type II error rate than REGWQ.


           Alpha                                      0.05
           Error Degrees of Freedom                     21
           Error Mean Square                       0.148869
           Critical Value of Studentized Range     3.56462
           Minimum Significant Difference           0.4863


       Means with the same letter are not significantly different.


       Tukey Grouping            Mean      N      powder

                      A        28.3875     8      KINGPOW
                      A
               B      A        28.0625     8      ZOOM
               B
               B                27.6375     8      BLASTO
```

You conclude that the mean muzzle velocity for **KINGPOW** is significantly different from **BLASTO**. The mean muzzle velocities for other pairs of types of gunpowder are not significantly different. These conclusions should make sense at an intuitive level when you look at the side-by-side box plots from **PROC UNIVARIATE**. However, now you know that the differences between groups are greater than could be expected by chance. Without the statistical test, you only knew that the groups seemed different, but you didn't know how likely the observed differences were.

Summary

Key Ideas

- To summarize data from more than two groups, separate the data into groups, and get summary information for each group using one or more SAS procedures. See "Syntax" for more detail.

- The steps for analyses are:
  1. Create a SAS data set.
  2. Check the data set for errors.
  3. Choose the significance level for the test.
  4. Check the assumptions for the test.
  5. Perform the test.
  6. Make conclusions from the test results.

- An analysis of variance (ANOVA) is a test for comparing the means of multiple groups at once. This test assumes that the observations are independent random samples from normally distributed populations, and that the groups have equal variances.

- The Welch's ANOVA compares the means of multiple groups when the assumption of equal variances does not seem reasonable.

- If the normality assumption does not seem reasonable, the Kruskal-Wallis test provides a nonparametric analogue.

- Compare the *p*-value for the test with the significance level.

  □ If the *p*-value is less than the significance level, then reject the null hypothesis. Conclude that the means for the groups are significantly different.

  □ If the *p*-value is greater than the significance level, then you fail to reject the null hypothesis. Conclude that the means for the groups are not significantly different. Do not conclude that the means for the groups are the same.

- Multiple comparisons tests show which means differ from other means. Use multiple comparison procedures only after an analysis of variance shows significant differences between groups. Multiple pairwise *t*-tests control the comparisonwise error rate, but increase the overall risk of incorrectly concluding that the means are different. The Tukey-Kramer test and Bonferroni approach control the experimentwise error rate. Dunnett's test is useful in the special case of a known control group.

Syntax

To summarize data from multiple groups

- To create a concise summary of multiple groups, use PROC MEANS with a CLASS statement.

- To create a detailed summary of multiple groups, use PROC UNIVARIATE with a CLASS statement. This summary contains output tables for each value of the *class-variable*. You can use the PLOT option to create line printer plots for each value of the *class-variable*.

- To create a detailed summary of multiple groups with comparative histograms:
  PROC UNIVARIATE DATA=*data-set-name* **NOPRINT;**
    CLASS *class-variable*;
    VAR *measurement-variables*;
    HISTOGRAM *measurement-variables* **/ NROWS=***value*;

  data-set-name   is the name of a SAS data set.

  class-variable   is the variable that classifies the data into groups.

  measurement-variables
       are the variables you want to summarize.

  The CLASS statement is required. The *class-variable* can be character or numeric and should be nominal. If you omit the *measurement-variables* from the VAR statement, then the procedure uses all numeric variables.

  You can use the options described in Chapters 4 and 5 with PROC UNIVARIATE and a CLASS statement. You can omit the NOPRINT option and create the detailed summary and histograms at the same time.

  The NROWS= option controls the number of histograms that appear on a single page. SAS automatically uses NROWS=2. For your data, choose a *value* equal to the number of groups.

  If any option is used in the HISTOGRAM statement, then the slash is required.

- To create a detailed summary of multiple groups with side-by-side bar charts, use PROC CHART. This is a low-resolution alternative to comparative histograms.

- To create side-by-side box plots:
    PROC BOXPLOT DATA=_data-set-name_**;**
      PLOT _measurement-variable*class-variable_
          / BOXWIDTHSCALE=1 BWSLEGEND;

    Items in italic were defined earlier in the chapter. Depending on your data, you might need to first sort by the *class-variable* before creating the plots. The asterisk in the PLOT statement is required.

    The BOXWIDTHSCALE=1 option scales the width of the box based on the sample size of each group. The BWSLEGEND option adds a legend to the bottom of the box plot. The legend explains the varying box widths. If either option is used, then the slash is required. You can use the options alone or together, as shown above.

To check the assumptions (step 4)

- To check the assumption of independent observations, you need to think about your data and whether this assumption seems reasonable. This is not a step where using SAS will answer this question.

- To check the assumption of normality, use PROC UNIVARIATE. For multiple groups, check each group separately.

- To test the assumption of equal variances:
     PROC ANOVA DATA=_data-set-name_**;**
      CLASS _class-variable_**;**
      MODEL _measurement-variable=class-variable_**;**
      MEANS _class-variable_ **/ HOVTEST;**

    Items in italic were defined earlier in the chapter.

    The PROC ANOVA, CLASS, and MODEL statements are required.

To perform the test (step 5)

- If the analysis of variance assumptions seem reasonable:
    PROC ANOVA DATA=_data-set-name_**;**
      CLASS _class-variable_**;**
      MODEL _measurement-variable=class-variable_**;**

    Items in italic were defined earlier in the chapter.

    The PROC ANOVA, CLASS, and MODEL statements are required.

- If the equal variances assumption does not seem reasonable:
    PROC ANOVA DATA=_data-set-name_**;**

CLASS *class-variable*;
MODEL *measurement-variable=class-variable*;
MEANS *class-variable* **/ WELCH**;

Items in italic were defined earlier in the chapter.

The PROC ANOVA, CLASS, and MODEL statements are required.

If the Welch ANOVA is used, then multiple comparison procedures are not appropriate.

- If you use PROC ANOVA interactively in the windowing environment mode, then end the procedure with the QUIT statement:

  QUIT;

- For non-normal data:

  PROC NPAR1WAY DATA=*data-set-name* **WILCOXON**;
    CLASS *class-variable*;
    VAR *measurement-variables*;

  Items in italic were defined earlier in the chapter.

  The CLASS statement is required.

  While the VAR statement is not required, you should use it. If you omit the VAR statement, then PROC NPAR1WAY performs analyses for every numeric variable in the data set.

For performing multiple comparison procedures

- First, compare groups and perform multiple comparison procedures only when you conclude that the groups are significantly different.

  PROC ANOVA DATA=*data-set-name*;
    CLASS *class-variable*;
    MODEL *measurement-variable=class-variable*;
    MEANS *class-variable* **/** *options*;

  Other items were defined earlier.

  The PROC ANOVA, CLASS, and MODEL statements are required.

The **MEANS** statement *options* can be one or more of the following:

T	multiple pairwise *t*-tests for pairwise comparisons.
BON	Bonferroni approach for multiple pairwise *t*-tests.
TUKEY	Tukey-Kramer test.
DUNNETT('*control***')**	Dunnett's test, where **'***control***'** identifies the control group. If you use formats in the data set, then use the formatted value in this option.
ALPHA=*level*	controls the significance level. Choose a *level* between 0.0001 and 0.9999. SAS automatically uses 0.05.
CLDIFF	shows differences with confidence intervals. SAS automatically uses confidence intervals for unequal group sizes.
LINES	shows differences with connecting letters. SAS automatically uses connecting letters for equal group sizes.

If any option is used, then the slash is required. You can use the options alone or together.

PROC ANOVA has options for many other multiple comparison procedures. See the SAS documentation or the references in Appendix 1, "Further Reading," for information.

Example

The program below produces the output shown in this chapter:

```
options nodate nonumber ps=60 ls=80;
proc format;
   value $subtxt
         'speced' = 'Special Ed'
         'mathem' = 'Mathematics'
         'langua' = 'Language'
         'music'  = 'Music'
         'scienc' = 'Science'
         'socsci' = 'Social Science';
run;
```

```
data salary;
   input subjarea $ annsal @@;
   label subjarea='Subject Area'
         annsal='Annual Salary';
   format subjarea $subtxt.;
   datalines;
speced 35584 mathem 27814 langua 26162 mathem 25470
speced 41400 mathem 34432 music 21827 music 35787 music 30043
mathem 45480 mathem 25358 speced 42360 langua 23963 langua 27403
mathem 29610 music 27847 mathem 25000 mathem 29363 mathem 25091
speced 31319 music 24150 langua 30180 socsci 42210 mathem 55600
langua 32134 speced 57880 mathem 47770 langua 33472 music 21635
mathem 31908 speced 43128 mathem 33000 music 46691 langua 28535
langua 34609 music 24895 speced 42222 speced 39676 music 22515
speced 41899 music 27827 scienc 44324 scienc 43075 langua 49100
langua 44207 music 46001 music 25666 scienc 30000 mathem 26355
mathem 39201 mathem 32000 langua 26705 mathem 37120 langua 44888
mathem 62655 scienc 24532 mathem 36733 langua 29969 mathem 28521
langua 27599 music 27178 mathem 26674 langua 28662 music 41161
mathem 48836 mathem 25096 langua 27664 music 23092 speced 45773
mathem 27038 mathem 27197 music 44444 speced 51096 mathem 25125
scienc 34930 speced 44625 mathem 27829 mathem 28935 mathem 31124
socsci 36133 music 28004 mathem 37323 music 32040 scienc 39784
mathem 26428 mathem 39908 mathem 34692 music 26417 mathem 23663
speced 35762 langua 29612 scienc 32576 mathem 32188 mathem 33957
speced 35083 langua 47316 mathem 34055 langua 27556 langua 35465
socsci 49683 langua 38250 langua 30171 mathem 53282 langua 32022
socsci 31993 speced 52616 langua 33884 music 41220 mathem 43890
scienc 40330 langua 36980 scienc 59910 mathem 26000 langua 29594
socsci 38728 langua 47902 langua 38948 langua 33042 mathem 29360
socsci 46969 speced 39697 mathem 31624 langua 30230 music 29954
mathem 45733 music 24712 langua 33618 langua 29485 mathem 28709
music 24720 mathem 51655 mathem 32960 mathem 45268 langua 31006
langua 48411 socsci 59704 music 22148 mathem 27107 scienc 47475
langua 33058 speced 53813 music 38914 langua 49881 langua 42485
langua 26966 mathem 31615 mathem 24032 langua 27878 mathem 56070
mathem 24530 mathem 40174 langua 27607 speced 31114 langua 30665
scienc 25276 speced 36844 mathem 24305 mathem 35560 music 28770
langua 34001 mathem 35955
;
run;

proc means data=salary;
   class subjarea;
   var annsal;
title 'Brief Summary of Teacher Salaries';
run;
```

```
proc univariate data=salary noprint;
   class subjarea;
   var annsal;
   histogram annsal / nrows=6;
title 'Comparative Histograms by Subject Area';
run;

proc sort data=salary;
   by subjarea;
run;

proc boxplot data=salary;
   plot annsal*subjarea / boxwidthscale=1 bwslegend;
title 'Comparative Box Plots by Subject Area';
run;

proc anova data=salary;
   class subjarea;
   model annsal=subjarea;
title 'ANOVA for Teacher Salaries';
run;
quit;

ods select means HOVFtest;
proc anova data=salary;
   class subjarea;
   model annsal=subjarea;
   means subjarea / hovtest;
title 'Testing for Equal Variances';
run;
quit;

ods select Welch;
proc anova data=salary;
   class subjarea;
   model annsal=subjarea;
   means subjarea / welch;
title 'Welch ANOVA for Salary Data';
run;

proc npar1way data=salary wilcoxon;
   class subjarea;
   var annsal;
title 'Nonparametric Tests for Teacher Salaries' ;
run;
```

```
proc anova data=salary;
   class subjarea;
   model annsal=subjarea;
   means subjarea / t;
title 'Multiple Comparisons with t tests';
run;

   means subjarea / bon ;
title 'Multiple Comparisons with Bonferroni Approach' ;
run;
   means subjarea / tukey;
title 'Means Comparisons with Tukey-Kramer Test';
run;
   means subjarea / tukey alpha=0.10 ;
title 'Tukey-Kramer Test at 90%';
run;
   means subjarea / dunnett('Mathematics') ;
title 'Means Comparisons with Mathematics as Control';
run;

data bullets;
   input powder $ velocity @@;
   datalines;
BLASTO 27.3 BLASTO 28.1 BLASTO 27.4 BLASTO 27.7 BLASTO 28.0
BLASTO 28.1 BLASTO 27.4 BLASTO 27.1 ZOOM 28.3 ZOOM 27.9
ZOOM 28.1 ZOOM 28.3 ZOOM 27.9 ZOOM 27.6 ZOOM 28.5 ZOOM 27.9
KINGPOW 28.4 KINGPOW 28.9 KINGPOW 28.3 KINGPOW 27.9
KINGPOW 28.2 KINGPOW 28.9 KINGPOW 28.8 KINGPOW 27.7
;
run;

proc means data=bullets;
   class powder;
   var velocity;
title 'Brief Summary of Bullets Data';
run;

proc boxplot data=bullets;
   plot velocity*powder / boxwidthscale=1 bwslegend;
title 'Comparative Box Plots by Gunpowder';
run;

ods select HOVFtest;
proc anova data=bullets;
   class powder;
   model velocity=powder;
   means powder / hovtest;
```

```
title 'Testing Equal Variances for Bullets Data';
run;
quit;

ods select TestsForNormality;
proc univariate data=bullets normal;
   class powder;
   var velocity;
title 'Testing Normality for Bullets Data';
run;

proc anova data=bullets;
   class powder;
   model velocity=powder;
title 'ANOVA for Bullets Data';
run;
   means powder / tukey;
title 'Tukey Kramer Test for Bullets Data';
run;
quit;

proc npar1way data=bullets wilcoxon;
   class powder;
   var velocity;
title 'Nonparametric Tests for Bullets Data' ;
run;
```

Part 4

Fitting Lines to Data

Chapter 10

Understanding Correlation and Regression

Can SAT scores be used to predict college grade point averages? How does the age of a house affect its selling price? Is heart rate affected by the amount of blood cholesterol? How much of an increase in sales results from a specific increase in advertising expenditures?

These questions involve looking at two continuous variables, and investigating whether the variables are related. As one variable increases, does the other variable increase, decrease, or stay the same? Because both variables contain quantitative measurements, comparing groups is not appropriate. This chapter discusses the following topics:

- summarizing continuous variables using scatter plots and statistics.

- using correlation coefficients to describe the strength of the linear relationship between two continuous variables.

- performing least squares regression analysis to develop equations that describe how one variable is related to another variable. Sections discuss fitting straight lines, fitting curves, and fitting an equation with more than two continuous variables.

- enhancing the regression analysis by adding confidence curves.

The methods in this chapter are appropriate for continuous variables. Regression analyses that handle other types of variables exist, but they are outside the scope of this book. See Appendix 1, "Further Reading," for suggested references.

Chapters 10 and 11 discuss the activities of regression analysis and regression diagnostics. Fitting a regression model and performing diagnostics are intertwined. In regression, you first fit a model. Then, you perform diagnostics to assess how well the model fits. You repeat this process until you find a suitable model. This chapter focuses on fitting models. Chapter 11 focuses on performing diagnostics.

Summarizing Multiple Continuous Variables

This section discusses methods for summarizing continuous variables. Chapter 4 described how to use summary statistics (such as the mean) and graphs (such as histograms and box plots) for a continuous variable. Use these graphs and statistics to check the data for errors before performing any statistical analyses.

For continuous variables, a *scatter plot* with one variable on the *y*-axis and another variable on the *x*-axis shows the possible relationship between the continuous variables. Scatter plots provide another way to check the data for errors.

Suppose a homeowner was interested in the effect that using the air conditioner had on the electric bill. The homeowner recorded the number of hours the air conditioner was used for 21 days, and the number of times the dryer was used each day. The homeowner also monitored the electric meter for these 21 days, and computed the amount of electricity used each day in kilowatt-hours.[1] Table 10.1 displays the data.

This data is available in the **kilowatt** data set in the sample data for the book. To create the data set in SAS, submit the following:

```
data kilowatt;
    input kwh ac dryer @@;
    datalines;
35 1.5 1 63 4.5 2 66 5.0 2 17 2.0 0 94 8.5 3 79 6.0 3
93 13.5 1 66 8.0 1 94 12.5 1 82 7.5 2 78 6.5 3 65 8.0 1
77 7.5 2 75 8.0 2 62 7.5 1 85 12.0 1 43 6.0 0 57 2.5 3
33 5.0 0 65 7.5 1 33 6.0 0
;
run;
```

---

[1] Data is from Dr. Ramon Littell, University of Florida. Used with permission.

Table 10.1 Kilowatt Data

AC	Dryer	Kilowatt-Hours
1.5	1	35
4.5	2	63
5	2	66
2	0	17
8.5	3	94
6	3	79
13.5	1	93
8	1	66
12.5	1	94
7.5	2	82
6.5	3	78
8	1	65
7.5	2	77
8	2	75
7.5	1	62
12	1	85
6	0	43
2.5	3	57
5	0	33
7.5	1	65
6	0	33

Creating Scatter Plots

SAS provides several ways to create scatter plots. **PROC CORR** produces both scatter plots and statistical reports. When you begin to summarize several continuous variables, using PROC CORR can be more efficient because you can perform all of the summary tasks with this one procedure.

```
ods graphics on;
ods select ScatterPlot;
proc corr data=kilowatt
          plots=scatter(noinset ellipse=none);
   var ac kwh;
run;
ods graphics off;
```

The **ODS** statement specifies that PROC CORR print only the scatter plot. The **ODS GRAPHICS** statements activate and deactivate the graphics for the procedure. See "Using ODS Statistical Graphics" in the next section.

In the PROC CORR statement, **PLOTS=SCATTER** specifies a scatter plot. **NOINSET** suppresses an automatic summary that shows the number of observations and the correlation coefficient (discussed later in this chapter). **ELLIPSE=NONE** suppresses the automatic prediction ellipse. The first variable in the **VAR** statement identifies the variable for the *x*-axis, and the second variable identifies the variable for the *y*-axis. When you use this combination of options, PROC CORR prints a simple scatter plot of the two variables. Figure 10.1 shows the results.

Figure 10.1 Scatter Plot of KWH by AC

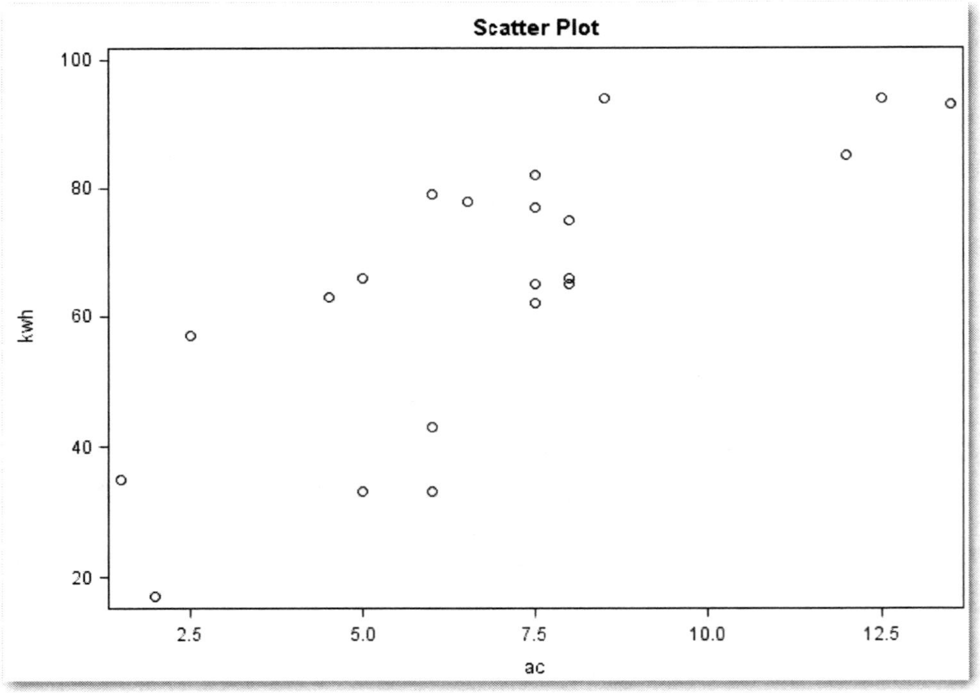

This simple scatter plot shows that higher values of **kwh** tend to occur with higher values of **ac**. However, the relationship is not perfect because some days have higher **ac** values with lower **kwh** values. This difference is because of other factors (such as how hot it was or how many other appliances were used) that affect the amount of electricity that was used on a given day.

Using ODS Statistical Graphics

As described in Chapter 4, you can use the ODS statement to control the output that is printed by a procedure. SAS includes ODS Statistical Graphics for some procedures.[2] ODS Statistical Graphics is a versatile and complex system that provides many features. This section discusses the basics.

The simplest way to use ODS Statistical Graphics is to use the automatic approach of including the graphs in SAS output. The key difference between ODS Statistical Graphics and plain ODS is that SAS displays the graphs in a separate window (when using ODS Statistical Graphics) rather than in the Graphics window (when using ODS). Figure 10.2 shows an example of how ODS Statistical Graphics output is displayed in the Results window when you are in the windowing environment mode. ODS Statistical Graphics created the graph labeled **kwh by ac**. For convenience, the rest of Chapter 10 and all of Chapter 11 use the term "ODS graphics" to refer to ODS Statistical Graphics.

Figure 10.2 Results When Using ODS Graphics

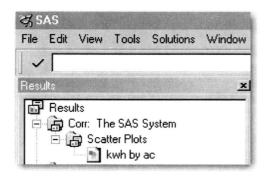

To use ODS graphics, you activate the feature before a procedure, and you deactivate the feature afterward. Look at the code before Figure 10.1. The first statement activates ODS graphics for **PROC CORR**, and the last statement deactivates ODS graphics.

---

[2] ODS Statistical Graphics are available in many procedures starting with SAS 9.2. For SAS 9.1, ODS Statistical Graphics were experimental and available in fewer procedures. For releases earlier than SAS 9.1, you need to use alternative approaches like traditional graphics or line printer plots, both of which Chapter 10 discusses. Also, to use ODS Statistical Graphics, you must have SAS/GRAPH software licensed.

The general form of the statements to activate and then deactivate ODS graphics in a SAS procedure is shown below:

ODS GRAPHICS ON;

ODS GRAPHICS OFF;

Creating a Scatter Plot Matrix

A simple scatter plot of two variables can be useful. When you have more than two variables, a *scatter plot matrix* can help you understand the possible relationships between the variables. A scatter plot matrix shows all possible simple scatter plots for all combinations of two variables:

```
ods graphics on;
ods select MatrixPlot;
proc corr data=kilowatt
          plots=matrix(histogram nvar=all);
   var ac dryer kwh;
run;
ods graphics off;
```

The ODS statement specifies that PROC CORR print only the scatter plot matrix. The VAR statement lists all of the variables that you want to plot.

In the PROC CORR statement, **PLOTS=MATRIX** specifies a scatter plot matrix. **HISTOGRAM** adds a histogram for each variable along the diagonal of the matrix.[3] **NVAR=ALL** includes all variables in the VAR statement in the scatter plot matrix. SAS automatically uses NVAR=5 for a 5×5 scatter plot matrix. For the **kilowatt** data, this option is not needed, but it is shown in the example to help you with your data.

Figure 10.3 shows the results.

---

[3] The **HISTOGRAM** option is available starting with SAS 9.2. For earlier releases of SAS, you can create the scatter plot matrix, but you cannot add histograms along the diagonal of the matrix.

Figure 10.3 Scatter Plot Matrix for Kilowatt Data

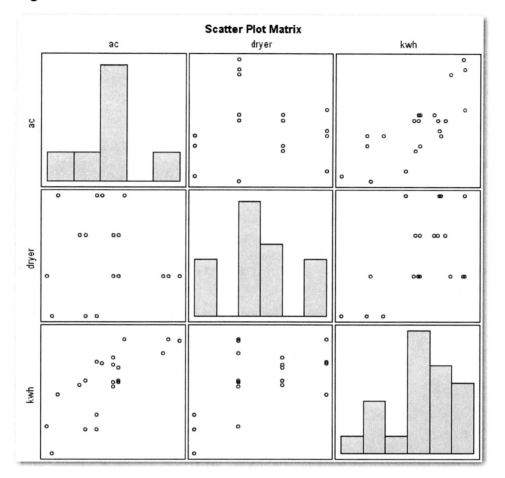

The scatter plot matrix shows all possible two-way scatter plots for the three variables. (For your data, the matrix will show all possible two-way scatter plots for the variables you choose.) The center diagonal automatically identifies the variables. When you use the HISTOGRAM option, the center diagonal displays a histogram for each variable. Figure 10.4 helps decode the structure of the scatter plot matrix.

Figure 10.4 Understanding the Structure of the Scatter Plot Matrix

ac	ac by dryer	ac by kwh
dryer by ac	dryer	dryer by kwh
kwh by ac	kwh by dryer	kwh

The ac by kwh scatter plot in the upper right corner shows ac on the *y*-axis and kwh on the *x*-axis. The lower left corner shows the mirror image of this scatter plot, with kwh on the *y*-axis and ac on the *x*-axis. kwh is the *response variable*, or the variable whose values are impacted by the other variables. Typically, scatter plots show the response variable on the vertical *y*-axis, and the other variables on the *x*-axis. PROC CORR shows the full scatter plot matrix, and you choose which scatter plot to use.

Look at the lower left corner of Figure 10.3. This is the same scatter plot as the simple scatter plot in Figure 10.1. This plot shows that higher values of kwh tend to occur with higher values of ac.

The kwh by dryer scatter plot shows that higher values of kwh tend to occur with higher values of dryer. By just looking at the scatter plot, this increasing relationship does not appear to be as strong as the increasing relationship between kwh and ac.

The histograms along the diagonal of Figure 10.3 help show the distribution of values for each variable. You can see that dryer has only a few values, and that kwh appears to be skewed to the right.

Figure 10.3 illustrates another benefit of a scatter plot matrix. It can highlight potential errors in the data by showing potential outlier points in two directions. For the kilowatt data, all of the data points are valid. Another example of a benefit of the scatter plot matrix is with blood pressure readings. A value of 95 for systolic blood pressure (the higher number of the two numbers) for a person is reasonable. Similarly, a value of 95 for diastolic blood pressure (the lower number of the two numbers) for a person is reasonable. However, a value of 95/95 for systolic/diastolic is not reasonable. Checking the two variables one at a time (with **PROC UNIVARIATE**, for example) would not highlight the value of 95/95 for systolic/diastolic as an issue. However, the scatter plot matrix would highlight this value as a potential outlier point that should be investigated.

Using Summary Statistics to Check Data for Errors

As a final task of summarizing continuous variables, think about checking the data for errors. The scatter plot matrix is an effective way to find potential outlier points. Chapter 4 discussed using box plots to find outlier points, and it recommended checking the minimum and maximum values to confirm that they seem reasonable. You could use PROC UNIVARIATE to produce a report of summary statistics for each variable, but using PROC CORR can be more efficient:

```
ods select SimpleStats;
proc corr data=kilowatt;
   var ac dryer kwh;
title 'Summary Statistics for KILOWATT Data Set';
run;
```

The ODS statement specifies that PROC CORR print only the simple summary statistics table. The VAR statement lists all of the variables that you want to summarize.

Figure 10.5 shows the results.

Figure 10.5 Simple Summary Statistics from PROC CORR

```
              Summary Statistics for KILOWATT Data Set

                         The CORR Procedure

                          Simple Statistics

Variable       N       Mean     Std Dev         Sum     Minimum     Maximum

ac            21    6.92857     3.13562   145.50000     1.50000    13.50000
dryer         21    1.42857     1.02817    30.00000           0     3.00000
kwh           21   64.85714    21.88444        1362    17.00000    94.00000
```

SAS identifies the variables under the **Variable** heading in the **Simple Statistics** table. See Chapter 4 for definitions of the other statistics in the other columns.

SAS uses all available values to calculate these statistics. Suppose the homeowner forgot to collect dryer information on one day, resulting in only 20 values for the dryer variable. The Simple Statistics table would show N as 20 for dryer, and N as 21 for the other two variables.

Reviewing PROC CORR Syntax for Summarizing Variables

The general form of the statements to summarize and plot variables using PROC CORR is shown below:

ODS GRAPHICS ON;

PROC CORR DATA=*data-set-name options*;

   VAR *variables*;

ODS GRAPHICS OFF;

data-set-name is the name of a SAS data set, and *variables* are the variables that you want to plot or summarize. If you omit the *variables* from the VAR statement, then the procedure uses all numeric variables.

The PROC CORR statement *options* can be:

 PLOTS=SCATTER(*scatter-options***)**   creates a simple scatter plot for the variables.

SAS uses the first variable in the VAR statement for the *x*-axis, and the second variable for the *y*-axis. If you use more than two variables, then SAS creates a scatter plot for each pair of variables. The *scatter-options* can be one or more of the following:

NOINSET	suppresses an automatic summary that shows the number of observations and the correlation coefficient.
ELLIPSE=NONE	suppresses the automatic prediction ellipse.

If you use *scatter-options*, then the parentheses are required.

And, the PROC CORR statement *options* can be:

 PLOTS=MATRIX(*matrix-options***)**   creates a scatter plot matrix for the variables.

The *matrix-options* can be one or more of the following:

HISTOGRAM	adds a histogram for each variable along the diagonal of the matrix.[4]
NVAR=ALL	includes all variables in the VAR statement in the scatter plot matrix. SAS automatically uses NVAR=5, so, if your data has fewer than 5 variables, you can omit this option.

If you use *matrix-options*, then the parentheses are required.

You can use both PLOTS=SCATTER and PLOTS=MATRIX in the same PROC CORR statement.

---

[4] The **HISTOGRAM** option is available starting with SAS 9.2. For earlier releases of SAS, you can create the scatter plot matrix, but you cannot add histograms along the diagonal of the matrix.

You can use the ODS statement to specify the output tables to print. (See "Special Topic: Using ODS to Control Output Tables" at the end of Chapter 4 for more detail.) Table 10.2 identifies the output tables for PROC CORR summary statistics and plots.

Table 10.2 ODS Table Names for PROC CORR

ODS Name	Output Table or Plot
ScatterPlot	Simple scatter plot of two variables
MatrixPlot	Scatter plot matrix
SimpleStats	Simple Statistics table

Calculating Correlation Coefficients

PROC CORR automatically produces correlation coefficients. In the previous sections, ODS statements have suppressed the correlation coefficients.

Understanding Correlation Coefficients

PROC CORR automatically produces *Pearson correlation coefficients*. Statistical texts represent the Pearson correlation coefficient with the letter "*r*." Values of *r* range from −1.0 to +1.0.

With positive correlation coefficients, the values of the two variables increase together. With negative correlation coefficients, the values of one variable increase, while the values of the other variable decrease. Values near 0 imply that the two variables do not have a linear relationship.

A correlation coefficient of 1 (*r*=+1) corresponds to a plot of points that fall exactly on an upward-sloping straight line. A correlation coefficient of −1 (*r*=−1) corresponds to a plot of points that fall exactly on a downward-sloping straight line. (Neither of these straight lines necessarily has a slope of 1.) The correlation of a variable with itself is always +1. Otherwise, values of +1 or −1 usually don't occur in real situations because plots of real data don't fall exactly on straight lines.

In reality, values of *r* are between −1 and +1. Relatively large positive values of *r*, such as 0.7, correspond to a plot of points that have an upward trend. Relatively large negative values of *r*, such as −0.7, correspond to a plot of points that have a downward trend.

What defines the terms "near 0" or "relatively large"? How do you know whether the correlation coefficient measures a strong or weak relationship between two continuous variables? SAS provides these answers in the **Pearson Correlation Coefficients** table:

```
proc corr data=kilowatt;
   var ac dryer kwh;
title 'Correlations for KILOWATT Data Set';
run;
```

The PROC CORR and VAR statements are the same as the statements used to produce simple statistics. The example above does not use an ODS statement, so the results contain all automatic reports from SAS.

Figure 10.6 shows the results.

Figure 10.6 Correlation Coefficients from PROC CORR

```
                    Correlations for KILOWATT Data Set

                           The CORR Procedure

              3  Variables:    ac        dryer      kwh


                            Simple Statistics

Variable        N         Mean      Std Dev        Sum     Minimum      Maximum

ac             21      6.92857      3.13562    145.50000     1.50000     13.50000
dryer          21      1.42857      1.02817     30.00000           0      3.00000
kwh            21     64.85714     21.88444         1362    17.00000     94.00000


                Pearson Correlation Coefficients, N = 21
                      Prob > |r| under H0: Rho=0

                            ac          dryer         kwh

            ac         1.00000       -0.02880      0.76528
                                      0.9014        <.0001

            dryer     -0.02880        1.00000      0.59839
                       0.9014                       0.0042

            kwh        0.76528        0.59839      1.00000
                       <.0001         0.0042
```

Figure 10.6 contains three tables. The **Variables Information** table lists the variables. The Simple Statistics table shows summary statistics, as described in the previous section. The Pearson Correlation Coefficients table shows the correlation coefficients and the results of statistical tests.

The **Pearson Correlation Coefficients** table displays the information for the variables in a matrix. This matrix has the same structure as the scatter plot matrix. See Figure 10.4 to help understand the structure. The heading for the table identifies the top number as the correlation coefficient, and the bottom number as a *p*-value. (See the next section, "Understanding Tests for Correlation Coefficients," for an explanation of the bottom number.) Because the data values are all nonmissing, the heading for the table also identifies the number of observations.

Figure 10.6 shows that the relationship between **ac** and **kwh** has a positive correlation of **0.76528**. Figure 10.6 shows that there is a weaker positive relationship between **dryer** and **kwh** with the correlation coefficient of **0.59839**. Figure 10.6 shows a very weak negative relationship between **ac** and **dryer** with the correlation coefficient of **–0.02880**. These results confirm the visual understanding from the scatter plot matrix.

The correlation coefficients along the diagonal of the matrix are **1.00000** because the correlation of a variable with itself is always +1.

Understanding Tests for Correlation Coefficients

The **Pearson Correlation Coefficients** table displays a *p*-value as the bottom number in most cells of the matrix. The table does not display a *p*-value for cells along the diagonal because the correlation of a variable with itself is always +1 and the test is not relevant.

The heading for the table identifies the test as **Prob > |r| under H0: Rho=0**. The true unknown population correlation coefficient is called Rho. (Chapter 5 discusses how population parameters are typically denoted by Greek letters.) For this test, the null hypothesis is that the correlation coefficient is 0. The alternative hypothesis is that the correlation coefficient is different from 0. This test provides answers to the questions of whether a correlation coefficient is relatively large or near 0.

Suppose you choose a 5% significance level. This level implies a 5% chance of concluding that the correlation coefficient is different from 0 when, in fact, it is not. This decision gives a reference probability of 0.05.

For the **kilowatt** data, focus on the two correlation coefficients involving **kwh**. Both of these correlation coefficients are significantly different from 0. The **ac** and **kwh** test has a *p*-value of **<.0001**. The **dryer** and **kwh** test has a *p*-value of **0.0042**. These results confirm the visual understanding from the scatter plot matrix.

In addition, the results for **ac** and **dryer** confirm the visual understanding from the scatter plot matrix. With a *p*-value of **0.9014**, you conclude that there is not a linear relationship between these two variables.

In general, if the *p*-value is less than the significance level, then reject the null hypothesis, and conclude that the correlation coefficient is significantly different from 0. If the *p*-value is greater than the significance level, then you fail to reject the null hypothesis.

You conclude that there is not enough evidence that the correlation coefficient is significantly different from 0.

Working with Missing Values

The **kilowatt** data has no missing values, so **PROC CORR** simply prints the number of observations in the heading for the **Pearson Correlation Coefficients** table.

For data sets with missing values, **PROC CORR** automatically uses as many observations as possible. Suppose the homeowner did not collect dryer information on the first day. The value for **dryer** would be missing for the first observation. (Suppose this data is the **kilowatt2** data set.) The statements below cause the automatic behavior to occur:

```
proc corr data=kilowatt2;
   var ac dryer kwh;
title 'Automatic Behavior with Missing Values for KILOWATT2';
run;
```

Figure 10.7 shows the results.

Figure 10.7 Correlation Coefficients with Missing Values

```
               Automatic Behavior with Missing Values for KILOWATT2

                            The CORR Procedure

               3  Variables:      ac        dryer      kwh


                              Simple Statistics

   Variable        N         Mean      Std Dev         Sum     Minimum     Maximum

   ac             21      6.92857      3.13562   145.50000     1.50000    13.50000
   dryer          20      1.45000      1.05006    29.00000           0     3.00000
   kwh            21     64.85714     21.88444        1362    17.00000    94.00000


                     Pearson Correlation Coefficients
                       Prob > |r| under H0: Rho=0
                          Number of Observations

                               ac          dryer          kwh

              ac          1.00000       -0.07298        0.76528
                                          0.7598         <.0001
                               21             20             21

              dryer      -0.07298        1.00000        0.60127
                           0.7598                         0.0050
                               20             20             20

              kwh         0.76528        0.60127        1.00000
                           <.0001          0.0050
                               21             20             21
```

Figure 10.7 indicates the missing value for **dryer** in the **Simple Statistics** table with an N of **20**. Figure 10.7 shows a third value in each cell in the **Pearson Correlation Coefficients** table. The heading for the table identifies this third value as **Number of Observations**. Compare the values in the cells. Any cell that involves **dryer** also involves only 20 observations because of the missing value.

In situations with several variables and missing values, you might want to use only the observations where all variables have data values. The **NOMISS** option in the PROC CORR statement limits analyses to the observations where all variables are nonmissing. The statements below use this option.

```
proc corr data=kilowatt2 nomiss;
   var ac dryer kwh;
title 'NOMISS Option for KILOWATT2';
run;
```

Figure 10.8 shows the results.

Figure 10.8 Correlation Coefficients with NOMISS Option

```
                       NOMISS Option for KILOWATT2

                          The CORR Procedure

              3  Variables:     ac        dryer     kwh


                           Simple Statistics

Variable        N          Mean     Std Dev         Sum      Minimum     Maximum

ac             20       7.20000     2.95314   144.00000      2.00000    13.50000
dryer          20       1.45000     1.05006    29.00000            0     3.00000
kwh            20      66.35000    21.32771        1327     17.00000    94.00000


              Pearson Correlation Coefficients,  N = 20
                     Prob > |r| under H0: Rho=0

                          ac           dryer          kwh

          ac          1.00000        -0.07298       0.73544
                                      0.7598         0.0002

          dryer       -0.07298        1.00000       0.60127
                       0.7598                        0.0050

          kwh         0.73544         0.60127       1.00000
                      0.0002          0.0050
```

Compare the results in Figure 10.7 and 10.8. Figure 10.8 shows an N of **20** for all variables. Figure 10.8 does not show the third value in each cell in the **Pearson Correlation Coefficients** table. Instead, the heading for the table identifies the number of observations as **20** for all correlations.

If your data has missing values, you might want to consider the **NOMISS** option. However, with many missing values, you might exclude a large number of observations, and you might receive misleading results. In this situation, consult a statistician because another analysis might be more appropriate for your data.

Reviewing PROC CORR Syntax for Correlations

The general form of the statements to create correlation coefficients and tests using PROC CORR is shown below:

PROC CORR DATA=*data-set-name options*;

   VAR *variables*;

The PROC CORR statement *options* can be as defined earlier and can also be:

      NOMISS   excludes observations with missing values.

If you omit the *variables* from the VAR statement, then the procedure uses all numeric variables. Specify at least two variables for the output to be meaningful because the correlation of a variable with itself is always +1.

Other items in italic were defined earlier.

You can use the ODS statement to specify the output tables to print. (See "Special Topic: Using ODS to Control Output Tables" at the end of Chapter 4 for more detail.) Table 10.4 identifies the output tables for PROC CORR.

Table 10.4 ODS Table Names for PROC CORR

ODS Name	Output Table or Plot
VarInformation	Variables Information
SimpleStats	Simple Statistics table
PearsonCorr	Pearson Correlation Coefficients

Cautions about Correlations

There are two important cautions to remember when interpreting correlations.

First, correlation is not the same as causation. When two variables are highly correlated, it does not necessarily imply that one variable causes the other. In some cases, there might be an underlying causal relationship, but you don't know this information from the correlation coefficient. In other cases, the relationship might be caused by a different variable. For example, consider a store owner who notices a high positive correlation between the sales of ice scrapers and hot coffee. Coffee sales don't cause ice scraper sales, and the reverse is not true, either. Perhaps the sales of both of these items are caused by weather—drivers purchasing more hot coffee and more ice scrapers on cold, snowy days. Proving cause-and-effect relationships is much more difficult than showing a high correlation.

Second, "shopping for significance" in a large group of correlations is not a good idea. Researchers sometimes make the mistake of measuring many variables, finding the correlations between all possible pairs of variables, and choosing the significant correlations to make conclusions. Recall the interpretation of a significance level, and you see the mistake: If a researcher performs 100 correlations, and tests their significance at the 0.05 significance level, then about 5 significant correlations will be found by chance alone.

Questions Not Answered by Correlation

From the value of *r* in Figure 10.6, you conclude that as the use of the air conditioner increases, so do the kilowatt-hours (**kwh**) that are consumed. This is not a surprise. Here are some other important questions:

- How many kilowatt-hours are consumed for each hour of using the air conditioner?

- What is a prediction of the kilowatt-hour consumption for a specific day when the air conditioner is used for a specified number of hours?

- What is an estimate of the average kilowatt-hour consumption on days when the air conditioner is used for a specified number of hours?

- What are the confidence limits for the predicted kilowatt-hour consumption?

A regression analysis of the data answers these questions. The next sections discuss fitting a straight line with regression analysis, fitting a curve, and regression analysis with multiple *x* variables.

Performing Straight-Line Regression

A correlation coefficient indicates that a linear relationship exists, and the *p*-value evaluates the strength of this linear relationship. However, a correlation coefficient does not fully describe the relationship between the two variables. Figure 10.9 shows a plot for two different *y* variables against the same *x* variable. The correlation coefficient between *y* and *x* is 1.0 in both cases. All of the points for each *y* variable fall exactly on a straight line. However, the relationships between *y* and *x* are quite different for the two *y* variables.

Figure 10.9 Correlation Coefficients of 1.0

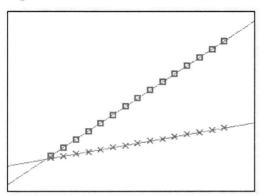

Look again at the kwh by ac scatter plot in Figure 10.1. You could draw a straight line through the data points. In a room of people, each person would draw a slightly different straight line. Which straight line is "best"? The next three sections discuss statistical methods and assumptions for fitting the "best" straight line to data.

Understanding Least Squares Regression

Least squares regression fits a straight line through the data points that minimizes the vertical differences between all of the points and the straight line. Figure 10.10 shows this approach. For each data point, calculate the vertical difference between the data point and the straight line. Then, square the difference. This gives points above and below the line the same importance (squared differences are all positive). Finally, sum the squared differences. The "best" straight line is the line that minimizes the sum of squared differences, which is the basis of the name "least squares regression."

Figure 10.10 Process of Least Squares Regression

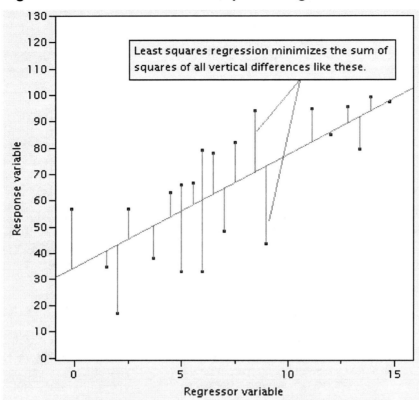

Explaining Regression Equations

The *regression equation* summarizes a straight-line relationship between two variables:

$$y = \beta_0 + \beta_1 x + \varepsilon$$

The *y* variable is the *dependent variable* or the *response variable*. The *x* variable is the *independent variable* or the *regressor variable*. The equation says that the dependent variable *y* is a straight-line function of *x*, with an *intercept* β_0 and a *slope* β_1. The intercept is the value of the *y*-axis when *x*=0. The slope is the amount of vertical increase for a one-unit horizontal increase. In algebra, this is called "the rise over the run." For the **kilowatt** data, the slope estimates the average number of kilowatt-hours for running the air conditioner for one hour.

Because a given *x* value does not always lead to the same *y* value, the data has random variation. The equation uses ε to denote the random variation. For the **kilowatt** data, the homeowner knows how long the air conditioner was used each day. But, other appliances and lights consume electricity in varying amounts from day to day. This variation is the "error" in the model equation. For straight-line regression, the "error" includes both the random variation in the data and other potential variables (such as **dryer**) that are not included in the model equation.

Statisticians use this equation when they have the entire population of values for *x* and *y*. In that case, the regression equation describes the entire population. In reality, you usually collect just a sample, and then use the sample data to estimate the straight line for the entire population. In this case, you use the following equation:

$$\hat{y} = b_0 + b_1\, x$$

b_0 and b_1 are the estimates of β_0 and β_1, respectively. The value b_0 gives the predicted value of *y* when *x* is 0. The value b_1 gives the increase in the *y* value that results from a one-unit increase in *x*. The values of b_0 and b_1 are called *parameter estimates* because they estimate the unknown population parameters β_0 and β_1, respectively. The value $\hat{y}$ indicates the predicted value of the response variable from the equation.

Assumptions for Least Squares Regression

Like all analyses, least squares regression requires some assumptions. Here they are:

- The data values are measurements. Both the *x* and *y* variables are continuous.

  For the **kilowatt** data, this assumption seems reasonable.

- Observations are independent. The values from one pair of *x-y* observations are not related to the values from another pair. To check the assumption, think about your data and whether this assumption seems reasonable. This is not a step where using SAS will answer this question.

  For the **kilowatt** data, the observations are independent because the measurements for one day are not related to the measurements for other days.

- Observations are a random sample from the population. You want to make conclusions about a larger population, not about just the sample. To check the assumption, think about your data and whether this assumption seems reasonable. This is not a step where using SAS will answer this question.

For the **kilowatt** data, you want to make conclusions about **kwh** and **ac** use in general, not for just the 21 days in the data set. Assuming that the days in the study are a random sample of all possible days seems reasonable.

- The values of x variables are known without error. To check the assumption, think about your data and whether this assumption seems reasonable. This is not a step where using SAS will answer this question.

  For the **kilowatt** data, assuming that the homeowner accurately measured the number of hours that the air conditioner was used seems reasonable.

- The errors in the data are normally distributed with a mean of 0 and a variance of σ^2. However, the regression equation has meaning even if the assumption does not seem reasonable. This assumption is needed for hypothesis tests and confidence intervals. (Chapter 11 discusses this assumption. For now, assume that it seems reasonable.)

Steps in Fitting a Straight Line

The steps for performing the regression analysis are basically the same as the steps for comparing groups:

1. Create a SAS data set.

2. Check the data set for errors.

3. Choose the significance level.

4. Check the assumptions.

5. Perform the test.

6. Make conclusions from the results.

For the **kilowatt** data, steps 1, 2, and 4 are done. For step 3, choose a 5% significance level, which requires a p-value less than 0.05 to conclude that the regression coefficients are different from 0. This level implies a 5% chance of concluding that the regression coefficients are different from 0 when, in fact, they are not.

Fitting a Straight Line with PROC REG

As with scatter plots, SAS provides more than one way to fit a straight line to the data. In fact, SAS includes more than 25 procedures that perform regression. This book focuses on **PROC REG**, which performs linear regression and includes diagnostic statistics and plots. If the topics in this chapter do not fit your situation, then check the SAS documentation. SAS very likely provides a procedure that meets your needs. To fit a straight line to kwh and ac in SAS:

```
proc reg data=kilowatt;
   model kwh=ac;
   plot kwh*ac / nostat cline=red;
title 'Straight-line Regression for KILOWATT Data';
run;
```

When performing regression, PROC REQ requires the **MODEL** statement. The MODEL statement above identifies kwh as the response (*y)* variable, and ac as the independent (*x*) variable. The **PLOT** statement is not required. Above, the PLOT statement produces a scatter plot with the fitted straight line.[5] **NOSTAT** suppresses several statistics that automatically appear to the right of the scatter plot. **CLINE=RED** specifies that the regression line is red.

Figure 10.11 shows the scatter plot. Figure 10.12 shows the printed results.

---

[5] To create this high-resolution plot, you must have SAS/GRAPH software licensed.

Figure 10.11 Straight-Line Regression Plot for Kilowatt Data

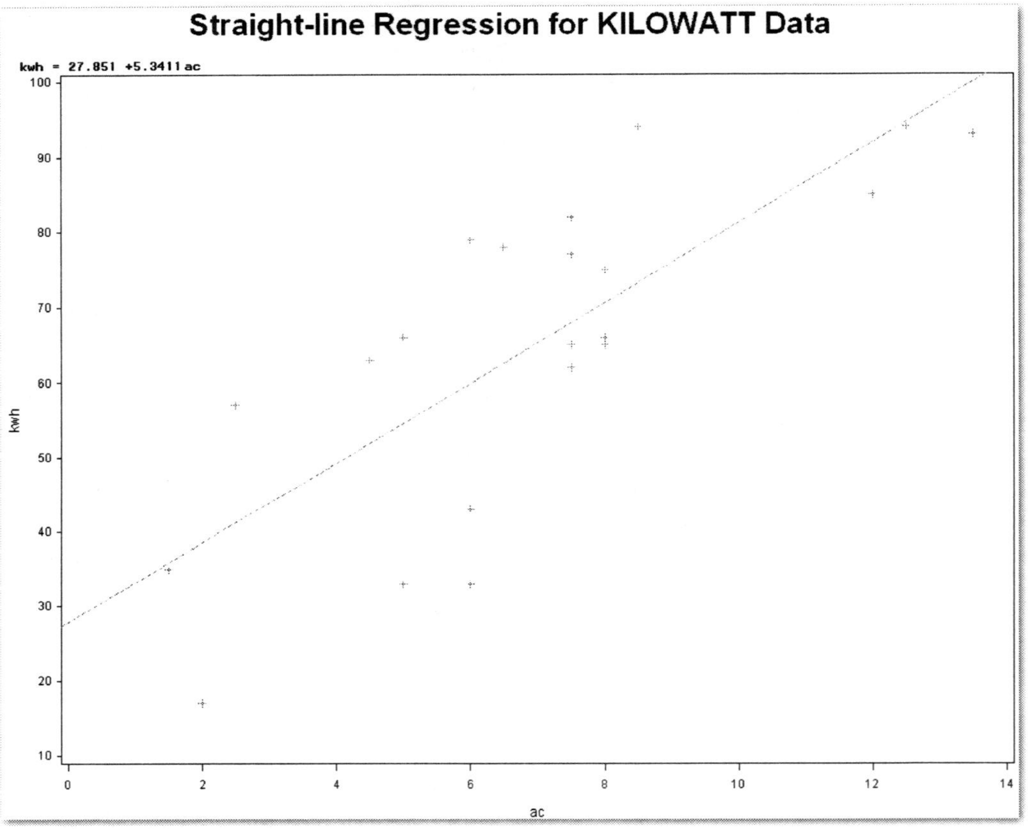

The scatter plot shows the fitted regression equation above the plot on the left.

Figure 10.12 Straight-Line Regression Results for Kilowatt Data

```
           Straight-line Regression for KILOWATT Data

                      The REG Procedure
                       Model: MODEL1
                    Dependent Variable: kwh

              Number of Observations Read        21
              Number of Observations Used        21

                      Analysis of Variance

                              Sum of         Mean
    Source            DF      Squares       Square    F Value    Pr > F

    Model              1   5609.66260   5609.66260      26.85    <.0001
    Error             19   3968.90883    208.88994
    Corrected Total   20   9578.57143


              Root MSE            14.45303    R-Square    0.5856
              Dependent Mean      64.85714    Adj R-Sq    0.5638
              Coeff Var           22.28440


                      Parameter Estimates

                         Parameter      Standard
    Variable      DF      Estimate         Error    t Value    Pr > |t|

    Intercept      1      27.85107       7.80654       3.57      0.0021
    ac             1       5.34108       1.03067       5.18      <.0001
```

The printed tables provide details for the regression analysis. The next five sections discuss these tables.

Finding the Equation for the Fitted Straight Line

First, answer the research question, "What is the relationship between the two variables?" Figure 10.11 shows the red fitted straight line in the scatter plot. Just above the scatter plot, on the left, SAS shows the equation for the fitted line:

kwh = 27.85 + 5.34*ac

This equation rounds the SAS parameter estimates to two decimal places.

The fitted line has an intercept value of 27.85. When the homeowner does not use the air conditioner at all, the fitted line predicts that 27.85 kilowatt-hours will be used that day.

The fitted line has a slope of 5.34. The homeowner uses 5.34 kilowatt-hours of electricity when the air conditioner runs for 1 hour. If the homeowner uses the air conditioner for 8 hours, the fitted line predicts that 42.72 (5.34*8) kilowatt-hours of electricity will be used.

For this entire day, the fitted line predicts the following:

kilowatt-hours $\quad$ = 27.85 + (5.34*8)
$\qquad\qquad\qquad\qquad$ = 27.85 + 42.72
$\qquad\qquad\qquad\qquad$ = 70.57

Correlation coefficients could not provide an answer to the question of how many kilowatt-hours are consumed for one hour's use of the air conditioner. The slope of the regression line gives the answer of 5.34 kilowatt-hours. (Because a kilowatt is 1000 watts, the air conditioner uses 5340 watts per hour. This is more kilowatt-hours than would be consumed by fifty-three 100-watt light bulbs.)

Another goal was to predict the kilowatt-hours that are consumed when using the air conditioner for a specified number of hours. Again, the regression line gives the answer, as shown in the equation for using the air conditioner for 8 hours. (See "Printing Predicted Values and Limits" and "Plotting Predicted Values and Limits" for more detail.)

Figure 10.12 shows the estimates for the intercept and slope in the **Parameter Estimates** table.

As with correlation coefficients, a natural question about the intercept and slope is whether they are significantly different from 0. The next section explains how to find the answer to this question.

Understanding the Parameter Estimates Table

Figure 10.12 shows the **Parameter Estimates** table, which answers the question about the significance of the estimates for the intercept and the slope. SAS tests the null hypothesis that the parameter estimate is 0. The alternative hypothesis is that the parameter estimate is different from 0.

Look under the heading **Pr > |t|** in Figure 10.12. The *p*-values for the intercept and the slope are **0.0021** and **<.0001**, respectively. Because these values are less than the significance level of 0.05, you reject the null hypothesis, and you conclude that the intercept and slope are significantly different from 0.

The statistical results make sense because the slope indicates that increasing the number of hours that the air conditioner is used produces an increase in the kilowatt-hours for that day. The intercept is significantly different from 0, which indicates that even if the homeowner does not use the air conditioner, some kilowatt-hours are consumed that day.

In general, to interpret SAS results, look at the *p*-values under the heading Pr > |t|. If the *p*-value is less than your significance level, reject the null hypotheses that the parameter estimate is 0. If the *p*-value is greater, you fail to reject the null hypothesis.

The list below describes items in the **Parameter Estimates** table:

Variable	identifies the parameter. SAS identifies the intercept as **Intercept,** and the slope of the line by the regressor variable (**ac** in this example).
DF	is the degrees of freedom for the parameter estimate. This should equal 1 for each parameter.
Parameter Estimate	lists the parameter estimate.
Standard Error	is the standard error of the estimate. It measures how much the parameter estimates would vary from one collection of data to the another collection of data.
t Value	gives the *t*-value for testing the hypothesis. SAS calculates the *t*-value as the parameter estimate divided by its standard error.
	For example, the *t*-value for the slope is 5.341/1.031, which is 5.18.

Understanding the Fit Statistics Table

Think about testing the parameter estimates. If neither the intercept nor the slope is significantly different from 0, then the straight line does not fit the data very well. The **Fit Statistics** table provides statistical tools to assess the fit.

Find the value for **R-Square** in Figure 10.12. This statistic is the proportion of total variation in the *y* variable due to the *x* variables in the model. R-Square ranges from 0 to 1 with higher values indicating better-fitting models. For the kilowatt data, R-Square is **0.5856**. Statisticians often express this statistic as a percentage, saying that the model explains about 59% of the variation in the data (for this example).

The Fit Statistics table provides other statistics. The **Dependent Mean** is the average for the *y* variable. Compare the Dependent Mean value in Figure 10.12 with the Mean value for **kwh** in Figure 10.5, and confirm that they are the same. See the SAS documentation for explanations of the other statistics in this table.

Understanding the Analysis of Variance Table

The **Analysis of Variance** table looks similar to tables in Chapter 9. For regression analysis, statisticians use this table to help assess whether the overall model explains a significant amount of the variation in the data. The table augments conclusions from the **R-Square** statistic. Statisticians typically evaluate the overall model before examining the details for individual terms in the model.

When performing regression, the null hypothesis for the **Analysis of Variance** table is that the variation in the *y* variable is not explained by the variation in the *x* variable (for straight-line regression). The alternative hypothesis is that the variation in the *y* variable is explained by the variation in the *x* variable. Statisticians often refer to these hypotheses as testing whether the model is significant or not.

Find the value for **Pr > F** in Figure 10.12. This is the *p*-value for testing the hypothesis. Compare this *p*-value with the significance level. For the **kilowatt** data, the *p*-value of **<.0001** is less than 0.05. You conclude that the variation in **ac** explains a significant amount of the variation in **kwh**.

In general, to interpret SAS results, look at the *p*-value under the heading **Pr > F**. If the *p*-value is less than your significance level, reject the null hypotheses that the variation in the *y* variable is not explained by the variation in the *x* variable. If the *p*-value is greater, you fail to reject the null hypothesis.

Chapter 9 describes the other information in the **Analysis of Variance** table. In Chapter 9, see "Understanding Results" in "Performing a One-Way ANOVA."

Figure 10.12 shows the **DF** for **Model** as **1**. When fitting a straight line with a single independent variable, the DF for **Model** will always be 1. In general, the degrees of freedom for the model is one less than the number of parameters estimated by the model. When fitting a straight line, **PROC REG** estimates two parameters—the intercept and the slope—so the model has 1 degree of freedom.

Figure 10.12 shows that the **Sum of Squares** for **Model** and **Error** add to be the **Corrected Total**. Here is the equation:

$$9578.57 = 5609.66 + 3968.91$$

Statisticians usually abbreviate sum of squares as SS. They describe the sum of squares with this equation:

$$\text{Total SS} = \text{Model SS} + \text{Error SS}$$

This is a basic and important equation in regression analysis. Regression partitions total variation into variation due to the variables in the model, and variation due to error.

The model cannot typically explain all of the variation in the data. For example, Table 10.1 shows three observations with **ac** equal to **8** and with **kwh** values of **66, 65,** and **75**. Because the three different **kwh** values occurred on days with the same value of **8** for **ac**, the straight-line regression model cannot explain the variation in these data points. This variation must be due to error (or due to factors other than the use of the air conditioner).

Understanding Other Items in the Results

PROC REG uses headings and the **Number of Observations** table to identify the data used in the regression.

The heading contains two lines labeled **Model** and **Dependent Variable**. The dependent variable is the measurement variable on the left side of the equation in the MODEL statement. PROC REG assigns a label to each MODEL statement. PROC REG automatically uses **MODEL1** for the model in the first MODEL statement, MODEL2 for the model in the second MODEL statement, and so on. This book uses the automatic labels.

The **Number of Observations** table lists the number of observations read from the data set, and the number of observations used in the regression. Because the **kilowatt** data does not have missing values, PROC REG uses all observations in the analysis. For your data, PROC REG uses all observations with nonmissing values for the variables in the regression.

Using PROC REG Interactively

PROC REG is an interactive procedure. If you use line mode or the windowing environment mode, you can perform an analysis of variance, and then perform other tasks without rerunning the regression. Type the PROC REG and MODEL statements, and add a **RUN** statement to see the results. Then, you can add other statements and a second RUN statement to perform other tasks. The following statements produce the output shown in Figures 10.11 and 10.12:

```
proc reg data=kilowatt;
   model kwh=ac;
title 'Straight-line Regression for KILOWATT Data';
run;
   plot kwh*ac / nostat cline=red;
run;
quit;
```

When you use these statements, the second RUN statement does not end PROC REG because the procedure is waiting to receive additional statements. The **QUIT** statement ends PROC REG. The form of the QUIT statement is simply the word QUIT followed by a semicolon.

Although you can end the procedure by starting another DATA or PROC step, you might receive an error. If you use an ODS statement before the next PROC step, and the table you specify in the ODS statement is not available in PROC REG (which is running interactively), SAS prints an error in the log and does not create output. To avoid this situation, use the QUIT statement.

Printing Predicted Values and Limits

The section "Finding the Equation for the Fitted Straight Line" showed how to predict how many kilowatt-hours are consumed when the air conditioner is used for 8 hours on a given day. You can manually use the equation to predict other values, or you can let SAS provide the predictions. You can also add two types of confidence limits.

Defining Prediction Limits

Remember that correlations could not provide confidence limits for an individual predicted kilowatt-hour consumption. Table 10.1 shows three days in the data where the homeowner used the air conditioner for 8 hours (**ac=8**). The regression equation predicts the same kilowatt-hours for these three days. But, the actual kilowatt-hours for these days range from 65 to 75.

Suppose you want to put confidence limits on the predicted value of 70.57 (calculated in the section "Finding the Equation for the Fitted Straight Line"). Just as confidence limits for the mean provide an interval estimate for the unknown population mean, confidence limits for a predicted value provide an interval estimate for the unknown future value. These *confidence limits on individual values* need to account for possible error in the fitted regression line. They need to account for variation in the *y* variable for observations that have the same value of the *x* variable. Statisticians often refer to confidence limits on individual values as *prediction limits*.

Defining Confidence Limits for the Mean

Suppose you want to estimate the average kilowatt-hours that are consumed on all days when the air conditioner is used for a given number of hours. This task is essentially the same task as estimating the average kilowatt-hours that are consumed on a given day. In other words, to predict a single future value or to estimate the mean response for a given value of the independent variable, you use the same value from the regression equation.

For example, suppose you want to estimate the mean kilowatt-hour consumption for all days when the air conditioner is used for 8 hours. To get an estimate for each day, insert the value **ac=8** into the fitted regression equation. Because **ac** is the same for all of these days, the equation produces the same predicted value for **kwh**. It makes sense to use this predicted value as an estimate of the mean kilowatt-hours consumed for all of these days.

Next, you can add confidence limits on the average kilowatt-hours that are consumed when the air conditioner is used for 8 hours. These confidence limits will differ from the prediction limits.

Because the mean is a population parameter, the mean value for all of the days with a given value for **ac** does not change. Because the mean does not vary and the actual values do vary, it makes sense that a *confidence interval for the mean* would be smaller than the confidence interval for a single predicted value. With a single predicted value, you need to account for the variation between actual values, as well as for any error in the fitted regression line. With the mean, you need to account for only the error in the fitted regression line.

Using PROC REG to Print Predicted Values and Limits

Because PROC REG is an interactive procedure, you can print predicted values, prediction limits, and confidence limits for the mean using one of two approaches. You can provide options in the MODEL statement, or you can use the PRINT statement. For both approaches, use the same option names for the statistics requested. For both approaches, the optional ID statement adds a column that identifies the observations. PROC REG requires that the ID statement appear before the first RUN statement. The two PROC REG activities below produce the same output.

```
proc reg data=kilowatt;
   id ac;
   model kwh=ac / p clm cli;
run;
quit;

proc reg data=kilowatt;
   id ac;
   model kwh=ac;
run;
   print p clm cli;
run;
quit;
```

The first **PROC REG** activity uses options in the **MODEL** statement and produces the results in Figures 10.12 and 10.13. The second **PROC REG** activity uses the **PRINT** statement and options, and produces the results in Figures 10.12 and 10.13. With the second approach, you can add results to the **PROC REG** output without re-running the model.

Both **PROC REG** activities use the same options. The **P** option requests predicted values, **CLI** requests prediction limits, and **CLM** requests confidence limits for the mean.

Figure 10.13 shows the same heading as Figure 10.12. It also shows the **Output Statistics** table.

Figure 10.13 Printing Predicted Values and Limits

```
              Straight-line Regression for KILOWATT Data

                        The REG Procedure
                        Model: MODEL1
                     Dependent Variable: kwh

                        Output Statistics

             Dependent    Predicted     Std Error
   Obs    ac   Variable      Value    Mean Predict       95% CL Mean

    1    1.5   35.0000     35.8627       6.4228      22.4197   49.3057
    2    4.5   63.0000     51.8859       4.0265      43.4585   60.3134
    3    5.0   66.0000     54.5565       3.7280      46.7536   62.3593
    4    2.0   17.0000     38.5332       5.9792      26.0186   51.0478
    5    8.5   94.0000     73.2503       3.5455      65.8295   80.6710
    6    6.0   79.0000     59.8976       3.2959      52.9991   66.7960
    7   13.5   93.0000     99.9557       7.4713      84.3181  115.5933
    8    8.0   66.0000     70.5797       3.3416      63.5856   77.5739
    9   12.5   94.0000     94.6146       6.5514      80.9023  108.3269
   10    7.5   82.0000     67.9092       3.2084      61.1939   74.6245
   11    6.5   78.0000     62.5681       3.1847      55.9025   69.2337
   12    8.0   65.0000     70.5797       3.3416      63.5856   77.5739
   13    7.5   77.0000     67.9092       3.2084      61.1939   74.6245
   14    8.0   75.0000     70.5797       3.3416      63.5856   77.5739
   15    7.5   62.0000     67.9092       3.2084      61.1939   74.6245
   16   12.0   85.0000     91.9441       6.1048      79.1666  104.7215
   17    6.0   43.0000     59.8976       3.2959      52.9991   66.7960
   18    2.5   57.0000     41.2038       5.5480      29.5916   52.8160
   19    5.0   33.0000     54.5565       3.7280      46.7536   62.3593
   20    7.5   65.0000     67.9092       3.2084      61.1939   74.6245
   21    6.0   33.0000     59.8976       3.2959      52.9991   66.7960

                        Output Statistics

         Obs    ac       95% CL Predict      Residual

          1    1.5      2.7597   68.9657     -0.8627
          2    4.5     20.4834   83.2884     11.1141
          3    5.0     23.3158   85.7971     11.4435
          4    2.0      5.7963   71.2702    -21.5332
          5    8.5     42.1028  104.3977     20.7497
          6    6.0     28.8704   90.9247     19.1024
          7   13.5     65.9024  134.0090     -6.9557
          8    8.0     39.5312  101.6283     -4.5797
          9   12.5     61.4013  127.8279     -0.6146
         10    7.5     36.9223   98.8961     14.0908
         11    6.5     31.5919   93.5443     15.4319
         12    8.0     39.5312  101.6283     -5.5797
         13    7.5     36.9223   98.8961      9.0908
         14    8.0     39.5312  101.6283      4.4203
         15    7.5     36.9223   98.8961     -5.9092
         16   12.0     59.1057  124.7824     -6.9441
         17    6.0     28.8704   90.9247    -16.8976
         18    2.5      8.8010   73.6065     15.7962
         19    5.0     23.3158   85.7971    -21.5565
         20    7.5     36.9223   98.8961     -2.9092
         21    6.0     28.8704   90.9247    -26.8976
```

The list below describes items in the **Output Statistics** table:

Obs	identifies the observations by observation number.
ac	identifies the observations by the value of **ac**. For your data, the heading and values will reflect the variable you use in the **ID** statement.
Dependent Variable	are actual values of the response variable **kwh**.
Predicted Value	are predicted values of the response variable **kwh**.
	Observations **8, 12,** and **14** have a predicted value of **70.5797**. These are the three observations with **ac=8**, so the predicted values are the same. The **Predicted Value** differs slightly from what was computed in "Finding the Equation for the Fitted Straight Line" because the equation in the book rounds off the parameter estimates.
Std Error Mean Predict	is the standard error of the predicted value. It is used in calculating confidence limits for the mean.
95% CL Mean	the left column gives the lower confidence limit for the mean, and the right column gives the upper confidence limit for the mean. For observation **8** with **ac=8**, the 95% confidence limits are **63.59** and **77.57**. This means that you can be 95% confident that the mean kilowatt-hour consumption on a day with **ac=8** will be somewhere between 63.59 and 77.57.
95% CL Predict	the left column gives the lower prediction limit, and the right column gives the upper prediction limit. For observation **8** with **ac=8**, the 95% prediction limits are **39.53** and **101.63**. This means that you can be 95% confident that the kilowatt-hour consumption on a day with **ac=8** will be somewhere between 39.53 and 101.62.

Residual gives the *residuals*, which are the differences between actual and predicted values. Residuals are calculated by subtracting the predicted value from the actual value. For the first observation, the residual is **35.0–35.86=–0.86.** You might find it helpful to think of the residual as the amount by which the fitted regression line "missed" the actual data point. Chapter 11 discusses using residuals for regression diagnostics.

Compare the prediction limits and the confidence limits for the mean for an observation in Figure 10.13. The confidence limits for the mean are narrower than the prediction limits. Especially for larger data sets, plotting the limits can be more helpful than looking at the table.

Plotting Predicted Values and Limits

PROC REG provides multiple ways to plot. You can use ODS graphics, similar to the approach used with PROC CORR earlier in the chapter. You can use the PLOT statement to produce high-resolution graphs, and you can also use the PLOT statement to produce line printer plots. This section discusses using ODS graphics and the PLOT statement for high-resolution graphs. See "Special Topic: Line Printer Plots" at the end of the chapter for information about line printer plots.

Using ODS Graphics

The statements below use ODS graphics to create a plot that shows the data, fitted line, prediction limits, and confidence limits for the mean:

```
ods graphics on;
proc reg data=kilowatt plots(only)=fit(stats=none);
   model kwh=ac;
title 'Straight-line Regression for KILOWATT Data';
run;
quit;
ods graphics off;
```

The first **ODS GRAPHICS** statement activates ODS graphics, as described earlier in this chapter. The **PLOTS=** option in the PROC REG statement uses several additional plotting options.[6] **ONLY** limits the plots to plots specified after the equal sign. When you use ODS graphics, PROC REG automatically produces several plots. (Chapter 11 discusses the other plots, which are useful for regression diagnostics.) **FIT** requests a plot with the fitted line, prediction limits, and confidence limits for the mean.
STATS=NONE suppresses a table of statistics that automatically appears on the right side of the plot. Figure 10.14 shows the plot.

---

[6] The **PLOTS=** option is available starting with SAS 9.2. For earlier releases of SAS, you can use traditional graphics to create a similar plot.

Figure 10.14 Plotting the Fitted Line and Limits with ODS Graphics

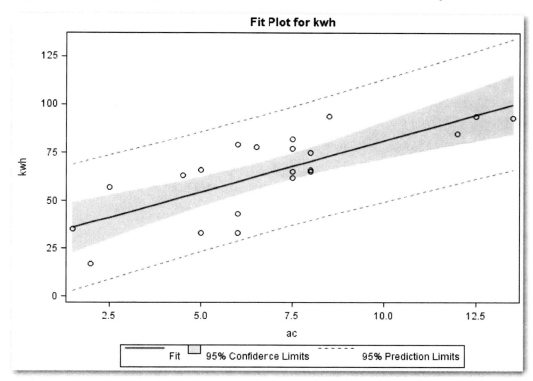

Figure 10.14 shows the data points as circles, and the fitted regression line as the thick, dark blue, solid line. The confidence limits for the mean are shown by the inner, blue-shaded area. The confidence limits for individual predicted values are shown by the outer, dashed, blue lines. SAS provides the legend below the plot as a helpful reminder.

You can use additional plotting options with ODS graphics to suppress the prediction limits, confidence limits, or both. The statements below use **NOCLI** to suppress the prediction limits shown in Figure 10.14:

```
ods graphics on;
proc reg data=kilowatt plots(only)=fit(nocli stats=none);
   model kwh=ac;
run;
quit;
ods graphics off;
```

The statements below use **NOCLM** to suppress the confidence limits for the mean shown in Figure 10.14:

```
ods graphics on;
proc reg data=kilowatt plots(only)=fit(noclm stats=none);
   model kwh=ac;
run;
quit;
ods graphics off;
```

The statements below use **NOLIMITS** to suppress both the prediction limits and the confidence limits for the mean shown in Figure 10.14:

```
ods graphics on;
proc reg data=kilowatt plots(only)=fit(nolimits stats=none);
   model kwh=ac;
run;
quit;
ods graphics off;
```

Using Traditional Graphics with the PLOT Statement

When you use the PLOT statement to produce high-resolution graphs, SAS documentation refers to these graphs as "traditional graphics." To create these graphs, you must have SAS/GRAPH software licensed. Figure 10.11 shows a scatter plot and fitted regression line created by the PLOT statement. By adding plotting options, the plot can include limits. When you add prediction limits and confidence limits for the mean, SAS automatically chooses a different color for each line. You can use **SYMBOL** statements to specify the same color for both the upper and lower prediction limits, and to specify another color for both the upper and lower confidence limits for the mean. This approach produces better graphs for presentations or reports.

The statements below assume that you have fit a straight line to the data, and that you are using PROC REG interactively. (If this is not the case, then add the PROC REG and MODEL statements.)

```
symbol1 color=blue;
symbol2 color=red line=1;
symbol3 color=green line=2;
symbol4 color=green line=2;
symbol5 color=purple line=3;
symbol6 color=purple line=3;
plot kwh*ac / pred conf nostat;
run;
quit;
```

The **SYMBOL1** statement controls the color of the data points. The **SYMBOL2** statement controls the color and appearance of the fitted line. The next two statements control the color and appearance of the confidence limits for the mean. The last two SYMBOL statements control the color and appearance of the prediction limits. SAS provides more than 45 line types, and SAS Help shows examples of each type. In general, the **COLOR=** option identifies the color, and the **LINE=** option specifies the line type.

In the PLOT statement, the NOSTAT option suppresses an automatic table of statistics as discussed earlier. The **PRED** option requests prediction limits, and the **CONF** option requests confidence limits for the mean. You can specify only one of these two options if you want. For example, you can use only the PRED option to create a plot with the fitted line and prediction limits.

Figure 10.15 shows the results.

Figure 10.15 Plotting the Fitted Line and Limits with the PLOT Statement

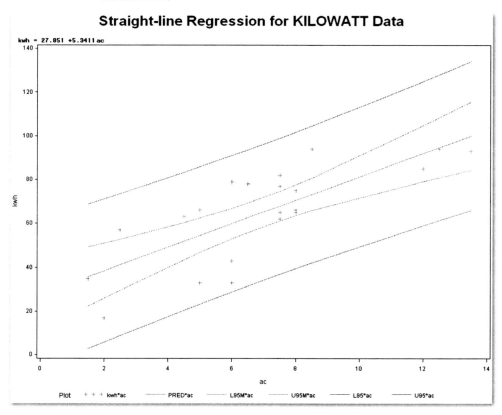

Figure 10.15 shows the data points as plus signs. The lines have the colors specified in the SYMBOL statements.

Summarizing Straight-Line Regression

Straight-line regression uses least squares to fit a line through a set of data points. Predicted values for the *y* variable for a given *x* value are computed by inserting the value of *x* into the equation for the fitted line. Prediction limits account for the error in fitting the regression line, and account for the variation between values for a given *x* value. Confidence limits on the mean account for the error in fitting the regression line because the mean does not change. Plots of the observed values, fitted regression line, prediction limits, and confidence limits on the mean are useful in summarizing the regression.

Because this chapter uses PROC REG for fitting lines, curves, and regression with multiple variables, the summary for the general form of the procedure appears only in the "Syntax" section at the end of the chapter. Similarly, the list of ODS tables for PROC REG appears only in the "Summary" section.

After fitting a regression line, your next step is to perform regression diagnostics. The diagnostic tools help you decide whether the regression equation is adequate, or whether you need to add more terms to the equation. Perhaps another variable needs to be added, or a curve fits the data better than a straight line. Chapter 11 discusses a set of basic tools for regression diagnostics.

The next section discusses fitting curves.

Fitting Curves

Understanding Polynomial Regression

For the kilowatt data, a straight line described the relationship between air conditioner use and kilowatt-hours. In other cases, though, the relationship between two variables is not represented well by a straight line. Sometimes, a curve does a better job of representing the relationship. A *quadratic polynomial* has the following equation:

$$\hat{y} = b_0 + b_1 x + b_2 x^2$$

With least squares regression, b_0, b_1, and b_2 are estimates of unknown population parameters. The squared term (x^2) is called the *quadratic term* when referring to the model.

If you could measure all of the values in the entire population, you would know the exact relationship between x and y. Because you can measure only values in a sample of the population, you estimate the relationship using regression. Least squares regression for curves minimizes the sum of squared differences between the points and the curve, as shown in Figure 10.16.

Figure 10.16 Least Squares Regression for a Curve

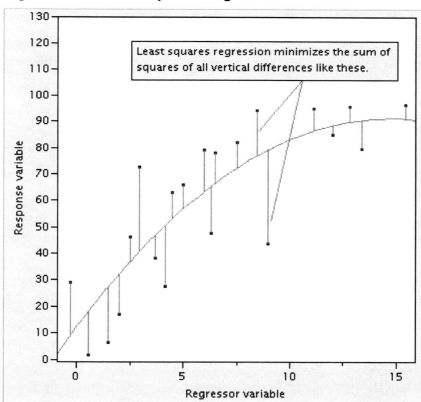

The assumptions for fitting a curve with least squares regression are the same as the assumptions for fitting a straight line. Similarly, the steps for fitting a curve are the same as the steps for fitting a straight line. As with fitting a straight line, **PROC REG** provides one way to fit a curve to data.

Fitting Curves with PROC REG

An engineer conducted an experiment to test the performance of an industrial engine. The experiment used a mixture of diesel fuel and gas derived from distilling organic materials. The engineer measured horsepower produced from the engine at several speeds, where speed is measured in hundreds of revolutions per minute (rpm × 100).[7] Table 10.2 shows the data.

The steps for fitting a curve are the following:

1. Create a SAS data set.

2. Check the data set for errors. Assume that you have performed this step and found no errors.

3. Choose the significance level. Choose a significance level of 0.05. This level implies a 5% chance of concluding that the regression coefficients are different from 0 when, in fact, they are not.

4. Check the assumptions.

5. Perform the analysis.

6. Make conclusions from the results.

Table 10.2 Engine Data

Speed	Horsepower
22.0	64.03
20.0	62.47
18.0	54.94
16.0	48.84
14.0	43.73
12.0	37.48
15.0	46.85
17.0	51.17
19.0	58.00
21.0	63.21
22.0	64.03

---

[7] Data is from Dr. Ramon Littell, University of Florida. Used with permission.

Speed	Horsepower
20.0	59.63
18.0	52.90
16.0	48.84
14.0	42.74
12.0	36.63
10.5	32.05
13.0	39.68
15.0	45.79
17.0	51.17
19.0	56.65
21.0	62.61
23.0	65.31
24.0	63.89

This data is available in the **engine** data set in the sample data for the book. The statements below create the data set, and then use PROC CORR to create a scatter plot of the data. Figure 10.17 shows the scatter plot.

```
data engine;
   input speed power @@;
   speedsq=speed*speed;
   datalines;
22.0 64.03 20.0 62.47 18.0 54.94 16.0 48.84 14.0 43.73
12.0 37.48 15.0 46.85 17.0 51.17 19.0 58.00 21.0 63.21
22.0 64.03 20.0 59.63 18.0 52.90 16.0 48.84 14.0 42.74
12.0 36.63 10.5 32.05 13.0 39.68 15.0 45.79 17.0 51.17
19.0 56.65 21.0 62.61 23.0 65.31 24.0 63.89
;
run;

ods graphics on;
ods select ScatterPlot;
proc corr data=engine plots=scatter(noinset ellipse=none);
   var speed power;
title 'Scatterplot for ENGINE Data';
run;
ods graphics off;
```

The DATA step includes a program statement that creates the new variable **SPEEDSQ**. (See Chapter 7 for an introduction to program statements.) PROC REG requires individual names for the independent variables in the MODEL statement. As a result, PROC REG does not allow terms like SPEED*SPEED. The solution is to create a new variable for the quadratic term in the regression model.

Figure 10.17 Scatter Plot Showing Curve for Engine Data

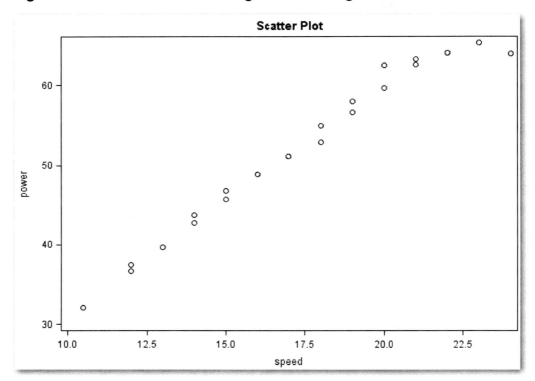

Looking at the scatter plot in Figure 10.17, it seems like a straight line would not fit the data well. **Power** increases with **Speed** up to about 21, and then the data levels off or curves downward. Instead of fitting a straight line, fit a curve to the data.

Checking the Assumptions for Regression

Think about the assumptions:

- Both of the variables are continuous, and the data values are measurements.

- The observations are independent because the measurement for one observation of speed and power is unrelated to the measurement for another observation.

- The observations are a random sample from the population. Because this is an experiment, the engineer selected the levels of **Speed** in advance, selecting levels that made sense in the industrial environment where the engine would be used.

- The engineer can set the value for **Speed** precisely, so assume that the x variable is known without error.

- Chapter 11 discusses the assumption that the errors in the data are normally distributed. For now, assume that this assumption seems reasonable.

Performing the Analysis

To perform the analysis, submit the following:

```
proc reg data=engine;
   id speed;
   model power=speed speedsq / p cli clm;
title 'Fitting a Curve to the ENGINE Data';
run;
```

The statements above use the same statements and options for fitting a straight line. The key difference is the SPEEDSQ term in the MODEL statement. Because SPEEDSQ is calculated as **speed**^{2}, this term is the quadratic term in the equation for fitting a curve. The ID statement identifies observations in the **Output Statistics** table. Figure 10.18 shows the results from the MODEL statement. Figure 10.19 shows the results of the options in the statement.

Figure 10.18 (for fitting a curve) shows the same SAS tables as Figure 10.12 (for fitting a straight line). Here are the key differences:

- The **Parameter Estimates** table now contains three parameters—Intercept, speed, and speedsq.

- The **Analysis of Variance** table shows a **Model** with 2 degrees of freedom. In general, the degrees of freedom for the model are one less than the number of parameters estimated by the model. When fitting a curve, PROC REG estimates three parameters, so the model has two degrees of freedom.

The next two sections focus on explaining the conclusions for the **engine** data.

Figure 10.18 Fitting a Curve with PROC REG

```
                Fitting a Curve to the ENGINE Data

                        The REG Procedure
                          Model: MODEL1
                     Dependent Variable: power

                 Number of Observations Read      24
                 Number of Observations Used      24


                        Analysis of Variance

                                  Sum of        Mean
        Source            DF     Squares       Square    F Value    Pr > F

        Model              2   2285.64535   1142.82268    749.84    <.0001
        Error             21     32.00598      1.52409
        Corrected Total   23   2317.65133


                Root MSE              1.23454    R-Square    0.9862
                Dependent Mean       52.19333    Adj R-Sq    0.9849
                Coeff Var             2.36533


                        Parameter Estimates

                          Parameter     Standard
        Variable    DF     Estimate        Error   t Value    Pr > |t|

        Intercept    1    -17.66377      5.43598     -3.25     0.0038
        speed        1      5.53776      0.64485      8.59     <.0001
        speedsq      1     -0.08407      0.01852     -4.54     0.0002
```

Understanding Results for Fitting a Curve

First, answer the research question, "What is the relationship between the two variables?" Using the estimates from the **Parameter Estimates** table, here is the fitted equation:

$$\text{power} = -17.66 + 5.54*\text{speed} - 0.08*\text{speed}^2$$

This equation rounds the SAS parameter estimates to two decimal places.

From the **Parameter Estimates** table, you conclude that all three coefficients are significantly different from 0. All three *p*-values are less than 0.05. At first, this might not make sense, because 0.08 seems close to 0. However, remember that the t **Value** divides the **Parameter Estimate** by **Standard Error**. Figure 10.18 shows a very small standard error for the coefficient of **speedsq**, and a correspondingly large t **Value**. This coefficient is multiplied by the **speedsq** value. Even though the coefficient is small, the impact of this term on the curve is important.

The **Fit Statistics** table shows **R-Square** is about 0.99, meaning that the fitted curve explains about 99% of the variation in the data. This is an excellent fit.

The **Analysis of Variance** table leads you to conclude that the fitted curve explains a significant amount of the variation in **Power**. The *p*-value for the *F* test is **<.0001**. Statisticians typically examine the overall test of the fitted model before reviewing the details of parameter estimates.

Printing Predicted Values and Limits

Just as when you fit a straight line, when you fit a curve, you can print confidence limits for the mean, and confidence limits on an individual predicted value. Figure 10.19 shows the first page of results from using the **P**, **CLI**, and **CLM** options in the **MODEL** statement. The second page shows the prediction limits and residuals for the last few observations in the data set.

Figure 10.19 for fitting a curve shows the same SAS tables as Figure 10.13. See the discussion after Figure 10.13 for an explanation of the columns in the output.

Figure 10.19 Printing Predicted Values and Limits for Engine Data

```
                    Fitting a Curve to the ENGINE Data

                             The REG Procedure
                             Model: MODEL1
                          Dependent Variable: power

                            Output Statistics

                Dependent   Predicted      Std Error
   Obs   speed  Variable      Value      Mean Predict        95% CL Mean

    1    22.0   64.0300      63.4763        0.4296        62.5829   64.3698
    2    20.0   62.4700      59.4628        0.3289        58.7788   60.1468
    3    18.0   54.9400      54.7767        0.3526        54.0435   55.5100
    4    16.0   48.8400      49.4181        0.3490        48.6924   50.1438
    5    14.0   43.7300      43.3869        0.3479        42.6633   44.1104
    6    12.0   37.4800      36.6831        0.5238        35.5937   37.7724
    7    15.0   46.8500      46.4865        0.3388        45.7820   47.1910
    8    17.0   51.1700      52.1815        0.3566        51.4398   52.9231
    9    19.0   58.0000      57.2039        0.3387        56.4995   57.9082
   10    21.0   63.2100      61.5537        0.3503        60.8251   62.2822
   11    22.0   64.0300      63.4763        0.4296        62.5829   64.3698
   12    20.0   59.6300      59.4628        0.3289        58.7788   60.1468
   13    18.0   52.9000      54.7767        0.3526        54.0435   55.5100
   14    16.0   48.8400      49.4181        0.3490        48.6924   50.1438
   15    14.0   42.7400      43.3869        0.3479        42.6633   44.1104
   16    12.0   36.6300      36.6831        0.5238        35.5937   37.7724
   17    10.5   32.0500      31.2138        0.8102        29.5290   32.8987
   18    13.0   39.6800      40.1190        0.4048        39.2773   40.9608
   19    15.0   45.7900      46.4865        0.3388        45.7820   47.1910
   20    17.0   51.1700      52.1815        0.3566        51.4398   52.9231
   21    19.0   56.6500      57.2039        0.3387        56.4995   57.9082
   22    21.0   62.6100      61.5537        0.3503        60.8251   62.2822
   23    23.0   65.3100      65.2309        0.5729        64.0394   66.4223
   24    24.0   63.8900      66.8173        0.7723        65.2112   68.4233


                            Output Statistics

            Obs   speed       95% CL Predict         Residual

             1    22.0     60.7580    66.1947         0.5537
             2    20.0     56.8059    62.1198         3.0072
             3    18.0     52.1067    57.4468         0.1633
             4    16.0     46.7501    52.0861        -0.5781
             5    14.0     40.7195    46.0542         0.3431
             6    12.0     33.8941    39.4720         0.7969
             7    15.0     43.8243    49.1488         0.3635
             8    17.0     49.5091    54.8538        -1.0115
             9    19.0     54.5416    59.8661         0.7961
            10    21.0     58.8849    64.2224         1.6563
            11    22.0     60.7580    66.1947         0.5537
            12    20.0     56.8059    62.1198         0.1672
            13    18.0     52.1067    57.4468        -1.8767
            14    16.0     46.7501    52.0861        -0.5781
            15    14.0     40.7195    46.0542        -0.6469
            16    12.0     33.8941    39.4720        -0.0531
            17    10.5     28.1430    34.2847         0.8362
            18    13.0     37.4172    42.8209        -0.4390
            19    15.0     43.8243    49.1488        -0.6965
```

Changing the Alpha Level

SAS automatically creates 95% prediction limits and confidence limits for the mean. You can specify a different confidence level with the **ALPHA=** option in the PROC REG statement:

```
ods select OutputStatistics;
proc reg data=engine alpha=0.1;
   model power=speed speedsq / p cli clm;
title '90% Limits for ENGINE Data';
run;
```

Because the regression fit for the model is the same as shown in Figure 10.18, the ODS statement limits the printed results to the Output Statistics table. The ALPHA= option specifies 90% confidence limits. Figure 10.20 shows the first page of the results.

Compare the results in Figures 10.19 and 10.20. The key difference is that Figure 10.20 shows 90% prediction limits and 90% confidence limits for the mean. When you use the ALPHA= option, PROC REG changes the alpha level for both types of limits.

You can also use the ALPHA= option when fitting a straight line or when performing multiple regression.

If you use the ALPHA= option and ODS graphics, then PROC REG applies the confidence level to the plots. For example, if you specify ALPHA=0.10, then ODS graphics display 90% prediction limits and 90% confidence limits for the mean.

Similarly, if you use the ALPHA= option and a PLOT statement, then PROC REG applies the confidence level to the plots.

Figure 10.20 Changing the Alpha Level

```
                        90% Limits for ENGINE Data

                            The REG Procedure
                             Model: MODEL1
                        Dependent Variable: power

                            Output Statistics

           Dependent      Predicted        Std Error
    Obs     Variable          Value     Mean Predict         90% CL Mean

     1       64.0300        63.4763           0.4296      62.7371      64.2156
     2       62.4700        59.4628           0.3289      58.8969      60.0288
     3       54.9400        54.7767           0.3526      54.1700      55.3835
     4       48.8400        49.4181           0.3490      48.8176      50.0186
     5       43.7300        43.3869           0.3479      42.7881      43.9856
     6       37.4800        36.6831           0.5238      35.7817      37.5845
     7       46.8500        46.4865           0.3388      45.9036      47.0695
     8       51.1700        52.1815           0.3566      51.5678      52.7952
     9       58.0000        57.2039           0.3387      56.6210      57.7867
    10       63.2100        61.5537           0.3503      60.9509      62.1565
    11       64.0300        63.4763           0.4296      62.7371      64.2156
    12       59.6300        59.4628           0.3289      58.8969      60.0288
    13       52.9000        54.7767           0.3526      54.1700      55.3835
    14       48.8400        49.4181           0.3490      48.8176      50.0186
    15       42.7400        43.3869           0.3479      42.7881      43.9856
    16       36.6300        36.6831           0.5238      35.7817      37.5845
    17       32.0500        31.2138           0.8102      29.8197      32.6079
    18       39.6800        40.1190           0.4048      39.4226      40.8155
    19       45.7900        46.4865           0.3388      45.9036      47.0695
    20       51.1700        52.1815           0.3566      51.5678      52.7952
    21       56.6500        57.2039           0.3387      56.6210      57.7867
    22       62.6100        61.5537           0.3503      60.9509      62.1565
    23       65.3100        65.2309           0.5729      64.2450      66.2167
    24       63.8900        66.8173           0.7723      65.4884      68.1462


                            Output Statistics

    Obs           90% CL Predict           Residual

     1       61.2271        65.7256           0.5537
     2       57.2644        61.6613           3.0072
     3       52.5675        56.9860           0.1633
     4       47.2105        51.6257          -0.5781
     5       41.1798        45.5939           0.3431
     6       34.3754        38.9907           0.7969
     7       44.2837        48.6894           0.3635
     8       49.9703        54.3927          -1.0115
     9       55.0010        59.4067           0.7961
    10       59.3455        63.7619           1.6563
    11       61.2271        65.7256           0.5537
    12       57.2644        61.6613           0.1672
    13       52.5675        56.9860          -1.8767
    14       47.2105        51.6257          -0.5781
    15       41.1798        45.5939          -0.6469
    16       34.3754        38.9907          -0.0531
    17       28.6729        33.7547           0.8362
    18       37.8834        42.3546          -0.4390
    19       44.2837        48.6894          -0.6965
```

Plotting Predicted Values and Limits

When fitting a curve, PROC REG has different choices for plotting. The ODS graphics FIT choice is not available. Also, the CONF and PRED options in the PLOT statement are not available. However, you can still use both ODS graphics and the PLOT statement to produce graphs. The next two sections show how.

Using ODS Graphics

The statements below use ODS graphics to create a plot that shows the data, fitted curve, prediction limits, and confidence limits for the mean.

```
ods graphics on;
ods select PredictionPlot;
proc reg data=engine alpha=0.10
        plots(only)=predictions(x=speed unpack);
   model power=speed speedsq;
run;
quit;
ods graphics off;
```

The first ODS GRAPHICS statement activates ODS graphics. The ODS statement limits the output to the plot. The PROC REG statement specifies ALPHA=0.10, so ODS graphics will create a plot with 90% confidence limits. The PLOTS option in the PROC REG statement uses several additional plotting options. ONLY limits the plots to plots specified after the equal sign. When you use ODS graphics, PROC REG automatically produces several plots. (Chapter 11 discusses the other plots, which are useful for regression diagnostics.) **PREDICTIONS** requests a plot with the fitted line, prediction limits, and confidence limits for the mean. Using **x=speed** specifies the *x* variable in the model. The PREDICTIONS option is appropriate only for models that have terms for a single variable. The models can be a straight line, curve, cubic model (with a linear, squared, and cubed term), and so on. The models cannot have more than a single *x* variable. PROC REG automatically produces two plots, and **UNPACK** separates these two plots. Figure 10.21 shows the first plot, which shows the fitted curve and limits. Chapter 11 discusses the second plot, which is useful for regression diagnostics.

Figure 10.21 Plotting the Fitted Curve and Limits with ODS Graphics

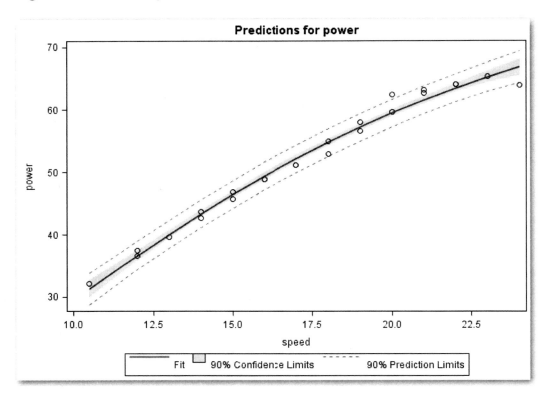

Figure 10.21 shows the data points as circles, and the fitted curve as the thick, dark blue, solid line. The confidence limits for the mean are shown by the inner, blue-shaded area. The confidence limits for individual predicted values are shown by the outer, dashed, blue lines. SAS provides the legend below the plot as a helpful reminder.

Figure 10.21 illustrates the narrow confidence curves for a model that fits very well. The curves for both the prediction limits and the confidence limits for the mean are very close to the fitted curve. Compare these limits for a model with an **R-Square** value of about 0.99 with the limits for a model that doesn't fit as well in Figure 10.12 (with an **R-Square** value of about 0.59). This comparison illustrates a general principle: poorly fitting models generate wider confidence limits.

You can use additional plotting options with ODS graphics to suppress the prediction limits, confidence limits, or both. (See the discussion for using ODS graphics to plot a fitted line and limits earlier in this chapter.) You can use the NOCLI, NOCLM, and NOLIMITS options with PREDICTIONS. The statement below provides one example:

```
proc reg data=engine
        plots(only)=predictions(x=speed noclm unpack);
```

To use NOCLI or NOLIMITS, replace NOCLM in the statement.

Using Traditional Graphics with the PLOT Statement

Although the CONF and PRED options in the PLOT statement are not available when fitting curves, you can use the PLOT statement and SYMBOL statements. The resulting plot does not display lines; it displays points instead. Here are the statements in SAS:[8]

```
symbol1 value=plus color=blue;
symbol2 value=dot color=red;
symbol3 value="I" color=purple;
symbol4 value="M" color=green;
symbol5 value="I" color=purple;
symbol6 value="M" color=green;
plot (power p. lcl. lclm. ucl. uclm.)*speed / overlay nostat;
run;
```

First, look at the PLOT statement. The statement identifies a single *x* variable—SPEED. The statement identifies several *y* variables, which are enclosed in parentheses. The first *y* variable is POWER, which plots the actual data values. The remaining *y* variables use statistical keywords and end with a dot (period). These keywords are defined in the table:

P.	predicted value
LCL.	lower prediction limit
LCLM.	lower confidence limit for the mean
UCL.	upper prediction limit
UCLM.	upper confidence limit for the mean

OVERLAY places all *y**x plot requests on a single plot. **NOSTAT** suppresses several automatic statistics that appear to the right of the graph.

Next, look at the SYMBOL statements. SAS uses these statements based on the order of the *y* variables in the PLOT statement. For the example, SYMBOL1 controls the

---

[8] These statements assume that you have fit a curve to the data with alpha=0.10, and that you are using PROC REG interactively. If this is not the case, then add the PROC REG and MODEL statements.

appearance of the actual data values, and SYMBOL2 controls the appearance of the predicted values. Because the prediction limits are the third and fifth y variables, the SYMBOL3 and SYMBOL5 statements control their appearance. Similarly, SYMBOL4 and SYMBOL6 control the appearance of the confidence limits for the mean.

The **VALUE=** option identifies the symbol used for the plotted point. SAS includes many automatic values. SYMBOL1 and SYMBOL2 use two of those automatic values. The remaining VALUE= options use text values. For your data, enter the single character enclosed in double quotation marks, as shown in the example.

Figure 10.22 shows the results.

Figure 10.22 Plotting Predicted Values and Limits for the Engine Data

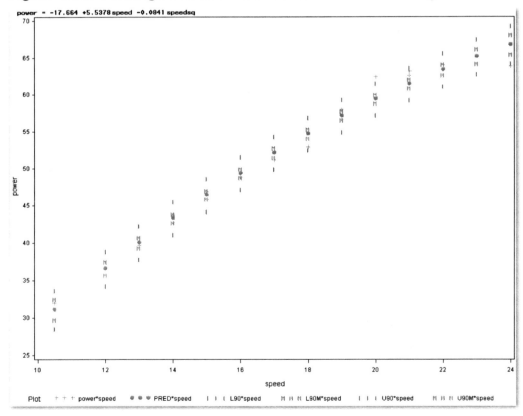

Data points in Figure 10.22 are difficult to see because either predicted values or points representing one of the limits overlay most of the data values.

Compare Figures 10.21 and 10.22. Figure 10.22 is more like a line printer plot. It does not show connected lines for the predicted values and limits. Instead, Figure 10.22 shows predicted values and limits for each *x* value in the data.

Summarizing Polynomial Regression

Fitting a curve to data involves a single dependent *y* variable, and a single *x* variable with a linear term and quadratic term. PROC REQ requires two separate variables for the linear term and quadratic term.

The same assumptions for fitting a straight line and the same analysis steps apply. Use PROC REG, and interpret the results in the same way. You can add confidence limits on the mean and on individual predicted values to the scatter plot. After you fit a curve, the next step is to perform diagnostics to check the fit of your model. Chapter 11 discusses a set of basic tools for regression diagnostics.

The next section discusses regression with more than one *x* variable.

Regression for Multiple Independent Variables

Understanding Multiple Regression

This chapter has discussed how to fit a straight line and a curve using regression. Both cases involved a single *y* variable and a single *x* variable. The curve included linear and quadratic (squared) terms for **speed** in the **engine** data, but the model included a single *x* variable.

Remember the homeowner-recorded dryer use for the **kilowatt** data? Suppose you want to add the **dryer** variable to the model. In this case, you are no longer fitting a straight line to the data because both **ac** and **dryer** are independent variables. *Multiple regression* is a regression model with multiple independent variables.

Multiple regression is difficult to picture in a simple scatter plot. For two independent variables, think of a three-dimensional picture. Figure 10.23 shows multiple regression, where the two independent variables are on the axes labeled **X1** and **X2**, and the dependent variable is on the axis labeled **Y**. Multiple regression is the process of finding the best-fitting plane through the data points. Here, the term "best-fitting" is defined as the plane that minimizes the squared distances between all of the data points and the plane.

Figure 10.23 Multiple Regression with Two Independent Variables

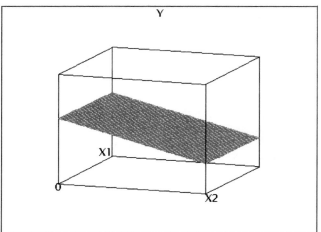

With two independent variables, here is the estimated regression equation:

$$\hat{y} = b_0 + b_1 x_1 + b_2 x_2$$

b_0, b_1, and b_2 estimate the population parameters β_0, β_1, and β_2. The two independent variables are x_1 and x_2.

The assumptions for multiple regression are the same as the assumptions for fitting a straight line. Similarly, the steps for multiple regression are the same as the steps for fitting a straight line.

Fitting Multiple Regression Models in SAS

For the **kilowatt** data, you use PROC REG with a different MODEL statement. Fit a multiple regression model with both **ac** and **dryer** as independent variables. You have already considered the assumptions for multiple regression because you considered these assumptions when fitting a straight line to the data. The only new assumption to consider for multiple regression is whether the homeowner measured **dryer** without error. Assume that the homeowner correctly recorded the number of times the dryer was used each day.

Also, assume that you want to continue to use an alpha level of 0.05.

```
proc reg data=kilowatt;
    model kwh=ac dryer / p clm cli;
title 'Multiple Regression for KILOWATT Data';
run;
```

The program uses statements and options discussed earlier in this chapter. The key difference is that the **MODEL** statement now includes terms for the two independent variables—**ac** and **dryer**. Figure 10.24 shows the results.

Figure 10.24 Multiple Regression Results for Kilowatt Data

```
             Multiple Regression for KILOWATT Data

                      The REG Procedure
                       Model: MODEL1
                   Dependent Variable: kwh

            Number of Observations Read          21
            Number of Observations Used          21


                        Analysis of Variance

                             Sum of         Mean
 Source            DF       Squares       Square    F Value    Pr > F

 Model              2    9299.80154   4649.90077     300.24    <.0001
 Error             18     278.76989     15.48722
 Corrected Total   20    9578.57143


              Root MSE             3.93538    R-Square     0.9709
              Dependent Mean      64.85714    Adj R-Sq     0.9677
              Coeff Var            6.06777


                        Parameter Estimates

                      Parameter      Standard
     Variable    DF    Estimate         Error    t Value    Pr > |t|

     Intercept    1     8.10539       2.48085       3.27      0.0043
     ac           1     5.46590       0.28076      19.47      <.0001
     dryer        1    13.21660       0.85622      15.44      <.0001
```

Figure 10.24 shows the same output tables as **PROC REG** shows for fitting a straight line or a curve. The next section explains how to interpret these results.

Understanding Results for Multiple Regression

Finding the Regression Equation

From the **Parameter Estimates** table in Figure 10.24, here is the fitted regression equation:

kwh = 8.11 + 5.47*ac + 13.22*dryer

The equation rounds the SAS parameter estimates to two decimal places. Interpret the coefficients as follows:

- The intercept is 8.11. It estimates the number of kilowatt-hours that were consumed on days when neither the air conditioner nor the dryer was used.

- The second term in the equation, **5.47*ac**, indicates that the air conditioner uses 5.47 kwh an hour.

- The third term in the equation, **13.22*dryer**, indicates that each dryer load uses 13.22 kwh.

You can specify values for **ac** and **dryer**, and then use the regression equation to predict the **kwh** that are consumed. PROC REG can create these predictions. See "Printing Predicted Values and Limits" later in this section.

Testing for Significance of Parameters

From the **Parameter Estimates** table in Figure 10.24, you conclude that all three coefficients are significantly different from 0. All three p-values are less than 0.05. The values for **ac** and **dryer** are both **<.0001**. These values provide overwhelming evidence of the significant effect of these variables on the kilowatt-hours that are consumed. The p-value for the intercept is **0.0043**. This value provides evidence that a significant (nonzero) number of kilowatt-hours are consumed even when neither the air conditioner nor the dryer is used.

A significant regressor in multiple regression indicates that the variation explained by the regressor is significantly larger than the random variation in the data. For the **kilowatt** data, the electricity used by the **ac** and **dryer** is significantly larger than the random variation. Suppose the homeowner measured the number of times a small appliance, such as a coffee maker, was used. The amount of electricity the coffee maker uses is so small that it would probably not be detected. If you added **coffeemaker** to the model, you would probably find a large p-value, indicating that the coffee maker is not a significant source of variation.

Checking the Fit of the Model

In Figure 10.24, the Fit Statistics table shows an R-Square value of about **0.97**, indicating that the fitted curve explains about 97% of the variation in the data. This is an excellent fit. Compare this fit with the R-Square value of 0.59 in Figure 10.12, which is for the model that contained only ac. Adding dryer to the model significantly improved the fit.

The Analysis of Variance table leads you to conclude that the fitted model with ac and dryer as regressors explains a significant amount of the variation in kwh. The *p*-value for the *F* test is **<.0001**. Statisticians typically examine the overall test of the fitted model before reviewing the details of parameter estimates.

Printing Predicted Values and Limits

Just as when you fit a straight line or a curve, you can print confidence limits for the mean, and confidence limits on an individual predicted value for multiple regression. Figure 10.25 shows the results from using the P, CLI, and CLM options in the MODEL statement.

Figure 10.25 for multiple regression shows the same SAS tables as Figure 10.19 for fitting a curve and Figure 10.13 for fitting a straight line.

Summarizing Multiple Regression

Multiple regression involves two or more independent variables. The same assumptions for fitting a straight line or a curve apply. In addition, the same analysis steps for fitting a straight line or a curve apply. Use PROC REG just as you use it for fitting a straight line or a curve, and interpret the results in the same way. You can print predicted values, confidence limits on the mean, and prediction limits.

After you fit the multiple regression, the next step is to perform diagnostics to check the fit of your model. Chapter 11 discusses a set of basic tools for regression diagnostics. The usual process of fitting a model is an iterative one. First, you fit a model, and then you look at the fit, revise the fit based on what you have learned from the diagnostic tools, look at the revised fit, and so on.

Figure 10.25 Predicted Values and Limits for Multiple Regression

```
                Multiple Regression for KILOWATT Data

                        The REG Procedure
                        Model: MODEL1
                     Dependent Variable: kwh

                        Output Statistics

          Dependent     Predicted      Std Error
   Obs     Variable        Value     Mean Predict        95% CL Mean

    1      35.0000       29.5208        1.7965        25.7466    33.2950
    2      63.0000       59.1351        1.1927        56.6294    61.6409
    3      66.0000       61.8681        1.1202        59.5147    64.2215
    4      17.0000       19.0372        2.0605        14.7082    23.3662
    5      94.0000       94.2154        1.6663        90.7145    97.7162
    6      79.0000       80.5506        1.6111        77.1658    83.9354
    7      93.0000       95.1117        2.0584        90.7871    99.4362
    8      66.0000       65.0492        0.9779        62.9947    67.1037
    9      94.0000       89.6458        1.8127        85.8375    93.4541
   10      82.0000       75.5329        1.0036        73.4245    77.6413
   11      78.0000       83.2836        1.5978        79.9267    86.6404
   12      65.0000       65.0492        0.9779        62.9947    67.1037
   13      77.0000       75.5329        1.0036        73.4245    77.6413
   14      75.0000       78.2658        1.0372        76.0867    80.4449
   15      62.0000       62.3163        0.9458        60.3293    64.3033
   16      85.0000       86.9128        1.6939        83.3540    90.4716
   17      43.0000       40.9008        1.5231        37.7008    44.1008
   18      57.0000       61.4199        1.9993        57.2195    65.6204
   19      33.0000       35.4349        1.6015        32.0702    38.7996
   20      65.0000       62.3163        0.9458        60.3293    64.3033
   21      33.0000       40.9008        1.5231        37.7008    44.1008

                        Output Statistics

           Obs        95% CL Predict          Residual

            1      20.4322      38.6095         5.4792
            2      50.4958      67.7745         3.8649
            3      53.2718      70.4644         4.1319
            4       9.7045      28.3699        -2.0372
            5      85.2368     103.1939        -0.2154
            6      71.6167      89.4845        -1.5506
            7      85.7811     104.4423        -2.1117
            8      56.5299      73.5686         0.9508
            9      80.5429      98.7486         4.3542
           10      67.0003      84.0654         6.4671
           11      74.3602      92.2070        -5.2836
           12      56.5299      73.5686        -0.0492
           13      67.0003      84.0654         1.4671
           14      69.7155      86.8161        -3.2658
           15      53.8129      70.8196        -0.3163
           16      77.9115      95.9141        -1.9128
           17      32.0352      49.7664         2.0992
           18      52.1462      70.6937        -4.4199
           19      26.5085      44.3613        -2.4349
           20      53.8129      70.8196         2.6837
           21      32.0352      49.7664        -7.9008
```

Summary

Key Ideas

- Scatter plots show the relationship between two variables. A scatter plot matrix shows multiple relationships between pairs of variables.

- The Pearson correlation coefficient, r, measures the strength of the relationship between two variables. However, it does not describe the form of the relationship, and it does not show that one of the variables causes the other.

- PROC CORR creates scatter plots, a scatter plot matrix, summary statistics, correlation coefficients, and tests for the coefficients.

- Regression analysis describes relationships between variables. The simplest model fits a straight line that shows the relationship between one dependent variable (y) and one independent variable (x). Fitting a curve shows the relationship between one y variable, an x variable, and the square of x. Multiple regression shows the relationship between one y variable and two or more x variables.

- Least squares regression provides the best-fitting models. Here, the term "best-fitting" means that the fitted model minimizes the sum of squared differences between all of the data points and the fitted line, fitted curve, or multidimensional plane.

- Once regression analysis has been performed, use the fitted equation to predict future values at a given value of the independent variable.

- Prediction limits or confidence limits for individual values put bounds around a future value.

- Confidence limits for the mean put bounds around the mean value of the dependent variable at a given value of the independent variable.

- Plots of the actual data points, the regression equation itself, and prediction or confidence limits help you see how the regression equation fits the data.

- Regression analysis and regression diagnostics are intertwined. First, fit a model. Next, perform diagnostics to assess how well the model fits. Repeat this process until you find a suitable model. (Chapter 11 discusses performing diagnostics.)

- SAS provides many procedures that perform regression analysis. PROC REG is the simplest way to perform linear regression for lines, curves, and multiple regression. The procedure performs a test for the overall significance of the model, provides parameter estimates and significance tests for the estimates, and provides the R-Square statistic. Using options with PROC REG, you can print predicted values, prediction limits, and confidence limits for the mean. You can also produce plots that show the fitted line or curve.

Syntax

To summarize data with scatter plots and statistics

ODS GRAPHICS ON;
  PROC CORR DATA=*data-set-name options*;
    VAR *variables*;
ODS GRAPHICS OFF;

data-set-name	is the name of a SAS data set.
variables	are the variables you want to summarize. If you omit the *variables* from the VAR statement, then the procedure uses all numeric variables.

The PROC CORR statement *options* can be:

PLOTS=SCATTER(*scatter-options***)**

creates a simple scatter plot for the variables.

SAS uses the first variable in the VAR statement for the *x*-axis, and the second variable for the *y*-axis. If you use more than two variables, then SAS creates a scatter plot for each pair of variables. The *scatter-options* can be one or more of the following:

NOINSET	suppresses an automatic summary that shows the number of observations and the correlation coefficient
ELLIPSE=NONE	suppresses the automatic prediction ellipse

If you use *scatter-options*, then the parentheses are required.

And, the PROC CORR statement *options* can be:

PLOTS=MATRIX(*matrix-options***)**

creates a scatter plot matrix for the variables.

The *matrix-options* can be one or more of the following:

HISTOGRAM	adds a histogram for each variable along the diagonal of the matrix. This option is available starting with SAS 9.2.
NVAR=ALL	includes all variables in the scatter plot matrix. SAS automatically uses NVAR=5, so if your data has fewer than 5 variables, you can omit this option.

If you use *matrix-options*, then the parentheses are required.

You can use both PLOTS=SCATTER and PLOTS=MATRIX in the same PROC CORR statement.

To create correlation coefficients and tests

> **PROC CORR DATA=***data-set-name options***;**
>   **VAR** *variables***;**

The PROC CORR statement *options* can be as defined earlier and can also be:

> **NOMISS**   excludes observations with missing values.

If you omit the *variables* from the VAR statement, then the procedure uses all numeric variables. Specify at least two variables for the output to be meaningful because the correlation of a variable with itself is always +1.

Other items in italic were defined earlier.

To perform regression to fit a line

PROC REG includes many other features and options that are not discussed in this chapter.

> **ODS GRAPHICS ON;**
>   **PROC REG DATA=***data-set-name* **ALPHA=***level plot-options***;**
>   **ID** *variable***;**
>   **MODEL** *y-variable=x-variable* **/ P CLI CLM;**
>   **PRINT P CLI CLM;**
>   **PLOT** *y-variable*x-variable* **/ *options***;**
> **ODS GRAPHICS OFF;**

data-set-name	is the name of a SAS data set.
variable	in the ID statement identifies observations.
y-variable	is the dependent variable.
x-variable	is the independent variable.

The PROC REG and MODEL statements are required. All other statements are optional.

PROC REG is an interactive procedure. The PROC REG, MODEL, and ID statements must be before the first RUN statement. The other statements can be before or after the first RUN statement. Use a QUIT statement to end the procedure when you are using PROC REG interactively.

In the PROC REG statement:

ALPHA=/level/	controls the confidence level for the limits. The value of **/level/** must be between 0 and 1. A **/level/** of 0.05 gives the automatic 95% limits.

The *plot-options* in the PROC REG statement require ODS graphics, and the *plot-options* discussed in this chapter require SAS 9.2 or later. The *plot-options* appropriate for a straight line are:

PLOTS(ONLY)=FIT(/fit-options/ **STATS=NONE)**

ONLY	limits the plots to plots specified after the equal sign. PROC REG automatically produces several plots. Chapter 11 discusses the other plots, which are useful for regression diagnostics. When ONLY is used, the parentheses are required.
FIT	requests a plot with the data, fitted line, prediction limits, and confidence limits for the mean.
STATS=NONE	suppresses a table of statistics that automatically appears on the right side of the plot. When STATS=NONE is used, the parentheses are required.

The *fit-options* require parentheses. *fit-options* can be one of the following:

NOCLI	suppresses prediction limits.
NOCLM	suppresses confidence limits for the mean.
NOLIMITS	suppresses both prediction limits and confidence limits for the mean.

You can use the P, CLI, or CLM options in the MODEL or PRINT statement. You do not need to use the options in both statements. If you know in advance that you want the predicted values and limits, then use the options in the MODEL statement. If you want to check the fit of the line first, then use PROC REG interactively, and use the options in the PRINT statement.

P	requests predicted values. If you use the CLM or CLI option, then using P is not necessary.
CLI	requests prediction limits, which are confidence limits for an individual value.
CLM	requests confidence limits for the mean.

The PLOT statement can create a simple scatter plot, add the fitted line, and add limits, depending on the values of *y-variable*. (SAS documentation refers to graphics created with the PLOT statement as "traditional graphics.") All plots should use the same *x-variable*. When you specify more than one *y-variable*, enclose the list of values for *y-variable* in parentheses. The *y-variable* can be the *y-variable* in the MODEL statement, or it can be one or more of the following statistical keywords:

P.   predicted value

LCL.   lower prediction limit

LCLM.   lower confidence limit for the mean

UCL.   upper prediction limit

UCLM.   upper confidence limit for the mean

The period after the statistical keyword is required.

The *options* in the PLOT statement can be:

NOSTAT   suppresses several automatic statistics that appear to the right of the graph.

CLINE=*color*   specifies the color of the regression line.

PRED   requests prediction limits.

CONF   requests confidence limits for the mean.

OVERLAY   overlays all *y-variable* requests onto a single plot.

To control the appearance of the plot, use **SYMBOL** statements:

   SYMBOL*n* **COLOR=***color* **LINE=***number;*

n   has values from 1 to 6 for the straight-line regression plot. For the example code in this book, **SYMBOL1** controls the color of the data points. **SYMBOL2** controls the color and appearance of the fitted line. **SYMBOL3** and **SYMBOL4** control the color and appearance of the confidence limits for the mean. **SYMBOL5** and **SYMBOL6** control the color and appearance of the prediction limits.

COLOR=*color*	controls the color. This option in the SYMBOL statement overrides the **CLINE=** option.
LINE=*number*	controls the appearance of the line. For better graphs, use the same value for *number* for the upper and lower limits. SAS provides more than 45 line types.

To perform regression to fit a quadratic curve

Before fitting a curve, use a program statement in the DATA step to create the quadratic term in the regression equation:

*quadratic-variable=independent-variable*independent-variable*;

To fit a quadratic curve, and then print the predicted values and limits:

```
ODS GRAPHICS ON;
  PROC REG DATA=data-set-name ALPHA=level plot-options;
    ID variable;
    MODEL y-variable=x-variable quadratic-variable / P CLI CLM;
    PRINT P CLI CLM;
    PLOT (y-variables)*x-variable / options;
ODS GRAPHICS OFF;
```

Items in italic were defined earlier. In the MODEL statement, the *quadratic-variable* is the new variable for the quadratic term. P, CLI, and CLM can be used in either the MODEL or PRINT statement.

The *plot-options* in the PROC REG statement require ODS graphics, and the *plot-options* discussed in this chapter require SAS 9.2 or later. The *plot-options* appropriate for a quadratic curve are:

PLOTS(ONLY)=PREDICTIONS(X=*x-variable fit-options* **UNPACK)**

ONLY	limits the plots to plots specified after the equal sign. PROC REG automatically produces several plots. Chapter 11 discusses the other plots, which are useful for regression diagnostics. When ONLY is used, the parentheses are required.
PREDICTIONS	requests a plot with the data, predicted curve, prediction limits, and confidence limits for the mean. The PREDICTIONS option is appropriate only for models that have terms for a single variable. The models can be a curve, cubic model, and so on. Models cannot have more than one *x-variable*.
X=*x-variable*	identifies the *x-variable* in the model.

UNPACK	separates the two plots that PROC REG automatically produces. Chapter 11 discusses the second plot, which is useful for regression diagnostics.

The *fit-options* require parentheses and are the same *fit-options* described for fitting a straight line.

The PLOT statement can create a simple scatter plot, add the fitted curve, and add limits, depending on the values of *y-variables*. (SAS documentation refers to graphics created with the PLOT statement as "traditional graphics.") All plots should use the same *x-variable*. When you specify more than one *y-variable*, enclose the list of values for the *y-variables* in parentheses. The *y-variables* can be the *y-variable* in the MODEL statement, or it can be one or more of the following statistical keywords:

P.	predicted value
LCL.	lower prediction limit
LCLM.	lower confidence limit for the mean
UCL.	upper prediction limit
UCLM.	upper confidence limit for the mean

The period after the statistical keyword is required.

The *options* in the PLOT statement can be:

OVERLAY	places all plot requests on a single plot.
NOSTAT	suppresses several automatic statistics that appear to the right of the plot.

To control the appearance of the plot, use **SYMBOL** statements:

> **SYMBOL***n* **VALUE=***symbol* **COLOR=***color* **LINE=***number.*

n	has values from 1 to 6. SYMBOL1 controls the appearance of the first *y-variable* in the PLOT statement. SYMBOL2 controls the appearance of the second *y-variable*, and so on.
VALUE=*symbol*	identifies the symbol used for the plotted point. SAS includes many automatic values, such as PLUS and DOT. You can also enter any single text character enclosed in double quotation marks.

COLOR= and **LINE=** are defined for fitting a line.

To perform multiple regression

To perform multiple regression, and then print the predicted values and limits:

PROC REG DATA=*data-set-name* **ALPHA**=*level options*;
    ID *variable*;
    MODEL *y-variable=x-variable1 x-variable2 /* **P CLI CLM**;
    PRINT P CLI CLM;

Items in italic were defined earlier in the chapter. In the MODEL statement, *x-variable1* is the first *x*-variable, *x-variable2* is the second *x*-variable, and so on. P, CLI, and CLM can be used in either the MODEL or PRINT statement. With the exception of creating a simple scatter plot, ODS graphics and the PLOT statement are not appropriate for multiple regression.

As described in Chapter 4, you can use the ODS statement to control the output table that is printed. The table below identifies the ODS names of the PROC REG output tables and plots that are discussed in this chapter.

ODS Name	Output Table or Plot
Nobs	Number of Observations (from MODEL statement)
ANOVA	Analysis of Variance (from MODEL statement)
FitStatistics	Fit Statistics (from MODEL statement)
ParameterEstimates	Parameter Estimates (from MODEL statement)
OutputStatistics	Output Statistics (from P, CLI, or CLM options)
FitPlot	Fitted line and limits (from PLOTS=FIT in the PROC REG statement)
PredictionPlot	Fitted curve and limits (from PLOTS=PREDICTION in the PROC REG statement)
REG*n*	ODS plots, or plots created by the PLOT statement, where the first plot in a SAS session is REG, the second plot is REG1, the third plot is REG2, and so on

Example

The program below produces the output shown in this chapter:

```
options ps=60 ls=80 nonumber nodate;

data kilowatt;
   input kwh ac dryer @@;
   datalines;
35 1.5 1 63 4.5 2 66 5.0 2 17 2.0 0 94 8.5 3 79 6.0 3
93 13.5 1 66 8.0 1 94 12.5 1 82 7.5 2 78 6.5 3 65 8.0 1
77 7.5 2 75 8.0 2 62 7.5 1 85 12.0 1 43 6.0 0 57 2.5 3
33 5.0 0 65 7.5 1 33 6.0 0
;
run;

ods select ScatterPlot;
ods graphics on;

proc corr data=kilowatt plots=scatter(noinset ellipse=none);
   var ac kwh;
run;
ods graphics off;

ods graphics on;
ods select MatrixPlot;
proc corr data=kilowatt plots=matrix(histogram nvar=all);
   var ac dryer kwh;
run;
ods graphics off;

ods select SimpleStats;
proc corr data=kilowatt;
   var ac dryer kwh;
   title 'Summary Statistics for KILOWATT Data Set';
run;

proc corr data=kilowatt;
   var ac dryer kwh;
   title 'Correlations for KILOWATT Data Set';
run;

data kilowatt2;
   input kwh ac dryer @@;
   datalines;
35 1.5 . 63 4.5 2 66 5.0 2 17 2.0 0 94 8.5 3 79 6.0 3
93 13.5 1 66 8.0 1 94 12.5 1 82 7.5 2 78 6.5 3 65 8.0 1
```

```
77 7.5 2 75 8.0 2 62 7.5 1 85 12.0 1 43 6.0 0 57 2.5 3
33 5.0 0 65 7.5 1 33 6.0 0
;
run;

proc corr data=kilowatt2;
   var ac dryer kwh;
title 'Automatic Behavior with Missing Values for KILOWATT2';
run;

proc corr data=kilowatt2 nomiss;
   var ac dryer kwh;
title 'NOMISS Option for KILOWATT2';
run;

proc reg data=kilowatt;
   id ac;
   model kwh=ac;
   plot kwh*ac / nostat cline=red;
title 'Straight-line Regression for KILOWATT Data';
run;
   print p clm cli;
run;
quit;

ods graphics on;
proc reg data=kilowatt plots(only)=fit(stats=none);
   model kwh=ac / p clm cli;
title 'Straight-line Regression for KILOWATT Data';
run;

quit;
ods graphics off;

proc reg data=kilowatt;
   model kwh=ac;
title 'Straight-line Regression for KILOWATT Data';
run;
   symbol1 color=blue;
   symbol2 color=red line=1;
   symbol3 color=green line=2;
   symbol4 color=green line=2;
   symbol5 color=purple line=3;
   symbol6 color=purple line=3;
   plot kwh*ac / pred conf nostat;
run;
quit;
```

```
data engine;
   input speed power @@;
   speedsq=speed*speed;
   datalines;
22.0 64.03 20.0 62.47 18.0 54.94 16.0 48.84 14.0 43.73
12.0 37.48 15.0 46.85 17.0 51.17 19.0 58.00 21.0 63.21
22.0 64.03 20.0 59.63 18.0 52.90 16.0 48.84 14.0 42.74
12.0 36.63 10.5 32.05 13.0 39.68 15.0 45.79 17.0 51.17
19.0 56.65 21.0 62.61 23.0 65.31 24.0 63.89
;
run;

ods graphics on;
ods select ScatterPlot;
proc corr data=engine plots=scatter(noinset ellipse=none);
   var speed power;
   title 'Scatterplot for ENGINE Data';
run;
ods graphics off;

proc reg data=engine;
   id speed;
   model power=speed speedsq / p cli clm;
title 'Fitting a Curve to the ENGINE Data';
run;
quit;

ods select OutputStatistics;
proc reg data=engine alpha=0.10;
   model power=speed speedsq / p cli clm;
title '90% Limits for ENGINE Data';
run;
quit;

title;
ods graphics on;
ods select PredictionPlot;
proc reg data=engine alpha=0.10 plots(only)=predictions(x=speed
unpack);
   model power=speed speedsq ;
run;
quit;
ods graphics off;
```

```
proc reg data=engine alpha=0.10;
   model power=speed speedsq ;
   symbol1 value=plus color=blue;
   symbol2 value=dot color=red;
   symbol3 value="I" color=purple;
   symbol4 value="M" color=green;
   symbol5 value="I" color=purple;
   symbol6 value="M" color=green;
   plot (power p. lcl. lclm. ucl. uclm.)*speed / overlay
nostat;
run;
quit;

proc reg data=kilowatt;
   model kwh=ac dryer  / p clm cli;
title 'Multiple Regression for KILOWATT Data';
run;
quit;
```

Special Topic: Line Printer Plots

Chapters 10 and 11 assume that you have the ability to create high-resolution graphics. These two chapters focus on graphs that you can create using either ODS graphics or high-resolution graphics. Each chapter includes a Special Topic section that discusses how to create line printer plots as an alternative. In some cases, you might not be able to create all of the plots in the chapter. Also, in some cases, SAS provides other procedures and other options. If you want line printer graphics, and the plots in this book do not meet your needs, then see the SAS documentation because graphic features that meet your needs might be available.

This section shows how to use PROC REG to create a line printer scatter plot of the data, a line printer scatter plot of the data and predicted values, and a line printer scatter plot that shows the data, predicted values, and limits.

To create a simple line printer scatter plot of the data with PROC REG:

```
options formchar="|----|+|---+=|-/\<>*";
proc reg data=kilowatt lineprinter;
   model kwh=ac;
   plot kwh*ac="+";
title 'Straight-line Regression for KILOWATT Data';
run;
quit;
```

The **OPTIONS** statement includes the **FORMCHAR=** option, which specifies characters that are used for various aspects of the plot. SAS documentation recommends this specific value of the FORMCHAR= option to ensure consistent output on different computers.

The **LINEPRINTER** option specifies line printer plots. The PLOT statement specifies a simple scatter plot, and specifies that the data values appear as plus signs. You can specify any single character. Enclose the character in double quotation marks.

Figure ST10.1 shows the results.

Figure ST10.1 Line Printer Scatter Plot

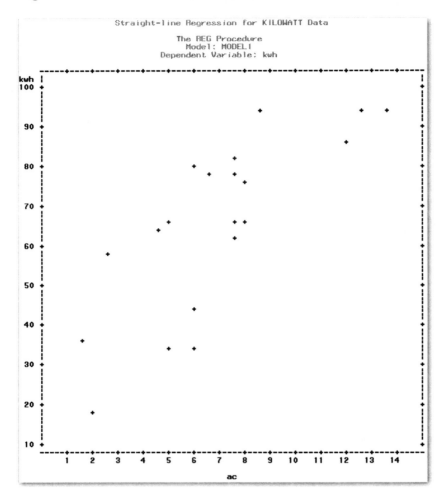

To create a line printer scatter plot of the data and predicted values with PROC REG:

```
options formchar="|----|+|---+=|-/\<>*";
proc reg data=kilowatt lineprinter;
   model kwh=ac;
   plot kwh*ac="+" p.*ac="p" / overlay;
title 'Straight-line Regression for KILOWATT Data';
run;
quit;
```

The PLOT statement includes two scatter plot requests. The OVERLAY option requests that the two scatter plots be placed on a single plot. P. specifies predicted values in the plot. For line printer plots, specify a symbol for each *y*x* request. Symbols will be helpful for understanding the plot. If symbols are omitted, then the plot shows the number of points at a given location as 1, 2, 3, and so on. The results can be confusing or difficult to interpret, especially for models where the predicted line or limits overlay the data values.

Figure ST10.2 shows the results.

Figure ST10.2 Line Printer Plot with Data and Predicted Values

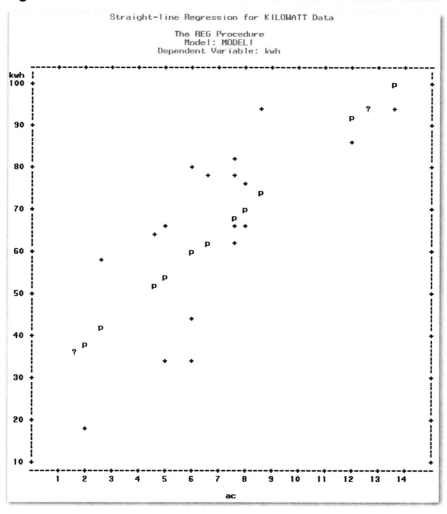

Figure ST10.2 shows two question marks. PROC REG uses a question mark on a line printer plot when more than one value should be plotted. For this plot, the question marks indicate two places where the data value and the predicted value overlap and cannot be distinctly plotted.

To create a line printer scatter plot of the data, predicted values, and limits with PROC REG:

```
options formchar="|----|+|---+=|-/\<>*";
proc reg data=kilowatt lineprinter;
   model kwh=ac;
   plot kwh*ac="+" p.*ac="p" lcl.*ac="I" lclm.*ac="M"
        ucl.*ac="I" uclm.*ac="M" / overlay ;
title 'Straight-line Regression for KILOWATT Data';
run;
quit;
```

The PLOT statement includes several scatter plot requests. The OVERLAY option requests that all scatter plots be placed on a single plot. All of the output statistics in the plot requests end with a period. The plot requests use the letter "M" to represent the confidence limits for the mean, and "I" to represent the prediction limits.

Figure ST10.3 shows the results. Figure ST10.3 shows several question marks that indicate places where more than one point from the multiple plot requests overlap.

Figure ST10.3 Line Printer Plot with Data, Predicted Values, and Limits

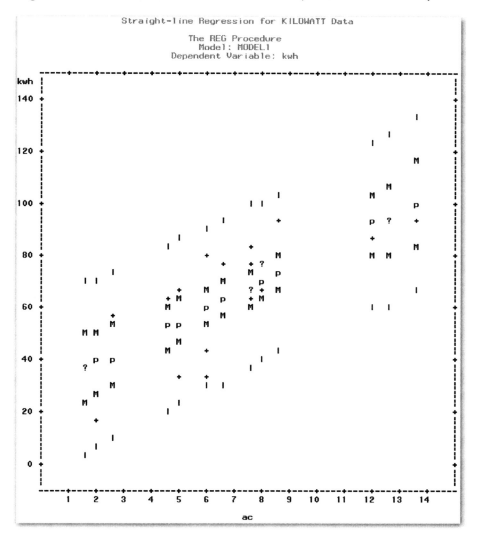

The general form of the statements to create line printer scatter plots in PROC REG is shown below:

OPTIONS FORMCHAR="|----|+|---+=|-/\<>*";
    PROC REG DATA=*data-set-name* **LINEPRINTER** *options*;
    MODEL statement;
PLOT *y*x="symbol" y*x="symbol"* ... **/ OVERLAY;**

Provide as many *y*x* plot requests as you want. The FORMCHAR= value ensures consistent output on different computers. OVERLAY requests that all of the *y*x* plot requests be placed on a single plot.

Chapter 11

Performing Basic Regression Diagnostics

How do you determine whether your regression model adequately represents your data?
How do you know when more terms should be added to the model? How do you identify
outlier points, where the model doesn't fit well? These questions can be answered using
regression diagnostics. This chapter discusses the following topics:

- understanding residuals plots

- using residuals plots and lack of fit tests to decide whether to add terms to the
  model

- using residuals plots to identify outlier points

- using residuals plots to detect a time sequence in the data

- using studentized residuals to check for possible outliers

- checking the regression assumption for normality of residuals

Chapters 10 and 11 discuss the activities of regression analysis and regression
diagnostics. Fitting a regression model and performing diagnostics are intertwined. In
regression, you first fit a model. Then, you perform diagnostics to assess how well the
model fits. You repeat this process until you find a suitable model. Chapter 10 focused on
fitting models. This chapter focuses on performing diagnostics.

Concepts in Plotting Residuals

A regression model fits an equation to an observed set of data points. Chapter 10 showed how to use the equation to get predicted values. The differences between the observed values and predicted values are *residuals* (residual=observed-predicted). Residuals can help show you whether your model fits the data well. The next four sections discuss plots using residuals. These plots use residuals on the *y*-axis. Statisticians often refer to this group of plots as *residuals plots*.

Plotting Residuals against Predicted Values

A plot of residuals against predicted values should look like a random scattering of points or a horizontal band. If your model needs another term, then a plot of residuals can suggest what type of term should be added to the model. Figure 11.1 shows some possible patterns in the plots.

Figure 11.1 Possible Patterns in Plots of Residuals

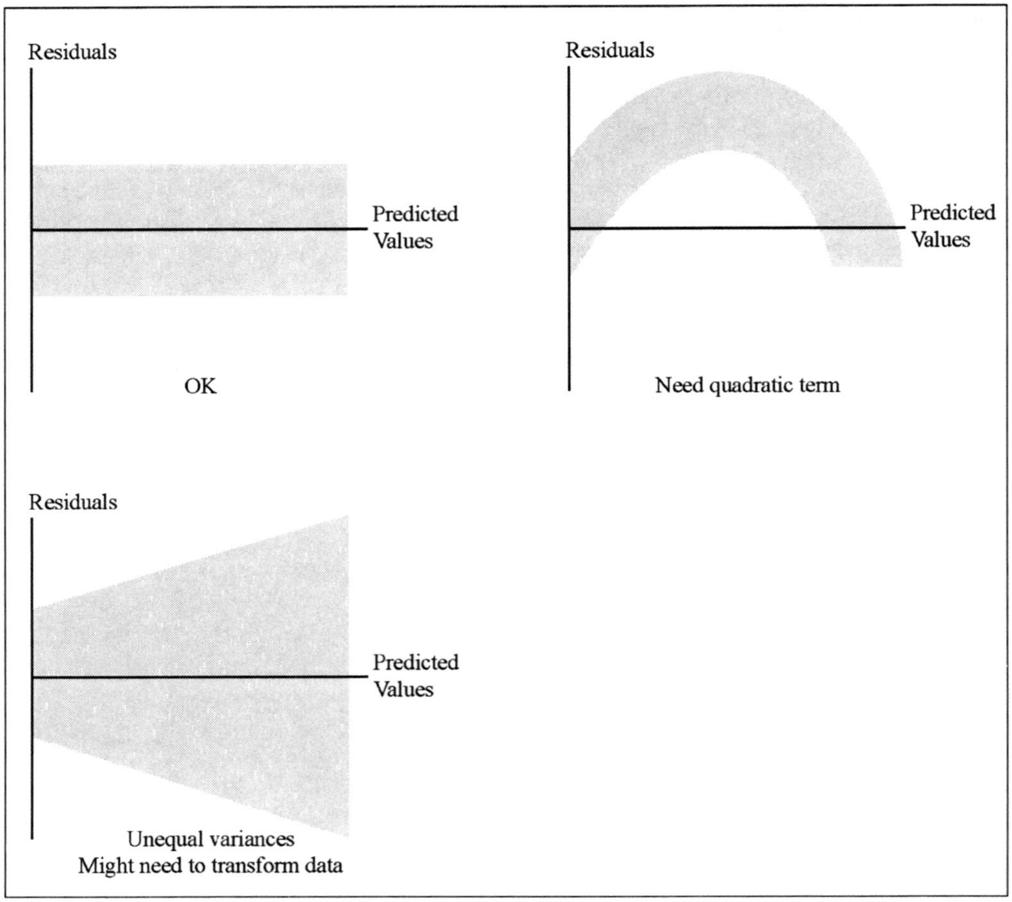

Residuals, Predicted Values, and Outlier Points

A plot of residuals against predicted values sometimes looks like a horizontal band, with the exception of one or two points. Figure 11.2 shows an example of a possible outlier point. All points except for one point fall inside the random pattern indicated by the gray horizontal band.

A large residual occurs when the predicted value is substantially different from the observed value. Here, the meaning of "large" refers to the absolute value of the residual because residuals can be either positive or negative. Observations with large residuals are possible outlier points. As with possible outlier points in the data, you need to investigate the residuals further. The possible outlier point might simply be due to chance or it might be due to a special cause.

Figure 11.2 Residuals Plot Showing Pattern for a Possible Outlier Point

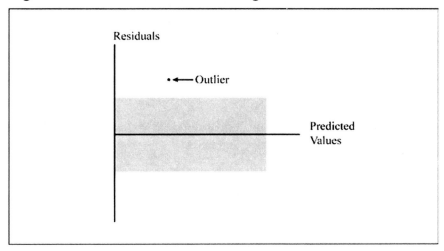

Plotting Residuals against Independent Variables

Plotting residuals against independent variables (*x* variables or regressors) helps show whether the model fits the data well. For a good-fitting model, plots of residuals against regressors show a random scattering of points.

Plots that show an obvious pattern indicate that another term needs to be added to the model. A curved pattern indicates that a quadratic term needs to be added to the model. A plot with a definite increasing or decreasing trend indicates that a regressor needs to be added to the model. For multiple regression, plot the residuals against regressors in the model and against independent variables not in the model.

Plotting Residuals in Time Sequence

When data is collected all at once, a time sequence is unlikely. For example, the homebuyer collected the mortgage **rates** data (in Chapter 6) on a single day. Typically, data is collected over time. This is especially true for data used in regression. For the **kilowatt** data, the homeowner collected data over 21 days. For the **engine** data, the engineer measured **power** and **speed** for 24 data points. When data is collected over time, the conditions that change over time might impact the model.

For data without a measurable time trend, the plot of residuals against the time sequence in which the data was collected shows a random pattern of points. For data with a measurable time trend, the plot often shows an up-and-down or wavy pattern. Figure 11.3 shows an example.

Here, the main focus is the relationship between dependent variables and regressors. You are not focused on time as a regressor. You are checking for a possible time sequence effect. In contrast, suppose you are interested in modeling changes in the stock market over the past five years. In that case, time is a regressor, and more advanced statistical techniques are required.

Figure 11.3 Residuals Plot Showing Pattern for a Time Trend

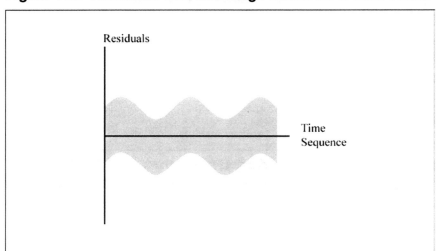

Creating Residuals Plots for the Kilowatt Data

Chapter 10 discussed the straight-line and multiple regression fits for the **kilowatt** data. This section shows the residuals plots for the straight-line model. It then shows the residuals plots for the multiple regression model.

Both topics use ODS graphics.[1] Both topics also request specific plots with ODS graphics. For a description of the automatic plots created by ODS graphics (including some plots that this chapter does not discuss in detail), see "Special Topic: Automatic ODS Graphics" at the end of the chapter. For traditional graphics created with a **PLOT** statement, or for line printer plots, see "Special Topic: Creating Diagnostic Plots with Traditional Graphics and Line Printer Plots" at the end of the chapter.

Because this chapter uses **PROC REG** for regression diagnostics for multiple data sets and multiple regression models, the summary for the general form of the procedure appears only in the "Syntax" section at the end of the chapter. Similarly, the list of ODS tables for **PROC REG** appears only in the "Syntax" section.

Residuals Plots for Straight-Line Regression

Chapter 10 used **PROC REG** to fit a straight line to the **kilowatt** data, and created the results in Figure 10.12. This section repeats the same **PROC REG** step for convenience, but it does not show the results.

Plotting Residuals against Independent Variables

The SAS statements below plot the residuals against **ac** for the straight-line model:

```
ods graphics on;
proc reg data=kilowatt plots(only)=residuals;
   model kwh=ac;
run;
quit;
ods graphics off;
```

The **PLOTS=RESIDUALS** option creates a plot of residuals against the independent variable.[2] As Chapter 10 discussed, **ONLY** limits the plots to plots specified after the equal sign. The program also produces the results shown in Figure 10.12 (but not shown in this chapter). Figure 11.4 shows the plot.

---

[1] ODS Statistical Graphics are available in many procedures starting with SAS 9.2. For SAS 9.1, ODS Statistical Graphics were experimental and available in fewer procedures. For releases earlier than SAS 9.1, you need to use alternative approaches like traditional graphics or line printer plots, both of which Chapter 11 discusses in the "Special Topic" sections. Also, to use ODS Statistical Graphics, you must have SAS/GRAPH software licensed.

[2] The **PLOTS=** option is available starting with SAS 9.2. For earlier releases of SAS, you can use traditional graphics to create a similar plot.

Figure 11.4 Residual by AC for Straight-Line Model

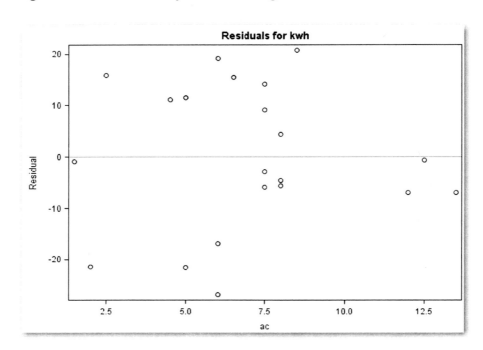

Figure 11.4 shows a random scatter of points, indicating that a quadratic term for **ac** is not needed.

To plot the residuals against variables not in the model, the simplest approach is to use PROC REG with a PLOT statement. In SAS,

```
ods graphics on;
proc reg data=kilowatt plots(only)=residuals;
   var dryer;
   model kwh=ac;
run;
   plot r.*dryer / nostat nomodel;
run;
quit;
ods graphics off;
```

The **VAR** statement identifies variables that do not appear in the **MODEL** statement and will be used in other statements. The VAR statement must appear before the first **RUN** statement. The PLOT statement uses the **R.** statistics keyword, which identifies residuals.

Also, the PLOT statement uses the **NOSTAT** option (discussed in Chapter 10), which suppresses several statistics that automatically appear to the right of the plot. **NOMODEL** suppresses the equation for the fitted model that automatically appears above the plot. (See Figure 10.11 for one example.) Since the model does not include **dryer**, NOSTAT and NOMODEL are especially appropriate.

The statements above produce the plots shown in Figures 11.4 and Figure 11.5, and produce the results in Figure 10.12.

Figure 11.5 Residuals by Dryer for the Straight-Line Model

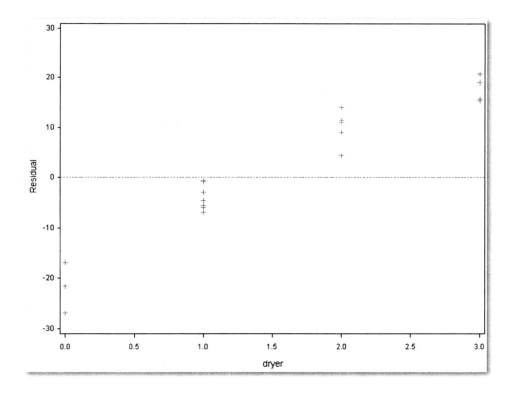

Figure 11.5 shows a definite increase in residuals with an increase in **dryer**. This pattern sends a message to add **dryer** to the model.

Plotting Residuals against Predicted Values

The SAS statements below plot the residuals against predicted values for the straight-line model:

```
ods graphics on;
proc reg data=kilowatt plots(only)=residualbypredicted;
   model kwh=ac;
run;
quit;
ods graphics off;
```

The **PLOTS=RESIDUALBYPREDICTED** option creates a plot of residuals against predicted values. ONLY suppresses other plots that PROC REG automatically creates.

Figure 11.6 shows the plot.

Figure 11.6 Residuals by Predicted Values for Straight-Line Model

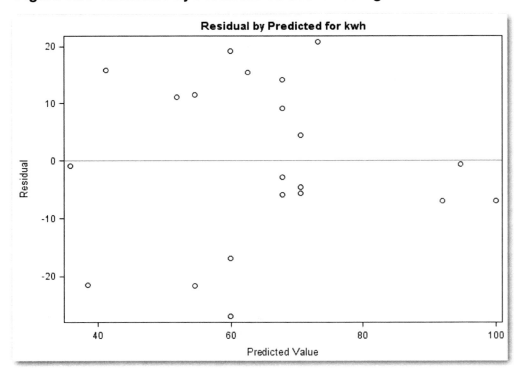

The points in Figure 11.6 appear in a horizontal band with no obvious outlier points. This plot indicates that the model fits the data reasonably well. From Figure 10.12 in Chapter 10, the results show that the model explains about 59% of the variation in the data. This shows why looking at more than one residuals plot is important. If you looked at only the plot of residuals against predicted values, you would probably miss the need to add **dryer** to the model.

Plotting Residuals in Time Sequence

The SAS statements below plot the residuals in time sequence for the straight-line model:

```
proc reg data=kilowatt;
   model kwh=ac;
   plot r.*obs. / nostat nomodel;
run;
quit;
```

The PLOT statement uses statistical keywords for automatic variables. The R. statistical keyword identifies residuals. The **OBS.** statistical keyword identifies the observation number. This approach assumes that the data set is sorted in time sequence. If the data set is not sorted, then use an *x*-variable that defines the time sequence. NOSTAT suppresses several statistics that automatically appear to the right of the plot. NOMODEL suppresses the equation for the fitted model that automatically appears above the plot. Figure 11.7 shows the results.

The points in Figure 11.7 appear in a horizontal band. Some people see a slightly curved pattern, indicating a possible time sequence effect. Looking carefully at the data in Table 10.1, notice that the days with the lowest air conditioner use are near the beginning and end of the 21-day period. Perhaps this is due to cooler days, or the homeowner's knowledge that data was being collected, or something else. This example shows the subjective nature of residuals plots. When the message is weak, the plots are more subjective than when the message is strong. Look at Figure 11.5 again. The message to add **dryer** to the model is obvious from the plot because it's a strong message.

Figure 11.7 Residuals in Time Sequence for Straight-Line Model

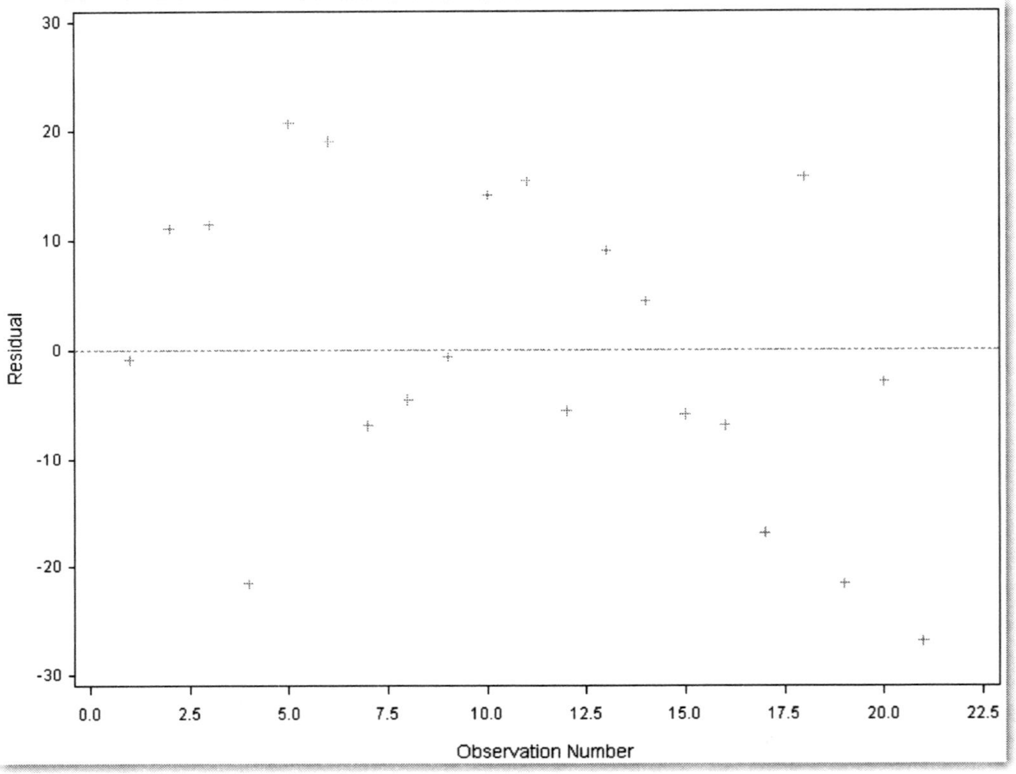

Summarizing Residuals Plots for the Straight-Line Model

For the straight-line model, residuals plotted against **ac** show that a quadratic term for **ac** is not needed. Plotting residuals against **dryer** shows a definite increasing pattern, which sends a message to add **dryer** to the model. Plotting residuals against predicted values shows a random pattern, indicating an adequate model. Plotting residuals in time sequence indicates no obvious pattern.

The previous SAS programs created each of the diagnostic plots separately. The statements below create all of the diagnostic plots in a single program:

```
ods graphics on;
proc reg data=kilowatt
        plots(only)=(residuals residualbypredicted);
   var dryer;
   model kwh=ac;
run;
   plot r.*dryer / nostat nomodel;
run;
   plot r.*obs. / nostat  nomodel;
quit;
ods graphics off;
```

Look at the PROC REG statement. When **PLOTS=** specifies more than one plot, enclose the plot names in parentheses. The program above creates the plots in Figures 11.4, 11.5, 11.6, and 11.7. (The program also creates the results shown in Figure 10.12.)

Residuals Plots for Multiple Regression

Chapter 10 used PROC REG for multiple regression with the kilowatt data, and created the results in Figure 10.24. To create diagnostic plots, use the same approach you used for the straight-line model. The statements below create all of the diagnostic plots discussed in this section:

```
ods graphics on;
proc reg data=kilowatt
        plots(only)=(residuals(unpack) residualbypredicted);
   model kwh=ac dryer;
run;
   plot r.*obs. / nostat  nomodel;
quit;
ods graphics off;
```

Plotting Residuals against Independent Variables

To check the fit of the model, plot the residuals against the independent variables in the model (ac and dryer). Using PLOTS=RESIDUALS creates both plots in a single window. Using **PLOTS=RESIDUALS(UNPACK)** creates two separate plots. Because the PROC REG statement also requests the plot of residuals by predicted values, the parentheses around the list of plots are required. Figures 11.8 and 11.9 show the plots.

Figure 11.8 Residuals by AC for Multiple Regression

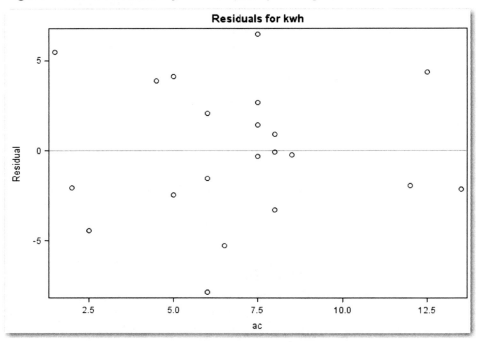

Figure 11.8 shows a random pattern. No additional term for **ac** (such as a quadratic term) is needed.

Figure 11.9 Residuals by Dryer for Multiple Regression

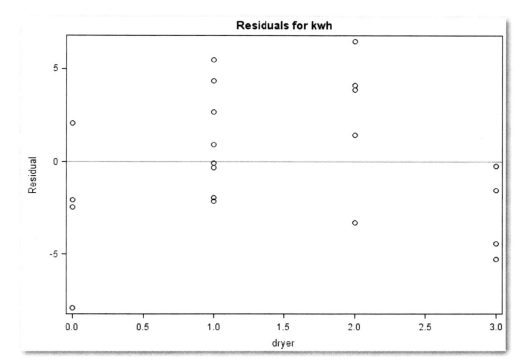

Figure 11.9 shows a curved pattern. Compare this pattern with Figure 11.1. The curvature in the plot indicates a need for a quadratic term for **dryer**.

These results might seem unexpected. If each use of the dryer consumes the same amount of electricity, then the response to **dryer** should be linear, and no quadratic term for **dryer** should be needed. Perhaps smaller loads of clothes were dried on days when several loads were washed, requiring less electricity per load on these days.

Plotting Residuals against Predicted Values

Using PLOTS=RESIDUALBYPREDICTED creates the plot of residuals by predicted values. Because the PROC REG statement also requests the plots of residuals against the independent variables, the parentheses around the list of plots are required. Figure 11.10 shows the plot.

Figure 11.10 Residuals by Predicted Values for Multiple Regression

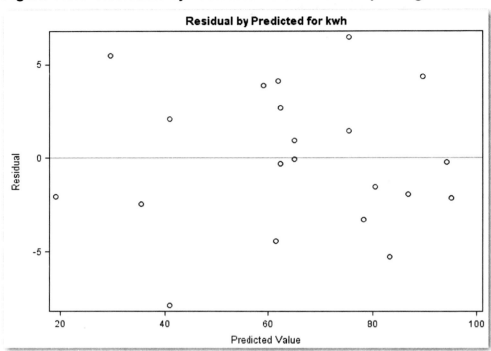

The points in Figure 11.10 appear in a horizontal band with no obvious outlier points. This plot indicates that the model fits the data well.

Plotting Residuals in Time Sequence

The PLOT statement uses statistical keywords for automatic variables to plot the residuals in time sequence. Figure 11.11 shows the plot.

Figure 11.11 Residuals in Time Sequence for Multiple Regression

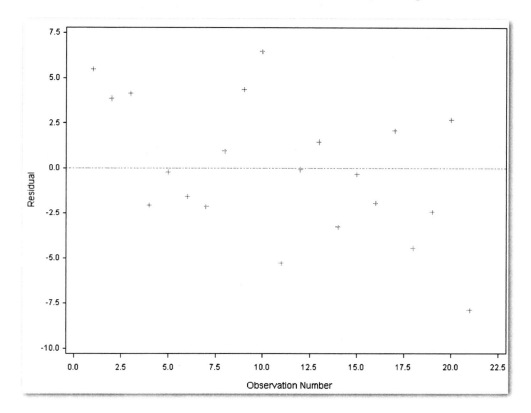

The points in Figure 11.11 appear in a horizontal band. Some people see a slightly decreasing trend. This decreasing trend indicates that the use of electricity might have decreased slightly with time. Perhaps the homeowner's knowledge that data was being collected affected the overall use of electricity. Or, perhaps the days near the end of the experiment were cooler and the air conditioner was used less. Just as with your own data, the true answers to these questions might be unknown. Or, careful sleuthing might reveal a cause for the time sequence effect.

Summarizing Residuals Plots for Multiple Regression

For the multiple regression model, plotting residuals against **ac** shows a horizontal band, indicating that there is no need for a quadratic term. Plotting residuals against **dryer** indicates the possible need for a quadratic term for **dryer**.[3] Plotting residuals by predicted values shows a horizontal band with no obvious outlier points, indicating that the model fits the data well. Plotting residuals in time sequence might show a slight time sequence effect in the data. This effect might be due to either a cooling trend during the time the data was collected, or some other cause, such as the homeowner's knowledge that the use of electricity was being monitored.

Creating Residuals Plots for the Engine Data

Chapter 10 discussed fitting a curve to the **engine** data. Suppose the engineer didn't see the slight curvature in the scatter plot and fit a straight line to the data. The residuals plots would help show that the model needed more terms. The next two sections show the plots for fitting a straight line and for fitting a curve.

Residuals Plots for Straight-Line Regression

Chapter 10 did not fit a straight line to the **engine** data. Figure 11.12 shows the results of a straight-line regression for the **engine** data. To generate these results, use a MODEL statement with **power** as the *y* variable, and **speed** as the *x* variable. Figure 11.13 shows the scatter plot with the fitted line.

---

[3] If you add a quadratic term for **dryer**, you will find that the term is significant. However, the model already provides an excellent fit that explains about 97% of the variation in the data. And, the model has a non-significant lack of fit test (as discussed later in this chapter). For these reasons, the homeowner decided not to add a quadratic term to the model.

Figure 11.12 Straight-Line Regression Results for Engine Data

```
                        The SAS System

                       The REG Procedure
                        Model: MODEL1
                    Dependent Variable: power

            Number of Observations Read          24
            Number of Observations Used          24

                      Analysis of Variance

                              Sum of         Mean
Source                 DF    Squares       Square    F Value   Pr > F

Model                   1  2254.23929   2254.23929    782.08   <.0001
Error                  22    63.41204      2.88237
Corrected Total        23  2317.65133


              Root MSE           1.69775    R-Square    0.9726
              Dependent Mean    52.19333    Adj R-Sq    0.9714
              Coeff Var          3.25282


                      Parameter Estimates

                      Parameter      Standard
     Variable    DF    Estimate        Error    t Value   Pr > |t|

     Intercept    1     6.38561      1.67426       3.81     0.0009
     speed        1     2.62697      0.09394      27.97     <.0001
```

Briefly, the model is significant. Both of the intercept and slope parameters are significantly different from 0. The model fits the data well, explaining about 97% of the variation in the data.

Figure 11.13 Fitted Line for Engine Data

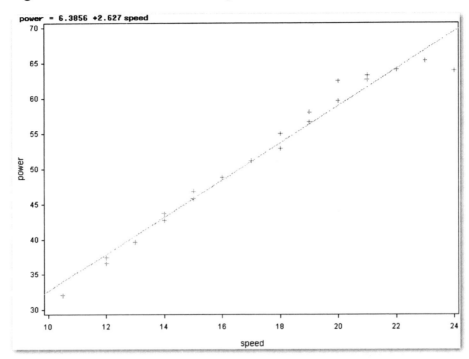

In the scatter plot, points near both ends of the fitted line appear below the line. Points in the middle appear above the line. This subtle curvature indicates that a quadratic term needs to be added to the model. Because the amount of curvature in the data is small, the indication is subtle. However, with a more dramatic curvature, this type of plot would clearly show that a quadratic term needs to be added to the model. The residuals plots can provide a clearer picture.

The statements below create all of the diagnostic plots discussed in this section:

```
ods graphics on;
proc reg data=engine
        plots(only)=(residuals residualbypredicted);
   model power=speed;
run;
   plot r.*obs. / nostat  nomodel;
run;
quit;
ods graphics off;
```

Plotting Residuals against the Independent Variable

The model has only one independent variable. Using PLOTS=RESIDUALS creates the plot of residuals against **speed**. Because the PROC REG statement also requests the plot of residuals by predicted values, the parentheses around the list of plots are required. Figure 11.14 shows the plot.

Figure 11.14 Residuals by Speed for Straight-Line Fit to Engine Data

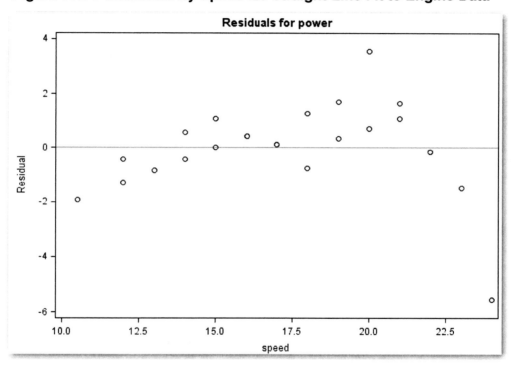

Figure 11.14 shows curvature. It starts on the left with negative values, changes to positive values, and then reverts to negative values. This pattern indicates that a quadratic term needs to be added to the model.

The statements do not plot residuals against the quadratic term (the quadratic term is not in the model). Because the quadratic term is simply a mathematical function (that is, the square) of the linear term, the plot would be the same. Compare this situation with the situation for the **kilowatt** data, where **ac** and **dryer** are not mathematically related. In that situation, plotting residuals against both variables makes sense.

Plotting Residuals against Predicted Values

Using PLOTS=RESIDUALBYPREDICTED creates the plot of residuals by predicted values. Because the PROC REG statement also requests the plots of residuals against the independent variable, the parentheses around the list of plots are required. Figure 11.15 shows the plot.

Figure 11.15   Residuals by Predicted Values for Straight-Line Fit to Engine Data

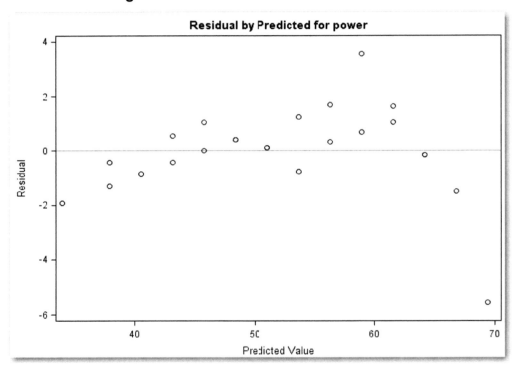

Figure 11.15 shows a much more obvious curvature than Figure 11.14. This pattern reinforces that a quadratic term for **speed** needs to be added to the model.

Plotting Residuals in Time Sequence

The PLOT statement uses statistical keywords for automatic variables to plot the residuals in time sequence. Figure 11.16 shows the plot.

Figure 11.16 Residuals in Sequence for Straight-Line Fit to Engine

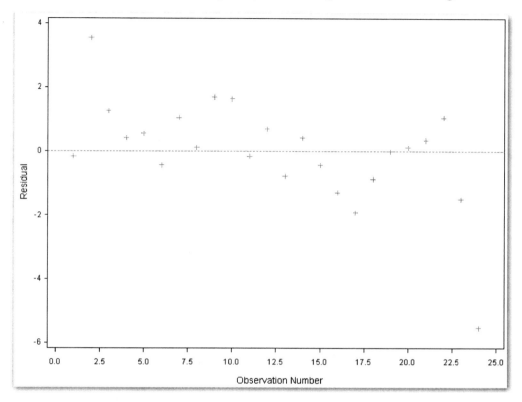

Figure 11.16 shows a wavy pattern similar to what was depicted in Figure 11.3. The wavy pattern is more obvious toward the end of the time period. This plot sends a message to add a term to the model.

Summarizing Residuals Plots for Straight-Line Fit

For the straight-line model, the scatter plot of the data with the fitted line gives a subtle signal that a quadratic term for **speed** needs to be added to the model. Plotting residuals against **speed** strengthens the signal. Plotting residuals against predicted values strengthens the signal even more. Plotting residuals in time sequence also indicates that the model needs another term.

The residuals plots all lead to the same conclusion—add a quadratic term to the model.

Residuals Plots for Fitting a Curve

Chapter 10 fit a curve to the **engine** data, and created the results in Figure 10.18. The statements below also produce these results, and create all of the diagnostic plots discussed in this section:

```
ods graphics on;
proc reg data=engine
         plots(only)=(residuals(unpack) residualbypredicted);
   model power=speed speedsq;
run;
   plot r.*obs. / nostat  nomodel;
quit;
ods graphics off;
```

Plotting Residuals against the Independent Variable

Using PLOTS=RESIDUALS creates the plot of residuals against **speed** and against **speedsq**. UNPACK creates two separate plots. Because the PROC REG statement also requests the plot of residuals by predicted values, the parentheses around the list of plots are required. These statements produce Figures 11.17 and 11.18.

Figure 11.17 shows a horizontal band with two possible outlier points. The plot shows a point higher than the other points at **speed** equal to **20**. It shows a point much lower than the other points at **speed** equal to **24**.

Figure 11.17 is the same as Figure 11.18. This situation occurs because **speedsq** is simply a mathematical function of **speed**. This example illustrates a general principle: when fitting a curve and looking at diagnostic plots, you can look at the residuals against the linear term. Looking at residuals against the quadratic term is not necessary. PROC REG produces both plots because the linear term and the quadratic term are defined by two different variables. In this situation, UNPACK is helpful because you can include only the plot for the linear term in presentations or reports.

Figure 11.17 Residuals by Speed for Fitting a Curve to Engine Data

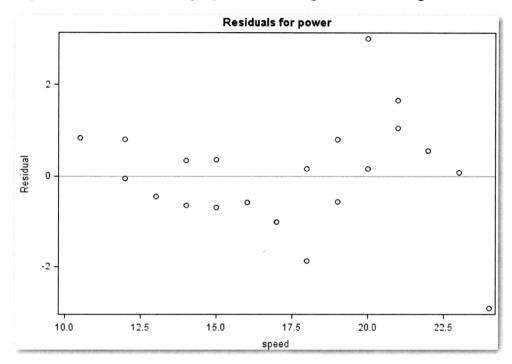

Figure 11.18 Residuals by Speedsq for Fitting a Curve to Engine Data

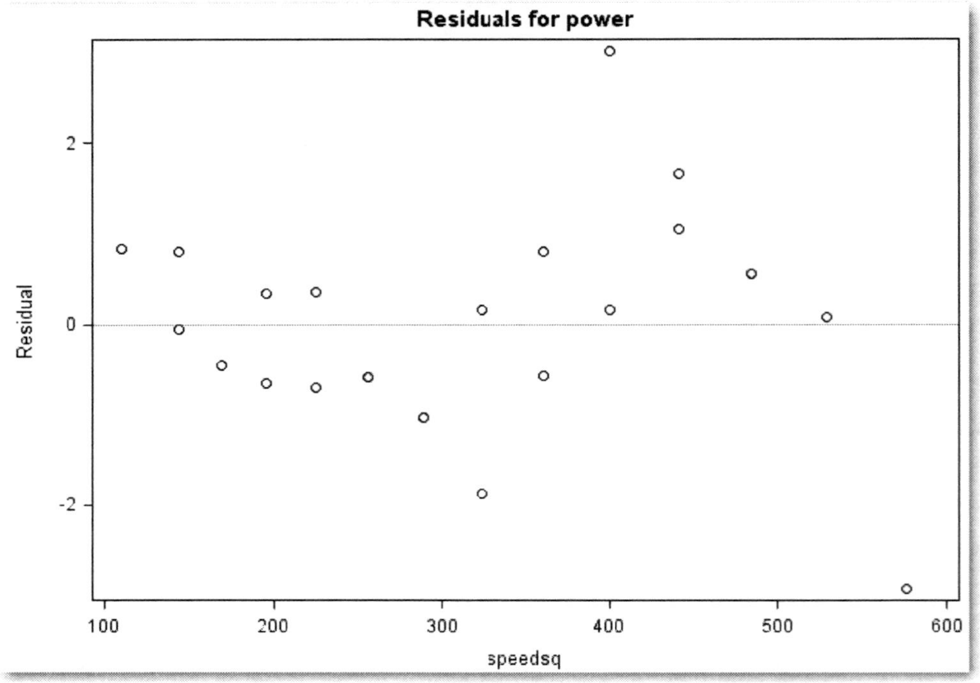

Plotting Residuals against Predicted Values

Using PLOTS=RESIDUALBYPREDICTED creates the plot of residuals by predicted
values. Because the PROC REG statement also requests the plot of residuals against
independent variables, the parentheses around the list of plots are required. Figure 11.19
shows the plot.

Figure 11.19 Residuals by Predicted Values for Fitting a Curve to Engine Data

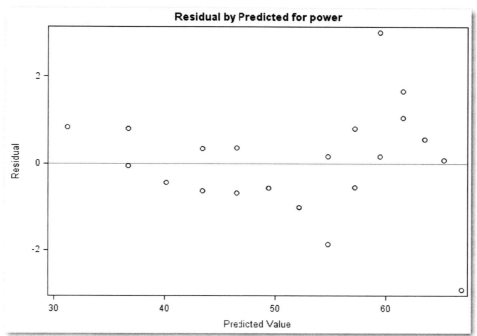

Figure 11.19 sends the same message as the Residuals by speed plot. You need to investigate the two possible outlier points.

Plotting Residuals in Time Sequence

The PLOT statement uses statistical keywords for automatic variables to plot the residuals in time sequence. Figure 11.20 shows the plot.

Figure 11.20 Residuals in Time Sequence for Fitting a Curve to Engine Data

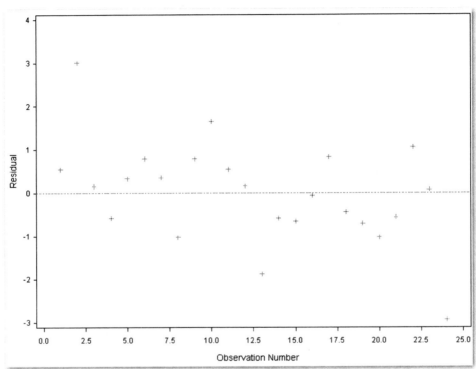

The points in Figure 11.20 form a rough horizontal band except for two possible outlier points. The large positive residual is the second point, and the large negative residual is the last point. This plot sends the same message as other residuals plots. You need to investigate the two possible outlier points.

Summarizing Residuals Plots for Fitting a Curve to Engine Data

For the curve model, plotting residuals against **speed** forms a mostly horizontal band with two possible outlier points. Plotting residuals against predicted values has the same results. Plotting residuals against a time sequence reinforces these results. The plot doesn't show an overall time sequence effect. However, it does show that one of the outlier points occurred near the beginning of the experiment, and the other outlier point occurred at the end of the experiment. These results highlight the subjective nature of residuals plots.

Looking for Outliers in the Data

Figure 11.17 shows a plot of residuals from the quadratic model for the **engine** data. The residuals fall mostly in a horizontal band except for two points that are away from the other points. These points are at the **speed** values of **20** and **24**.

To find out if these residuals are too large to have occurred reasonably by chance alone, you can use *studentized residuals*. Studentized residuals are obtained by dividing residuals by their standard errors. They are also called *standardized residuals*. PROC REG produces both a list and plot of studentized residuals. Reviewing the studentized residuals augments the residuals plots, and interpretation is less subjective.

For a model that fits the data well and has no outliers, most of the studentized residuals should be close to 0. In general, studentized residuals that are less than or equal to 2.0 in absolute value ($-2.0 \le$ studentized residual ≤ 2.0) can easily occur by chance. Values between 2.0 and 3.0 in absolute value occur infrequently. Values larger than 3.0 in absolute value occur very rarely by chance alone. If the studentized residual is between 2.0 and 3.0 in absolute value, then you might consider the observation suspicious. If it is 3.0 or larger in absolute value, then you can consider the observation a probable outlier. Some statisticians use different cutoff values to decide whether an observation is an outlier.

In general, look at the printed residuals, and see whether you can find a systematic pattern in the distribution of positive and negative values.

When there is an obvious need to add variables to the model, you often observe a pattern of negative residuals, then positive residuals, and then negative residuals again.

Data with Outliers (Engine)

The residuals plots from the quadratic model for the **engine** data appear to be a random scatter of points with a couple of exceptions. The studentized residuals can show whether these points are more likely to be a result of chance, or whether they point out special situations.

```
proc reg data=engine;
   model power=speed speedsq / r;
   plot student.*obs. / vref=-2 2 cvref=red nostat;
run;
quit;
```

The **R** option (for residuals) in the MODEL statement creates the **Output Statistics** table and simple plot shown in Figure 11.21. Figure 11.22 shows the plot created by the PLOT statement. This program also produces the regression analysis. (See Figure 10.18.)

The PLOT statement uses the **STUDENT.** statistical keyword to specify studentized residuals as the *y*-variable for the plot. It uses the OBS. statistical keyword to specify observation numbers as the *x*-variable. NOSTAT suppresses several statistics that automatically appear to the right of the plot. **VREF=** specifies reference lines at −**2** and **2**, which are helpful in interpreting the plot. **CVREF=RED** specifies that the reference lines be red.

Figure 11.21 shows the residuals in the **Residual** column, and the studentized residuals in the **Student Residual** column.

You can look at the column of studentized residuals in Figure 11.21 and identify observations that might be outliers. However, the plot makes this easier. This plot, labeled −**2** −**1** −**0 1 2**, is a simple plot of studentized residuals. Each asterisk (*****) corresponds to one half of a unit. For example, observations with four or five asterisks have studentized residuals between 2.0 and 3.0. By reviewing the plot, you can identify observations **2** and **24** as possible outliers.

Most of the remaining columns in the Output Statistics table are discussed in Chapter 10. Here are two new columns:

Std Error Residual is the standard error of the residual, used in calculating the studentized residual. Student Residual=Residual/Std Error Residual.

Cook's D is Cook's D statistic, which measures the change in predicted values when the observation is deleted from the data. Observations with large values of D are considered influential. (See "Special Topic: Automatic ODS Graphs" for more discussion.)

Figure 11.21 Studentized Residuals for Fitting a Curve to Engine Data

```
                       The REG Procedure
                        Model: MODEL1
                    Dependent Variable: power

                        Output Statistics
```

Obs	Dependent Variable	Predicted Value	Std Error Mean Predict	Residual	Std Error Residual	Student Residual
1	64.0300	63.4763	0.4296	0.5537	1.157	0.478
2	62.4700	59.4628	0.3289	3.0072	1.190	2.527
3	54.9400	54.7767	0.3526	0.1633	1.183	0.138
4	48.8400	49.4181	0.3490	-0.5781	1.184	-0.488
5	43.7300	43.3869	0.3479	0.3431	1.184	0.290
6	37.4800	36.6831	0.5238	0.7969	1.118	0.713
7	46.8500	46.4865	0.3388	0.3635	1.187	0.306
8	51.1700	52.1815	0.3566	-1.0115	1.182	-0.856
9	58.0000	57.2039	0.3387	0.7961	1.187	0.671
10	63.2100	61.5537	0.3503	1.6563	1.184	1.399
11	64.0300	63.4763	0.4296	0.5537	1.157	0.478
12	59.6300	59.4628	0.3289	0.1672	1.190	0.140
13	52.9000	54.7767	0.3526	-1.8767	1.183	-1.586
14	48.8400	49.4181	0.3490	-0.5781	1.184	-0.488
15	42.7400	43.3869	0.3479	-0.6469	1.184	-0.546
16	36.6300	36.6831	0.5238	-0.0531	1.118	-0.0475
17	32.0500	31.2138	0.8102	0.8362	0.932	0.898
18	39.6800	40.1190	0.4048	-0.4390	1.166	-0.376
19	45.7900	46.4865	0.3388	-0.6965	1.187	-0.587
20	51.1700	52.1815	0.3566	-1.0115	1.182	-0.856
21	56.6500	57.2039	0.3387	-0.5539	1.187	-0.467
22	62.6100	61.5537	0.3503	1.0563	1.184	0.892
23	65.3100	65.2309	0.5729	0.0791	1.094	0.0723
24	63.8900	66.8173	0.7723	-2.9273	0.963	-3.039

```
                        Output Statistics

                                                 Cook's
              Obs      -2-1  0  1  2                D

                1      |       |           |      0.011
                2      |       |*****      |      0.163
                3      |       |           |      0.001
                4      |       |           |      0.007
                5      |       |           |      0.002
                6      |       |*          |      0.037
                7      |       |           |      0.003
                8      |      *|           |      0.022
                9      |       |*          |      0.012
               10      |       |**         |      0.057
               11      |       |           |      0.011
               12      |       |           |      0.001
               13      |    ***|           |      0.074
               14      |       |           |      0.007
               15      |      *|           |      0.009
               16      |       |           |      0.000
               17      |       |*          |      0.203
               18      |       |           |      0.006
               19      |      *|           |      0.009
               20      |      *|           |      0.022
               21      |       |           |      0.006
               22      |       |*          |      0.023
               23      |       |           |      0.000
               24      |******|           |      1.980
```

This plot shows the studentized residuals. It adds reference lines at −2 and 2. Points that appear outside the reference lines should be investigated as outliers. Figure 11.22 indicates two such points, which correspond to observations **2** and **24**.

Figure 11.22 Studentized Residuals for Fitting a Curve to Engine Data

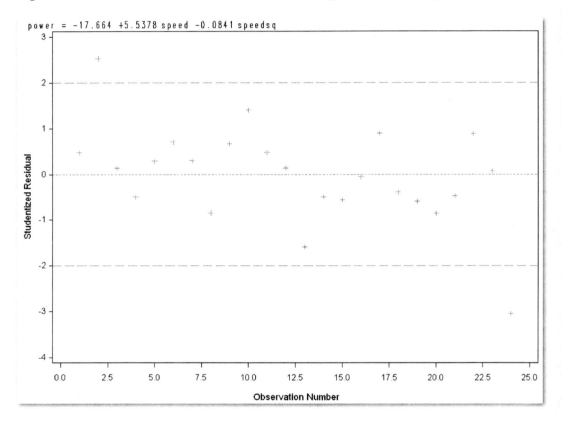

Data without Outliers (Kilowatt)

The residuals plots for the multiple regression model for the **kilowatt** data appeared to be a random scatter of points. The studentized residuals can confirm this initial conclusion, or identify possible outliers that need to be investigated.

```
proc reg data=kilowatt;
   model kwh=ac dryer / r;
   plot student.*obs. / vref=-2 2 cvref=red nostat;
run;
quit;
```

The statements use the same approach that was used for the **engine** data, and produce Figures 11.23 and 11.24. The statements also produce the regression analysis. (See Figure 10.24.)

Figure 11.21 shows the residuals and a simple plot. Figure 11.21 shows no observation with more than four asterisks. Observation 21 has a studentized residual of **–2.177** and is only mildly suspicious. In a data set with 21 observations, you are likely to find one or two residuals in the suspicious range due to chance alone. This is likely to occur, even if there are no real outliers.

Figure 11.23 Studentized Residuals for Multiple Regression for Kilowatt Data

```
                        The REG Procedure
                         Model: MODEL1
                      Dependent Variable: kwh

                        Output Statistics

       Dependent   Predicted      Std Error              Std Error    Student
Obs    Variable       Value    Mean Predict    Residual   Residual   Residual

  1     35.0000     29.5208        1.7965       5.4792      3.501      1.565
  2     63.0000     59.1351        1.1927       3.8649      3.750      1.031
  3     66.0000     61.8681        1.1202       4.1319      3.773      1.095
  4     17.0000     19.0372        2.0605      -2.0372      3.353     -0.608
  5     94.0000     94.2154        1.6663      -0.2154      3.565     -0.0604
  6     79.0000     80.5506        1.6111      -1.5506      3.590     -0.432
  7     93.0000     95.1117        2.0584      -2.1117      3.354     -0.630
  8     66.0000     65.0492        0.9779       0.9508      3.812      0.249
  9     94.0000     89.6458        1.8127       4.3542      3.493      1.247
 10     82.0000     75.5329        1.0036       6.4671      3.805      1.700
 11     78.0000     83.2836        1.5978      -5.2836      3.596     -1.469
 12     65.0000     65.0492        0.9779      -0.0492      3.812     -0.0129
 13     77.0000     75.5329        1.0036       1.4671      3.805      0.386
 14     75.0000     78.2658        1.0372      -3.2658      3.796     -0.860
 15     62.0000     62.3163        0.9458      -0.3163      3.820     -0.0828
 16     85.0000     86.9128        1.6939      -1.9128      3.552     -0.538
 17     43.0000     40.9008        1.5231       2.0992      3.629      0.579
 18     57.0000     61.4199        1.9993      -4.4199      3.390     -1.304
 19     33.0000     35.4349        1.6015      -2.4349      3.595     -0.677
 20     65.0000     62.3163        0.9458       2.6837      3.820      0.703
 21     33.0000     40.9008        1.5231      -7.9008      3.629     -2.177

                        Output Statistics

                                                    Cook's
             Obs     -2-1 0 1 2                        D

               1     |        |***   |               0.215
               2     |        |**    |               0.036
               3     |        |**    |               0.035
               4     |       *|      |               0.046
               5     |        |      |               0.000
               6     |        |      |               0.013
               7     |       *|      |               0.050
               8     |        |      |               0.001
               9     |        |**    |               0.139
              10     |        |***   |               0.067
              11     |      **|      |               0.142
              12     |        |      |               0.000
              13     |        |      |               0.003
              14     |       *|      |               0.018
              15     |        |      |               0.000
              16     |       *|      |               0.022
              17     |        |*     |               0.020
              18     |      **|      |               0.197
              19     |       *|      |               0.030
              20     |        |*     |               0.010
              21     |    ****|      |               0.278
```

Figure 11.24 Studentized Residuals for Multiple Regression for Kilowatt Data

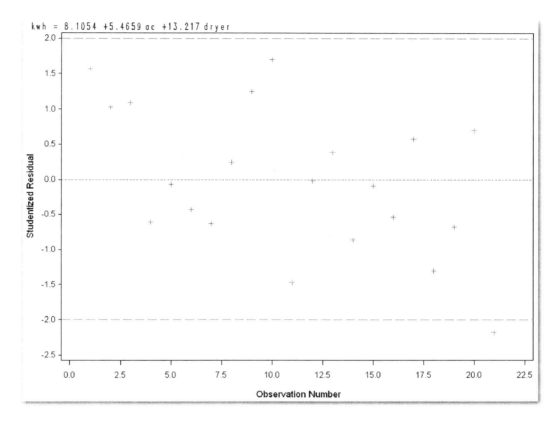

This plot shows the studentized residuals. It adds reference lines at −2 and 2. Figure 11.23 indicates only one possible point to investigate, which corresponds to observation 21.

Investigating Lack of Fit

This section first explains lack of fit. It then shows how to use PROC REG to create tables that analyze the fit of a model.

Concepts for Lack of Fit

Chapter 10 discussed the concepts for least squares regression. It explained how regression partitions total variation in the data into variation due to variables in the model, and variation due to error.

The error variation can be separated even further. When the data contains observations with the same *x* value and different *y* values, the variation in these observations is *pure error*. No matter what model you fit, these data points associated with pure error will still vary.

For example, Chapter 10 introduced this issue for the **kilowatt** data, which has three data points with **ac** equal to **8**. These three points have the same *x* value and different observed *y* values. The straight-line regression equation predicts the same *y* value for these three points.

Lack of fit tests are a statistical way of separating the error variation into pure error and *lack of fit* error. The null hypothesis for the lack of fit test is that the lack of fit error is 0. The alternative hypothesis is that the lack of fit error is different from 0. With a significant *p*-value, you conclude that the model has significant lack of fit, and other models should be considered. Statisticians usually perform lack of fit tests with a reference probability of 0.05. This gives a 5% chance of incorrectly concluding that the model has significant lack of fit when, in fact, it does not.

Checking Lack of Fit for the Kilowatt Data

Straight-Line Regression

Figure 10.12 shows the results of the straight-line regression for the **kilowatt** data, but it does not request a lack of fit analysis.

```
proc reg data=kilowatt;
   model kwh=ac / lackfit;
run;
```

The **LACKFIT** option in the **MODEL** statement requests a lack of fit analysis.[4] SAS performs the lack of fit analysis only when the data has points with duplicate *x* values. If your data has unique *x* values for every data point, SAS cannot perform the analysis.

Figure 11.25 shows the results.

---

[4] The **LACKFIT** option is available starting with SAS 9.2.

Figure 11.25 Lack of Fit from Straight-Line Fit for Kilowatt Data

```
                       The REG Procedure
                        Model: MODEL1
                   Dependent Variable: kwh

              Number of Observations Read          21
              Number of Observations Used          21


                       Analysis of Variance

                              Sum of         Mean
Source                DF      Squares       Square    F Value   Pr > F

Model                  1   5609.66260   5609.66260     26.85   <.0001
Error                 19   3968.90883    208.88994
   Lack of Fit        11   1920.07549    174.55232      0.68   0.7281
   Pure Error          8   2048.83333    256.10417
Corrected Total       20   9578.57143


              Root MSE              14.45303    R-Square     0.5856
              Dependent Mean       64.85714    Adj R-Sq     0.5638
              Coeff Var            22.28440


                       Parameter Estimates

                       Parameter      Standard
     Variable    DF     Estimate         Error    t Value   Pr > |t|

     Intercept    1     27.85107       7.80654       3.57     0.0021
     ac           1      5.34108       1.03067       5.18     <.0001
```

In the **Analysis of Variance** table, look at the row for **Error**. This term is further divided into a row for **Lack of Fit** and a row for **Pure Error**. Find the value for **Pr > F** for the Lack of Fit row. This gives the *p*-value for the lack of fit test. Figure 11.25 shows a *p*-value of **0.7281**. Because this value is greater than the reference probability of 0.05, you conclude that the model does not have significant lack of fit.

In general, to perform the lack of fit test at the 5% significance level, you conclude that the lack of fit is significant if the *p*-value is less than 0.05.

Look at the row for Pure Error. It shows **8** for the degrees of freedom (**DF**). Looking at the data shows how these 8 degrees of freedom are derived:

AC Value	Number of Rows	Degrees of Freedom
5	2	1
6	3	2
7.5	4	3
8	3	2
		Total=8

For each case with duplicate values of the *x* variable, the degrees of freedom are one less than the number of duplicate rows. (Table 10.1 in Chapter 10 shows the **kilowatt** data.)

When looking at lack of fit results, check the degrees of freedom for pure error. The lack of fit test is less useful when the data has only a few duplicate *x* values. With only a few duplicate *x* values, the estimate for pure error is not as good as for data that has many duplicate *x* values. The straight-line fit for the **kilowatt** data has enough duplicate *x* values.

From the lack of fit results, you conclude that the model is adequate. This example shows why multiple diagnostics are important. If you review only this report, instead of the report and the residuals plots, you miss the need to add **dryer** to the model.

Multiple Regression

Figure 10.24 shows the results from multiple regression for the **kilowatt** data, but it does not request a lack of fit analysis:

```
proc reg data=kilowatt;
   model kwh=ac dryer / lackfit;
run;
```

Figure 11.26 shows the results.

Figure 11.26 shows a *p*-value of **0.6130** for the lack of fit test. This value leads you to conclude that the model is adequate.

Look at the row for **Pure Error**. It shows **4** for the degrees of freedom (DF). Looking at the data shows how these 4 degrees of freedom are derived:

AC Value	Dryer Value	Number of Rows	Degrees of Freedom
6	0	2	1
7.5	1	2	1
7.5	2	2	1
8	1	2	1
			Total=4

This might be a case where the data does not have enough duplicate *x* values. If this were a designed experiment instead of an observational study, the design could be improved by adding more duplicate *x* values.

Figure 11.26 Lack of Fit from Multiple Regression for Kilowatt Data

```
                          The REG Procedure
                           Model: MODEL1
                       Dependent Variable: kwh

                 Number of Observations Read          21
                 Number of Observations Used          21

                          Analysis of Variance

                                  Sum of        Mean
     Source              DF       Squares      Square   F Value   Pr > F

     Model                2    9299.80154  4649.90077    300.24   <.0001
     Error               18     278.76989    15.48722
       Lack of Fit       14     211.26989    15.09071      0.89   0.6130
       Pure Error         4      67.50000    16.87500
     Corrected Total     20    9578.57143

                 Root MSE             3.93538   R-Square     0.9709
                 Dependent Mean      64.85714   Adj R-Sq     0.9677
                 Coeff Var            6.06777

                          Parameter Estimates

                         Parameter    Standard
     Variable      DF     Estimate       Error    t Value   Pr > |t|

     Intercept      1      8.10539     2.48085       3.27     0.0043
     ac             1      5.46590     0.28076      19.47     <.0001
     dryer          1     13.21660     0.85622      15.44     <.0001
```

Checking Lack of Fit for the Engine Data

Straight-Line Regression

Figure 11.12 shows the results of the straight-line regression for the **engine** data, but it does not request a lack of fit analysis.

```
proc reg data=engine;
   model power=speed / lackfit;
run;
```

Figure 11.27 shows the results.

Figure 11.27 Lack of Fit from Straight-Line Fit for Engine Data

```
                      The REG Procedure
                        Model: MODEL1
                   Dependent Variable: power

            Number of Observations Read          24
            Number of Observations Used          24


                        Analysis of Variance

                                 Sum of          Mean
Source                  DF      Squares        Square    F Value    Pr > F

Model                    1   2254.23929    2254.23929     782.08    <.0001
Error                   22     63.41204       2.88237
  Lack of Fit           12     54.79409       4.56617       5.30    0.0064
  Pure Error            10      8.61795       0.86179
Corrected Total         23   2317.65133


              Root MSE              1.69775   R-Square     0.9726
              Dependent Mean      52.19333    Adj R-Sq     0.9714
              Coeff Var            3.25282


                        Parameter Estimates

                      Parameter      Standard
       Variable   DF    Estimate        Error    t Value    Pr > |t|

       Intercept   1     6.38561      1.67426       3.81      0.0009
       speed       1     2.62697      0.09394      27.97      <.0001
```

Figure 11.27 shows a *p*-value of **0.0064** for the lack of fit test. This value leads you to conclude that the model is inadequate.

Table 10.2 shows the data. Looking at the data shows where the **10** degrees of freedom are derived. Most of the *x* values appear in the data table twice. The 10 degrees of freedom come from duplicate *x* values for the **speed** values of 12, 14, 15, 16, 17, 18, 19, 20, 21, and 22. This planned experiment was designed with enough duplicate *x* values to test for lack of fit.

Fitting a Curve

Figure 10.18 shows the results of fitting a curve to the **engine** data, but it does not request a lack of fit analysis.

```
proc reg data=engine;
   model power=speed speedsq / lackfit;
run;
```

Figure 11.28 shows the results.

Figure 11.28 Lack of Fit from Fitting a Curve to Engine Data

```
                         The REG Procedure
                          Model: MODEL1
                     Dependent Variable: power

              Number of Observations Read        24
              Number of Observations Used        24


                       Analysis of Variance

                               Sum of         Mean
Source                 DF      Squares       Square    F Value    Pr > F

Model                   2   2285.64535   1142.82268     749.84    <.0001
Error                  21     32.00598      1.52409
   Lack of Fit         11     23.38803      2.12618       2.47    0.0830
   Pure Error          10      8.61795      0.86179
Corrected Total        23   2317.65133


              Root MSE              1.23454    R-Square     0.9862
              Dependent Mean      52.19333    Adj R-Sq     0.9849
              Coeff Var            2.36533


                       Parameter Estimates

                      Parameter     Standard
       Variable   DF   Estimate       Error    t Value    Pr > |t|

       Intercept   1   -17.66377     5.43598      -3.25      0.0038
       speed       1     5.53776     0.64485       8.59      <.0001
       speedsq     1    -0.08407     0.01852      -4.54      0.0002
```

Figure 11.28 shows a *p*-value of **0.0830** for the lack of fit test. This value leads you to conclude that the model is adequate.

Testing the Regression Assumption for Errors

Chapter 10 defined the last assumption for regression as that the errors in the data are normally distributed with a mean of 0 and a variance of σ. Chapter 10 did not test this assumption because testing involves residuals from the regression model. Specifically, testing this assumption involves testing that residuals are normally distributed with a given mean and variance. In practice, many statisticians rely on the theory behind regression, and they often don't test this assumption. The theory behind least squares regression ensures that the mean of the residuals is 0. The least squares regression theory also ensures that the **Root MSE** from the regression fit is the best estimate of the standard deviation of the residuals. As a result, most statisticians test this last assumption for regression by checking the normality of the residuals. Many statisticians simply review plots of the residuals, rather than performing a formal statistical test.

Chapter 5 discussed testing for normality, and Chapter 7 discussed testing that the mean difference is 0. Testing the errors assumption uses the same approaches used in those chapters. However, the approaches in Chapters 5 and 7 require creating an output data set that contains the residuals, and then performing the analysis on the residuals. This chapter does not use those approaches.

Instead, this chapter shows diagnostic plots available in PROC REG, which help you check the normality of the residuals. These plots are available with ODS graphics. The next two sections discuss checking the normality assumption for the final models for the **kilowatt** and **engine** data.

Checking Normality of Errors for the Kilowatt Data

Linear regression assumes that the errors are normally distributed. To create plots that help you check this assumption, submit the following:

```
ods graphics on;
proc reg data=kilowatt
     plots(only)=(residualhistogram residualboxplot qqplot);
   model kwh=ac dryer;
run;
quit;
ods graphics off;
```

The PLOTS= option in the PROC REG statement identifies three plots. The parentheses around the three plots are required. ONLY suppresses other plots that PROC REG automatically creates. **RESIDUALHISTOGRAM** creates a histogram of residuals. **RESIDUALBOXPLOT** creates a box plot of residuals. **QQPLOT** creates a normal quantile plot of residuals. (Chapter 5 discussed using these three types of plots in addition to the formal test for normality. See the discussion in Chapter 5 for details about these types of plots.) Figure 11.29 shows the plots.

Figure 11.29 Checking for Normality of Residuals for Kilowatt Data

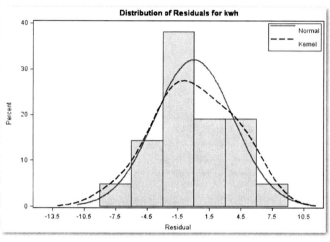

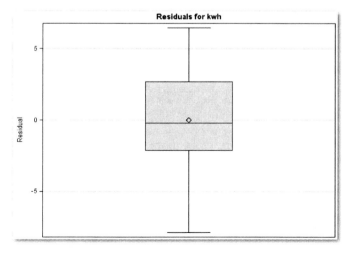

Figure 11.29 Checking for Normality of Residuals for Kilowatt Data (continued)

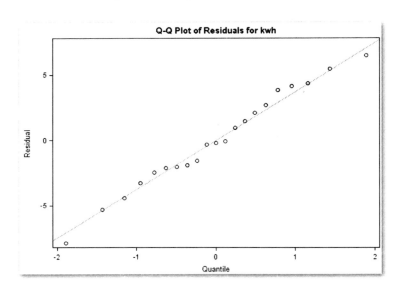

Figure 11.29 shows a histogram with an overlaid normal curve (solid blue line). The red dotted line is the kernel curve, which this book does not discuss. The histogram is roughly mound-shaped, and the assumption of normality seems reasonable.

Figure 11.29 also shows a box plot for the residuals. The median and mean are close together (the center line and the diamond). This matches the expected behavior for a normal distribution. The distribution appears to be slightly skewed to the left.

Figure 11.29 shows a normal quantile plot where most of the points fall close to the reference line for a normal distribution.

Combining information from the three plots, proceeding with the assumption that the residuals are normally distributed is reasonable.

Checking Normality of Errors for the Engine Data

Linear regression assumes that the errors are normally distributed. To create plots that help you check the assumption, submit the following:

```
ods graphics on;
proc reg data=engine
      plots(only)=(residualhistogram residualboxplot qqplot);
    model power=speed speedsq;
run;
quit;
ods graphics off;
```

The options are the same as the options used with the **kilowatt** data.

Figure 11.30 shows the plots.

Figure 11.30 Checking for Normality of Residuals for Engine Data

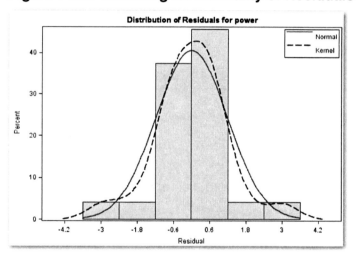

Figure 11.30 shows a histogram with an overlaid normal curve (solid blue line). The red dotted line is the kernel curve, which this book does not discuss. The histogram is mound-shaped, and the assumption of normality seems reasonable.

Figure 11.30 also shows a box plot for the residuals. (See next page.) The median and mean are close together (the center line and the diamond). This matches the expected behavior for a normal distribution. The median is very close to the center of the box, which also matches the expected behavior for a normal distribution. The box plot shows two outliers, which correspond to the measurements for observations 20 and 24. The other regression diagnostic plots identified these observations as potential outliers to investigate.

Figure 11.30 shows a normal quantile plot where most of the points fall close to the reference line for a normal distribution.

Figure 11.30 Checking for Normality of Residuals for Engine Data (continued)

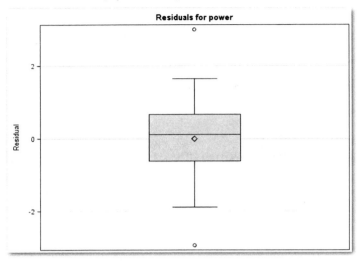

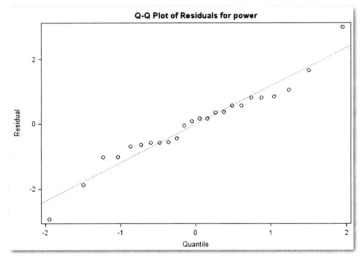

Combining information from the three plots, proceeding with the assumption that the residuals are normally distributed is reasonable.

Summary

Key Ideas

- Regression analysis is an iterative process of fitting a model, and then performing diagnostics. Residuals plots show inadequacies in models. Lack of fit tests indicate whether a different model would fit the data better.

- In general, in plots of residuals, a random scatter of points or a horizontal band indicates a good model. Definite patterns indicate different situations, depending on the plot. Stronger signals have more obvious patterns than weak signals.

- Plot residuals against independent variables in the model, and against independent variables not in the model. Curved patterns indicate that a quadratic term needs to be added to the model. Linear patterns indicate that a variable needs to be added to the model.

- Plot residuals against predicted values. A random scatter of points or a horizontal band indicates a good model. Investigate obvious outlier points to determine whether they are due to a special cause.

- Plot residuals against the time sequence in which the data was collected. An obvious pattern shows a possible time sequence effect in the data.

- Look at a list of the residuals for definite patterns. This approach is most useful for small data sets where there is a very obvious need for another variable.

- Use studentized residuals to check for outliers. For a model that fits the data well, most of the studentized residuals should be close to 0.

- Lack of fit tests separate error variation into pure error and lack of fit error. A significant p-value for a lack of fit test indicates that the model has significant lack of fit and that you need to consider a different regression model.

- Use residuals from the regression model to check the regression assumption that errors in the data are normally distributed.

Syntax

This section shows the general form for **PROC REG** to perform regression diagnostics. See the "Syntax" section in Chapter 10 for the general form of statements to fit regression equations. This chapter uses ODS Statistical Graphics, which are available in many procedures starting with SAS 9.2. For SAS 9.1, ODS Statistical Graphics were experimental and available in fewer procedures. For releases earlier than SAS 9.1, you need to use alternative approaches like traditional graphics or line printer plots, both of which are discussed in the "Special Topic" sections. Also, to use ODS Statistical Graphics, you must have SAS/GRAPH software licensed.

To create diagnostic plots

```
ODS GRAPHICS ON;
PROC REG DATA=data-set-name plot-options;
    VAR variables;
    MODEL statement
    PLOT y-variable*x-variable / options;
RUN;
QUIT;
ODS GRAPHICS OFF;
```

plot-options for a plot of residuals against variables in the model can be:

PLOTS(ONLY)=RESIDUALS(UNPACK);

plot-options for a plot of residuals against predicted values can be:

PLOTS(ONLY)=RESIDUALBYPREDICTED;

For all *plot-options*, the items in parentheses are:

ONLY   suppresses other plots that PROC REG automatically creates.

UNPACK   creates separate plots for the *x*-variables in the model. This is helpful for models with more than one *x*-variable. When fitting a straight line, UNPACK is not needed.

You can combine the *plot-options* and enclose the request in parentheses:

PLOTS(ONLY)=(RESIDUALS(UNPACK)
                            RESIDUALBYPREDICTED);

The PLOT statement for plotting residuals against *x*-variables that are not in the model is:

PLOT R.*x-variable / NOSTAT NOMODEL;

R.	is the statistical keyword for residuals.
x-variable	is a variable that is listed in the VAR statement, but it is not in the model.

The period after the statistical keyword is required. The items after the slash are:

NOSTAT	suppresses several statistics that automatically appear to the right of the plot.
NOMODEL	suppresses the equation for the fitted model that automatically appears above the plot.

The PLOT statement for plotting residuals in time sequence is:

PLOT R.*OBS. / NOSTAT NOMODEL;

OBS.	is the statistical keyword for the observation number.
	The period after the statistical keyword is required. This approach assumes that the data set is sorted in time sequence. If the data set is not sorted, then use an *x*-variable that defines the time sequence.

Other items in italic were defined earlier in the chapter.

To print and plot studentized residuals

```
PROC REG DATA=data-set-name;
  MODEL y-variable=x-variables / R;
  PLOT STUDENT.*OBS. / VREF=-2 2 CVREF=RED NOSTAT ;
RUN;
QUIT;
```

R	in the MODEL statement creates the Output Statistics table that shows the studentized residuals and a simple plot.
STUDENT.	in the PLOT statement is the statistical keyword to specify studentized residuals.

The period after the statistical keyword is required. The items after the slash are not required and are:

VREF=-2 2   specifies reference lines at −2 and 2, which are helpful in interpreting the plot.

CVREF=RED   specifies that the reference lines be red.

Other items were defined earlier.

To test for lack of fit

```
PROC REG DATA=data-set-name;
  MODEL y-variable=x-variables / LACKFIT;
RUN;
QUIT;
```

LACKFIT   in the MODEL statement requests a lack of fit analysis. SAS performs the lack of fit analysis only when the data has points with duplicate x values. If your data has unique x values for every data point, SAS cannot perform the analysis. The LACKFIT option is available starting with SAS 9.2.

To check normality of errors from the regression analysis

```
ODS GRAPHICS ON;
  PROC REG DATA=data-set-name
            PLOTS(ONLY)=(RESIDUALHISTOGRAM
            RESIDUALBOXPLOT QQPLOT);
      MODEL statement
RUN;
QUIT;
ODS GRAPHICS OFF;
```

The parentheses are required.

RESIDUALHISTOGRAM   creates a histogram of residuals.

RESIDUALBOXPLOT creates a box plot of residuals.

QQPLOT   creates a normal quantile plot of residuals.

You can use the ODS statement to choose the output tables to print. Table 11.1 identifies the names of the PROC REG output tables and plots discussed in this chapter:

Table 11.1 ODS Names for Selected PROC REG Tables and Plots

ODS Name	Output Table or Plot
ResidualPlot	ODS plot of residuals by x-variables in the model
ResidualByPredicted	ODS plot of residuals by predicted values
REG*n*	ODS plots, or plots created by the PLOT statement, where the first plot in a SAS session is REG, the second plot is REG1, the third plot is REG2, and so on
ResidualHistogram	ODS histogram of residuals
QQPlot	ODS normal quantile plot of residuals
ResidualBoxPlot	ODS box plot of residuals
PrintPlot	Line printer plots (discussed in "Special Topic: Creating Diagnostic Plots with Traditional Graphics and Line Printer Plots"); all line printer plots have the same ODS name
DiagnosticPanel	ODS set of nine diagnostic plots (discussed in "Special Topic: Automatic ODS Graphics")
OutputStatistics	Output Statistics (from R option) and the simple plot of studentized residuals
ANOVA	Analysis of Variance (from MODEL statement), which contains the lack of fit test when the LACKFIT option is used

Example

The program below produces the output shown in this chapter. This program includes the output in the two "Special Topic" sections. The program is designed for ease of use in following along with the book. Many of the **PROC** steps could be combined for efficiency.

```
options ps=60 ls=80 nonumber nodate;

data kilowatt;
   input kwh ac dryer @@;
   datalines;
35 1.5 1 63 4.5 2 66 5.0 2 17 2.0 0 94 8.5 3 79 6.0 3
93 13.5 1 66 8.0 1 94 12.5 1 82 7.5 2 78 6.5 3 65 8.0 1
77 7.5 2 75 8.0 2 62 7.5 1 85 12.0 1 43 6.0 0 57 2.5 3
33 5.0 0 65 7.5 1 33 6.0 0
;
run;

data engine;
   input speed power @@;
   speedsq=speed*speed;
   datalines;
22.0 64.03 20.0 62.47 18.0 54.94 16.0 48.84 14.0 43.73
12.0 37.48 15.0 46.85 17.0 51.17 19.0 58.00 21.0 63.21
22.0 64.03 20.0 59.63 18.0 52.90 16.0 48.84 14.0 42.74
12.0 36.63 10.5 32.05 13.0 39.68 15.0 45.79 17.0 51.17
19.0 56.65 21.0 62.61 23.0 65.31 24.0 63.89
;
run;

ods graphics on;
proc reg data=kilowatt
         plots(only)=(residuals residualbypredicted) ;
   var dryer;
   model kwh=ac / lackfit;
run;
   plot r.*dryer / nostat nomodel;
run;
   plot r.*obs. / nostat nomodel;
run;
quit;
ods graphics off;
```

```
ods graphics on;
proc reg data=kilowatt
        plots(only)=(residuals(unpack) residualbypredicted) ;

   model kwh=ac dryer;
run;
   plot r.*obs. / nostat nomodel;
run;
quit;
ods graphics off;

ods graphics on;
proc reg data=engine
        plots(only)=(residuals residualbypredicted) ;
   model power=speed;
run;
   plot power*speed / nostat cline=red;
run;
   plot r.*obs. / nostat nomodel;
run;
quit;
ods graphics off;

ods graphics on;
proc reg data=engine
        plots(only)=(residuals(unpack) residualbypredicted) ;
   model power=speed speedsq;
run;
   plot r.*obs. / nostat nomodel;
run;
quit;
ods graphics off;

options ps=70;
proc reg data=engine;
   model power=speed speedsq / r;
   plot student.*obs. / vref=-2 2 cvref=red nostat;
run;
quit;

options ps=60;
proc reg data=kilowatt;
   model kwh=ac dryer / r;
   plot student.*obs. / vref=-2 2 cvref=red nostat;
run;
quit;
```

```
proc reg data=kilowatt;
   model kwh=ac / lackfit;
run;
quit;

proc reg data=kilowatt;
   model kwh=ac dryer / lackfit;
run;
quit;

proc reg data=engine;
   model power=speed / lackfit;
run;
quit;

proc reg data=engine;
   model power=speed speedsq / lackfit;
run;
quit;

ods graphics on;
proc reg data=kilowatt
         plots(only)=(residualhistogram residualboxplot qqplot);
   model kwh=ac dryer;
run;
quit;
ods graphics off;

ods graphics on;
proc reg data=engine
        plots(only)=(residualhistogram residualboxplot qqplot);
   model power=speed speedsq;
run;
quit;
ods graphics off;

proc reg data=kilowatt ;
   model kwh=ac dryer;
run;
   symbol v='A';
   plot r.*ac/ nostat;
run;
   symbol v='D';
   plot r.*dryer / nostat;
run;
```

```
   symbol v=star;
   plot r.*p. / nostat;
run;
   plot r.*obs./ nostat;
run;
   plot student.*obs. / vref=-2 2 cvref=red nostat;
run;
   plot nqq.*obs. / nostat;
quit;

options formchar="|----|+|---+=|-/\<>*";
proc reg data=kilowatt lineprinter;
   model kwh=ac dryer;
run;
   plot r.*ac="*" ;
run;
   plot r.*dryer="+";
run;
   plot r.*p.;
run;
   plot r.*obs. ;
run;
   plot student.*obs. ;
run;
quit;

ods graphics on;
proc reg data=kilowatt plots=diagnostics(stats=none);
   model kwh=ac dryer / lackfit;
run;
quit;
ods graphics off;

ods graphics on;
proc reg data=engine plots=diagnostics(stats=none);
   model power=speed speedsq / lackfit;
run;
quit;
ods graphics off;
```

Special Topic: Creating Diagnostic Plots with Traditional Graphics and Line Printer Plots

Chapters 10 and 11 assume that you have the ability to create high-resolution graphics. This chapter has used ODS graphics for diagnostic plots whenever possible. It has used traditional graphics (high-resolution) in other situations.[5] For all of the diagnostic plots discussed in this chapter, PROC REG also provides high-resolution graphics with the PLOT statement. Also, if you do not have SAS/GRAPH software licensed, you can create most of the diagnostic plots with line printer plots. This section provides sample syntax for creating diagnostic plots.

Traditional Graphics

For plotting studentized residuals and for plotting residuals in time sequence, this chapter used the PLOT statement in PROC REG. You can also use the PLOT statement for other diagnostic plots. The statements below provide an example for multiple regression for the **kilowatt** data:

```
proc reg data=kilowatt ;
   model kwh=ac dryer;
run;
   symbol v='A';
   plot r.*ac/ nostat;
run;
   symbol v='D';
   plot r.*dryer / nostat;
run;
   symbol v=star;
   plot r.*p. / nostat;
run;
   plot r.*obs./ nostat;
run;
   plot student.*obs. / vref=-2 2 cvref=red nostat;
run;
```

---

[5] ODS Statistical Graphics are available in many procedures starting with SAS 9.2. For SAS 9.1, ODS Statistical Graphics were experimental and available in fewer procedures. For releases earlier than SAS 9.1, you need to use alternative approaches like traditional graphics or line printer plots, which are summarized in this section. Also, to use ODS Statistical Graphics, you must have SAS/GRAPH software licensed.

```
     plot nqq.*obs. / nostat;
run;
quit;
```

The first three plots use a **SYMBOL** statement to specify the plotting symbol. Once you use a **SYMBOL** statement, SAS continues to use a symbol specification until you provide another **SYMBOL** statement. The last three plots use a star (asterisk) as the plotting symbol.

All of the **PLOT** statements use the **NOSTAT** option to suppress statistics that automatically appear to the right of the plot. All of the plots use statistical keywords, which are summarized in the table below. The plot of studentized residuals (not shown here) adds reference lines and colors the reference lines (at −2 and 2).

Figure ST11.1 shows the graph of residuals by predicted values as an example.

Figure ST11.1 Traditional Graphic—Residuals by Predicted

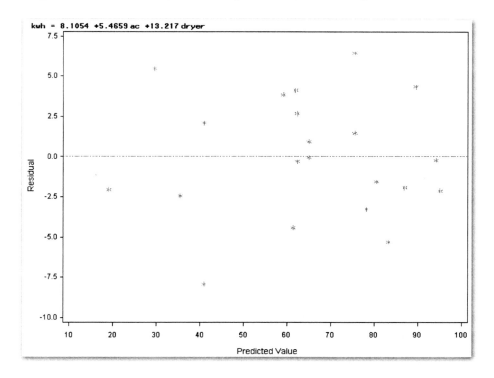

The table below identifies statistical keywords used in the PLOT statements:

R.	residual
P.	predicted value
OBS.	observation number
STUDENT.	studentized residual
NQQ.	normal quantiles for residuals

Line Printer Plots

When you do not have SAS/GRAPH software licensed, you can create most of the diagnostic plots as line printer plots.

```
options formchar="|----|+|---+=|-/\<>*";

proc reg data=kilowatt lineprinter;
   model kwh=ac dryer;
run;
   plot r.*ac="*" ;
run;
   plot r.*dryer="+";
run;
   plot r.*p.;
run;
   plot r.*obs. ;
run;
   plot student.*obs. ;
run;
quit;
```

The **OPTIONS** statement includes the **FORMCHAR=** option, which specifies characters that are used for various aspects of the plot. SAS documentation recommends this specific value of the FORMCHAR= option to ensure consistent output on different computers.

The **LINEPRINTER** option specifies line printer plots. Each PLOT statement specifies a scatter plot. The first two plots specify the appearance of the plotted points. You can specify any single character. Enclose the character in double quotation marks. For a plot where a plotting symbol is omitted, the plot shows the number of points at a given location as 1, 2, 3, and so on.

The NOSTAT, VREF=, and CVREF= options are not available for line printer plots. The SYMBOL statement is not available for line printer plots. Also, the normal quantile plot of residuals is not available for line printer plots.

Figure ST11.2 shows the graph of residuals by predicted values as an example.

Figure ST11.2 Line Printer Plot—Residuals by Predicted

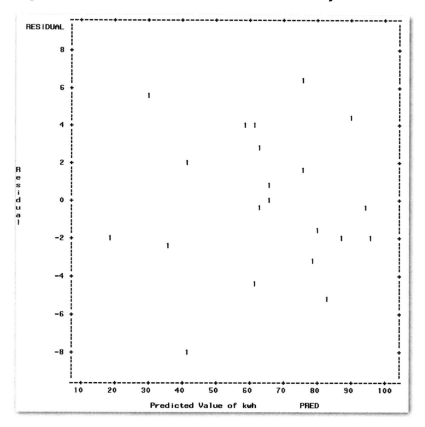

Special Topic: Automatic ODS Graphics

You can use many diagnostic tools to decide whether your regression model fits the data well. This chapter concentrates on residuals plots and lack of fit tests. Where possible, this chapter used ODS graphics and created specific plots for each topic.[6] SAS provides a panel of automatic ODS graphics. Some of these graphics provide more advanced diagnostics. This section briefly discusses the panel of automatic ODS graphics. It briefly introduces the more advanced diagnostics. See SAS documentation or the references in Appendix 1, "Further Reading" for more discussion.

The statements below create the automatic ODS graphics for multiple regression for the **kilowatt** data:

```
ods graphics on;
proc reg data=kilowatt plots=diagnostics(stats=none);
   model kwh=ac dryer;
run;
quit;
ods graphics off;
```

Using **PLOTS=DIAGNOSTICS** produces the diagnostics panel of plots. **STATS=NONE** suppresses a table of statistics that automatically appears on the lower right corner of the panel, and replaces the table with the box plot of residuals.

These statements produce the diagnostics panel in Figure ST11.3. The statements also produce a panel of residuals plots, with both residuals plots in a single panel. The panel combines the plots shown in Figures 11.8 and 11.9.

---

[6] ODS Statistical Graphics are available in many procedures starting with SAS 9.2. For SAS 9.1, ODS Statistical Graphics were experimental and available in fewer procedures. For releases earlier than SAS 9.1, you need to use alternative approaches like traditional graphics or line printer plots. Also, to use ODS Statistical Graphics, you must have SAS/GRAPH software licensed.

Figure ST11.3 Diagnostics Panel for Kilowatt Data

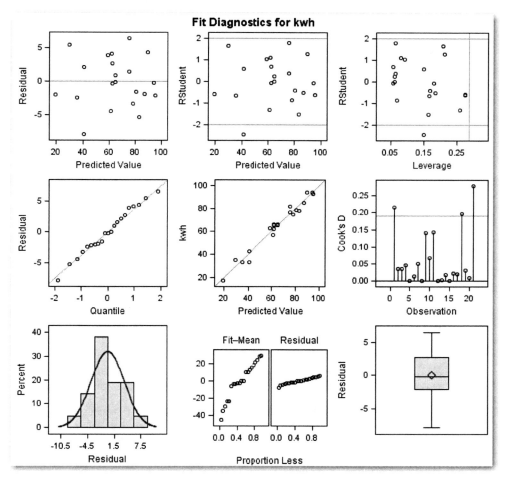

The diagram identifies the nine plots in the diagnostics panel:

Residuals by Predicted	RSTUDENT by Predicted	Leverage
QQ plot of residuals	Observed by Predicted	Cook's D by OBS
Histogram of residuals	Residual-Fit	Box plot of residuals

Four of these plots were discussed earlier. The histogram of residuals shows the normal curve, but not the kernel curve. Here are the five plots that haven't been discussed:

- **RSTUDENT by Predicted.** This plot is similar to the plots of studentized residuals against observation numbers produced earlier in the chapter. This plot uses the **RSTUDENT.** statistical keyword, which is the studentized residual calculated with the current observation deleted. RSTUDENT helps to identify observations that have a large influence on the fit of the model. As a general guideline, investigate points that appear outside the reference lines at −2 and 2. For the **kilowatt** data, this point is the possible outlier that has been discussed (observation 21).

- **Leverage.** This plot shows RSTUDENT by leverage, and is another way to identify possible outliers. As with the RSTUDENT by Predicted plot, this plot shows reference lines at −2 and 2. Points that appear outside the reference lines should be investigated as potential outliers. For the **kilowatt** data, only one point appears as a possible outlier. This is observation **21**, which is the last value in the data set.

  The plot also shows a vertical reference line for the leverage. Points to the right of the reference line are defined as influential, and they have a large impact on the fitted model. For more advanced analyses, statisticians fit the regression model with all of the observations, and then they fit the model without the influential points, and then they compare the results. The vertical reference line is placed at $2p/n$, where p is the number of parameters including the intercept, and n is the number of observations in the model. For multiple regression for the **kilowatt** data, the line is at $(2*3)/21=6/21=0.2857$. For the **kilowatt** data, no points appear to the right of the line, which indicates that there are no influential points.

- **Observed by Predicted.** The line on this plot represents a perfect fit of the model to the data. The circles on the plot are the data values. For a good model, the points (circles) on the plot are close to the line. Figure ST11.3 shows a good-fitting model. This result is what you would expect, given the other diagnostics and the R-Square of about 0.97.

- **Cook's D by OBS.** This plot shows the Cook's D statistic by observation number, and is another way to identify possible influential points. Points that appear above the reference line are defined as influential. The line is placed at $4/n$, where n is the number of observations in the model. For the **kilowatt** data, the line is at $4/21=0.1905$. For the **kilowatt** data, three points appear above the line. These are the observations for days 1, 18, and 21. Point 21 has the most influence.

- **Residual-Fit.** This plot shows two side-by-side normal quantile plots. The left plot shows the centered fit, and the right plot shows the residuals. In theory, when the spread in the two plots is similar, then the model provides a good fit.

The statements below create the automatic ODS graphics for multiple regression for the **engine** data:

```
ods graphics on;
proc reg data=engine plots=diagnostics(stats=none);
   model power=speed speedsq;
run;
quit;
ods graphics off;
```

Figure ST11.4 shows the results.

Figure ST11.4 Diagnostics Panel for Engine Data

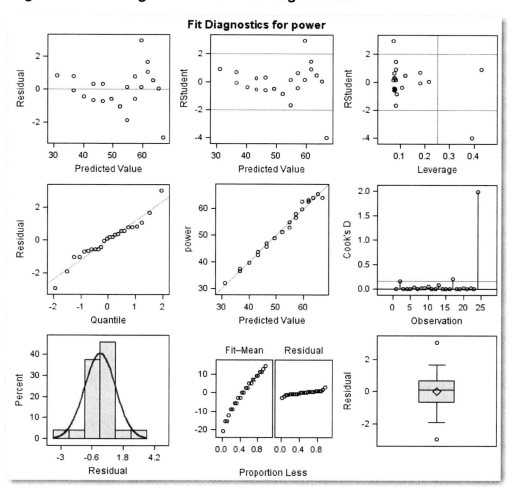

Four of these plots were discussed earlier. The histogram of residuals shows the normal curve, but not the kernel curve. Here are the five plots that have not been discussed:

- **RSTUDENT by Predicted.** This plot shows two possible outliers. These are observations **2** and **24**, which were previously identified as possible outliers.

- **Leverage.** This plot shows two possible outliers for observations **2** and **24**. The plot also shows two possible influential points, for observation **17** (which is not also a possible outlier) and **24**.

- **Observed by Predicted.** The plot shows a good-fitting model. This result is what you would expect, given the other diagnostics and the R-Square of about 0.98.

- **Cook's D by OBS.** This plot shows a strongly influential point for observation **24**. The plot also shows observation **17** just above the reference line.

- **Residual-Fit.** This plot shows that the model provides a good fit.

P a r t **5**

Data in Summary Tables

Chapter **12**

Creating and Analyzing Contingency Tables

Do Democrats and Republicans have the same responses to a survey question asking about campaign reforms? Do rural and urban residents own different types of vehicles (sports cars, sedans, SUVs, and trucks)? Do graduate and undergraduate students differ in whether they share an apartment, have their own apartment, or live in a dorm?

These questions involve looking at two classification variables, and testing whether the variables are related. Neither variable contains quantitative measurements. Instead, both variables identify the respondent as belonging to a group. For example, one respondent might be "rural, truck", and another respondent might be "urban, sedan". The two variables can be character or numeric, and are either nominal or ordinal. This chapter discusses the following topics:

- summarizing classification data using tables

- testing for independence between classification variables

- using measures of association between classification variables

The methods in this chapter are appropriate for both nominal and ordinal variables.

Defining Contingency Tables

Suppose you have nominal or ordinal variables that classify the data into groups. You want to summarize the data in a table. To make discussing summary tables easier, this section introduces some notation.

Tables that summarize two or more classification variables are called *contingency tables*. These tables are also called *crosstabulations*, *summary tables*, or *pivot tables* (in Microsoft Excel). Tables that summarize two variables are called *two-way tables*, tables that summarize three variables are called *three-way tables*, and so on. A special case of the two-way table occurs when both variables have only two levels. This special case is called a *2×2 table*. Although this chapter shows how to create contingency tables involving several variables, the analyses are appropriate only for two-way tables. Figure 12.1 shows the parts of a contingency table.

Figure 12.1 Parts of a Contingency Table

		Columns		
	column 1	**column 2**	**...**	**column c**
row 1	$cell_{11}$	$cell_{12}$	...	$cell_{1c}$
row 2	$cell_{21}$	$cell_{22}$	...	$cell_{2c}$
...	...	...	...	...
row r	$cell_{r1}$	$cell_{r2}$	...	$cell_{rc}$

(**Rows** label at left of the row entries)

The table consists of rows and columns. Figure 12.1 contains r rows and c columns, and is an $r{\times}c$ table. The rows and columns form *cells*. Each cell of a table is uniquely indexed by its row and column. For example, the cell in the second row and first column is $cell_{21}$. A contingency table usually shows the number of observations in each cell, also known as the *cell frequency*. The total number of observations in the table is n. The number of observations in each cell follows the same notation pattern as the cells. For example, the number of observations in $cell_{21}$ is n_{21}.

The contingency table in Figure 12.1 is a two-way table because it summarizes two variables. **The phrase "two-way" does not refer to the number of rows or columns. It refers to the number of variables that are included in the table.**

Sometimes, you have the raw data, and you want to summarize the data in a table and analyze it. Other times, you already have a summary table, and you want to analyze it. The next three sections discuss summarizing data in tables. The rest of the chapter discusses analyses for two-way tables.

Summarizing Raw Data in Tables

Suppose you have raw data that you want to summarize in a table. Table 12.1 shows data from an introductory statistics class. The instructor collected data on the gender of each student, and whether the student was majoring in Statistics.[1]

Table 12.1 Class Data

Student	Gender	Major
1	Male	Statistics
2	Male	Other
3	Female	Statistics
4	Male	Other
5	Female	Statistics
6	Female	Statistics
7	Male	Other
8	Male	Other
9	Male	Statistics
10	Female	Statistics
11	Male	Other
12	Female	Statistics
13	Male	Statistics
14	Male	Statistics
15	Male	Other
16	Female	Statistics
17	Male	Statistics
18	Male	Other
19	Female	Other
20	Male	Statistics

---

[1] Data is from Dr. Ramon Littell, University of Florida. Used with permission.

This data is available in the **statclas** data set in the sample data for the book. To create and summarize the data set in SAS, submit the following:

```
proc format;
   value $gentxt 'M' = 'Male'
                 'F' = 'Female';
   value $majtxt 'S' = 'Statistics'
                 'NS' = 'Other';
run;

data statclas;
   input student gender $ major $ @@;
   format gender $gentxt.;
   format major $majtxt.;
   datalines;
1 M S 2 M NS 3 F S 4 M NS 5 F S 6 F S 7 M NS 8 M NS 9 M S
10 F S 11 M NS 12 F S 13 M S 14 M S 15 M NS 16 F S
17 M S 18 M NS 19 F NS 20 M S
;
run;

proc freq data=statclas;
   tables gender*major;
   title 'Major and Gender for Students in Statistics Class';
run;
```

The **TABLES** statement identifies the *row-variable*, and then the *column-variable*. To create a contingency table, the asterisk is required. Figure 12.2 shows the results. The title identifies the *column-variable* (**major**) and the *row-variable* (**gender**).

Figure 12.2 Contingency Table

```
Major and Gender for Students in Statistics Class

               The FREQ Procedure

           Table of gender by major

     gender        major

     Frequency│
     Percent  │
     Row Pct  │
     Col Pct  │Other    │Statisti│  Total
              │         │cs      │
     ─────────┼─────────┼────────┼
     Female   │    1    │    6   │     7
              │ 5.00    │ 30.00  │ 35.00
              │14.29    │ 85.71  │
              │12.50    │ 50.00  │
     ─────────┼─────────┼────────┼
     Male     │    7    │    6   │    13
              │35.00    │ 30.00  │ 65.00
              │53.85    │ 46.15  │
              │87.50    │ 50.00  │
     ─────────┼─────────┼────────┼
     Total         8        12        20
                40.00     60.00    100.00
```

Understanding the Results

Figure 12.2 summarizes the data. The top row shows the values for the *column-variable*, and the leftmost column shows the values for the *row-variable*.

The outside edges of the table give *row totals* and *column totals*. Looking at the row totals, 7 students are **Female**, and 13 are **Male**. Looking at the column totals, 12 students are majoring in **Statistics**, and 8 are majoring in something else (**Other**). The outside edges also give the *row percentages* and *column percentages*. Looking at the row percentages, 35% (calculated from 7/20) of the students are **Female**. Looking at the column percentages, 60% (calculated from 12/20) of the students are majoring in **Statistics**.

The lower right outside edge of the table gives the overall total for the table, which is **20**. Because there are no missing values, this number matches the number of observations in the data table. However, if one observation in the data table contained the gender, but not the major, the contingency table would contain only 19 observations.

The top left corner cell gives a key to understanding the main body of the table. The list below gives details:

Frequency   Number of observations in each cell. This class has 1 female student who is majoring in **Other**.

Percent   Percentage of the total number of observations represented by the cell count. The single **Female** student majoring in **Other** represents 5% of the class. The class has 20 students, so 1/20=5%.

Row Pct   Percentage of observations in the row represented by the cell frequency. The single **Female Other** student represents 14.29% of the females in the class. The class has 7 females, so 1/7=14.29%.

The **Row Pct** values sum to 100% for each row.

Col Pct   Percentage of observations in the column represented by the cell frequency. The single **Female Other** student represents 12.5% of the **Other** majors in the class. The class has 8 students majoring in **Other**, so 1/8=12.5%.

The **Col Pct** values sum to 100% for each column.

Suppressing Statistics

Figure 12.2 shows four statistics in each cell. In some cases, you want only the frequency counts, and not the percentages. The TABLES statement includes four options to control the statistics that are displayed. To suppress all percentages, type the following:

```
proc freq data=statclas;
   tables gender*major / norow nocol nopercent;
title 'Counts for Students in Statistics Class';
run;
```

Figure 12.3 shows the results.

Compare the results in Figures 12.2 and 12.3. Figure 12.3 shows only the frequency counts. The TABLES statement also provides an option, **NOFREQ**, which suppresses the frequency counts.

Figure 12.3 Contingency Table with Frequency Counts Only

```
Counts for Students in Statistics Class

            The FREQ Procedure

        Table of gender by major

 gender        major

 Frequency│Other    │Statisti│  Total
          │         │cs      │

 Female       1          6         7

 Male         7          6         13

 Total        8         12         20
```

Summarizing PROC FREQ for Tables from Raw Data

The general form of the statements to create a contingency table from raw data using PROC FREQ is shown below:

PROC FREQ DATA=_data-set-name_**;**

   TABLES _row-variable*column-variable I options_**;**

data-set-name is the name of a SAS data set, _row-variable_ is the variable that defines rows, and _column-variable_ is the variable that defines columns. The asterisk is required.

The **TABLES** statement _options_ can be one or more of the following:

NOFREQ   suppresses frequency counts.

NOCOL   suppresses column percentages.

NOROW   suppresses row percentages.

NOPERCENT suppresses overall percentages.

If any option is used, the slash is required. You can use the options one at a time or together.

You can use the **ODS** statement to choose the output tables to print. Table 12.2 identifies the name of the contingency table for PROC FREQ:

Table 12.2 ODS Table Names for PROC FREQ

ODS Name	Output Table
CrossTabFreqs	Contingency table

Creating Contingency Tables from an Existing Summary Table

Sometimes, you already have a summary table of the data. You can create a SAS data set that matches the summary table, and then analyze it. You do not need to expand the summary table into a data set that has one row for every observation in the summary table. Table 12.3 shows frequency counts for several court cases. The columns show the defendant's race, and the rows identify whether the death penalty was imposed after the defendant was convicted of homicide.[2]

Table 12.3 Death Penalty Data

	Defendant's Race	
Decision	**Black**	**White**
No	149	141
Yes	17	19

This data is available in the **penalty** data set in the sample data for the book.

To create the data set in SAS, follow the steps in Chapter 2. Your data set will have four observations, one for each cell in the table. It will have three columns, one for decision, one for race, and one for count.

Table 12.3 already summarizes the data. However, the results from PROC FREQ provide more features, including the percentages and statistical tests.

---

[2] Data is from A. Agresti, *Analysis of Ordinal Categorical Data* (New York: John Wiley & Sons, Inc., 1984). Reprinted with permission of John Wiley & Sons, Inc.

```
data penalty;
    input decision $ defrace $ count @@;
    datalines;
Yes White 19 Yes Black 17
No White 141 No Black 149
;
run;

proc freq data=penalty;
    tables decision*defrace;
    weight count;
    title 'Table for Death Penalty Data';
run;
```

The statements use PROC FREQ with a **WEIGHT** statement. The *weight-variable* provides the frequency count for each cell. Without the *weight-variable*, SAS shows a single observation in each cell. Figure 12.4 shows the results.

SAS creates the same reports with a *weight-variable* as it does when creating a summary table from raw data. The numbers in each cell have the same meaning as they have when you create a summary table from raw data.

Figure 12.4 Contingency Table with a WEIGHT Variable

```
              Table for Death Penalty Data

                  The FREQ Procedure

            Table of decision by defrace

    decision        defrace

    Frequency|
    Percent  |
    Row Pct  |
    Col Pct  |Black    |White    |  Total

    No          149       141       290
              45.71     43.25     88.96
              51.38     48.62
              89.76     88.13

    Yes          17        19        36
               5.21      5.83     11.04
              47.22     52.78
              10.24     11.88

    Total       166       160       326
              50.92     49.08    100.00
```

Summarizing PROC FREQ for Tables from Summary Data

The general form of the statements to create a contingency table from summary data using PROC FREQ is shown below:

PROC FREQ DATA=*data-set-name*;

    TABLES *row-variable*column-variable I options*;

    WEIGHT *weight-variable*;

weight-variable is the variable that provides the frequency count for each cell.

Other items were defined earlier.

Creating Contingency Tables for Several Variables

To create contingency tables for several variables, use the **DATA** step and PROC FREQ as described in the previous two sections. Create the data set, and then use PROC FREQ with a TABLES statement. Use a WEIGHT statement if it is needed. In the TABLES statement, use asterisks to join all of the variables that form the table. The final two variables define the rows and columns of a two-way table that is produced for each combination of the levels of the other variables.

For example, suppose you conducted an opinion survey about welfare reform in seven cities. In addition to asking the opinion question and recording the city for each person, you ask for their political affiliation (Democrat, Republican, or Independent), and for their employment status (Employed, Unemployed, Student, or Retired). The following statements produce a two-way table of political affiliation by opinion for each of the 28 (=7×4) combinations of city and employment status:

```
proc freq data=welfare;
   tables city*employ*politic*opinion;
```

You should use caution when you enter a long list of variables joined by asterisks, and, as a result, create an enormous number of tables. For the example above, 28 tables are created. If **city** and **employ** each had 10 levels, 100 tables would have been created!

Printing Only One Table per Page

PROC FREQ automatically fits as many tables as possible on each page of output. This approach uses less paper.

You can override this default, and print only one table on each page. To do so, use the **PAGE** option in the PROC FREQ statement. For example, think about the welfare reform survey discussed in the previous section. The statements produce 28 tables, each on its own page:

```
proc freq data=welfare page;
   tables city*employ*politic*opinion;
```

The PAGE option does not affect tables that can't fit on one page. If you have two variables with many levels for each variable, PROC FREQ might not be able to fit the entire table on one page. Before trying the PAGE option, you might want to first try increasing your line size with the **LINESIZE=** option in the **OPTIONS** statement.

The general form for using the PAGE option is simply to add PAGE to the PROC FREQ statement.

Performing Tests for Independence

When you collect classification data, you want to know whether the variables are related in some way. For the **penalty** data, is the defendant's race related to the verdict? Does knowing the defendant's race tell you anything about the likelihood that the defendant will receive the death penalty?

In statistical terms, the null hypothesis is that the row and column variables are independent. The alternative hypothesis is that the row and column variables are not independent. To test for independence, you compare the observed cell frequencies with the *expected cell frequencies*, which are the cell frequencies that would occur in the situation where the null hypothesis is true. (See "Technical Details: Expected Cell Frequencies" at the end of this section.)

One commonly used test is a chi-square test, which tests the hypothesis of independence. A test statistic is calculated, and then compared with a critical value from a chi-square distribution. Suppose you want to test the hypothesis of independence at the 10% significance level for the **penalty** data. (For more information, see the general discussion about building hypothesis tests in Chapter 5.)

The steps for performing the analysis are the same steps for other analyses:

1. Create a SAS data set.

2. Check the data set for errors.

3. Choose the significance level.

4. Check the assumptions.

5. Perform the test.

6. Make conclusions from the results.

To check the data set for errors, use **PROC UNIVARIATE** and the plots, graphs, and tables from Chapter 4.

To test for independence with two classification variables, here are the assumptions:

- Data values are counts. Practically, this assumption requires that the variables are nominal or ordinal.

- Observations are independent. The values for one observation are not related to the values for another observation. To check this assumption, think about your data and whether this assumption seems reasonable. This is not a step where using SAS will answer this question.

- Observations are a random sample from the population. You want to make conclusions about a larger population, not about just the sample. For the statistics class, you want to make conclusions about statistics classes in general, not about just this single class. To check this assumption, think about your data and whether this assumption seems reasonable. This is not a step where using SAS will answer this question.

- Sample size is large enough for the tests. As a general rule, the sample size should be large enough to expect five responses in each cell of the summary table. SAS prints warning messages when this assumption is not met.

Before performing the test for independence for the **penalty** data, think about the assumptions.

- The data values are counts, so the first assumption is reasonable.

- The observations are independent, because one defendant's race and decision are unrelated to the race and decision of another defendant. (This assumes that the defendants are not being tried for the same crime.) The second assumption is reasonable.

- The observations are a random sample from the population of defendants convicted of homicide. The third assumption is reasonable.

To check the fourth assumption, you can add the **EXPECTED** option. Or, you can simply review the SAS log. When this assumption is not reasonable, SAS prints a warning message. To check the fourth assumption, and then perform the statistical test in SAS, type the following:

```
proc freq data=penalty;
    tables decision*defrace / expected chisq
           norow nocol nopercent;
    weight count;
    title 'Death Penalty Data: Statistical Tests';
run;
```

The WEIGHT statement is needed because the data points represent counts instead of individual values. The EXPECTED option displays the expected values for each cell in the table, assuming the null hypothesis of independence. The **CHISQ** option performs the test. The other three options in the TABLES statement suppress percentages. Figure 12.5 shows the results.

Understanding Chi-Square Test Results

First, answer the research question, "Is the defendant's race related to the verdict?"

SAS displays multiple chi-square tests. The Pearson test (**Chi-Square** test) and **Likelihood Ratio** test have the same assumptions. The Pearson test uses the observed and expected cell frequencies. The Likelihood Ratio test uses a more complex formula. Some statisticians prefer the Likelihood Ratio test. Although the *p*-values for the two tests are identical in Figure 12.5, this is not always true. Typically, the *p*-values are similar, but are not identical.

Look in the table of tests in Figure 12.5, and find the row with a **Statistic** of Chi-Square. Look at the number in the **Prob** column in this row. This value is **0.6379**, which is greater than the significance level of 0.10. You conclude that there is not enough evidence to reject the null hypothesis of independence between the defendant's race and the verdict. (Refer to Agresti in Appendix 1, "Further Reading," for an additional analysis of this data that considers the race of the victim.)

In general, to interpret SAS results, look at the *p*-value under Prob in the row for Chi-Square. If the *p*-value is less than the significance level, reject the null hypothesis that the two variables are independent. If the *p*-value is greater, you fail to reject the null hypothesis.

Figure 12.5 Testing for Independence with the Penalty Data

```
            Death Penalty Data: Statistical Tests

                    The FREQ Procedure

              Table of decision by defrace

        decision        defrace

        Frequency│
        Expected │Black    │White    │    Total

        No       │    149  │    141  │      290
                 │ 147.67  │ 142.33  │
                 ├─────────┼─────────┤
        Yes      │     17  │     19  │       36
                 │ 18.331  │ 17.669  │

        Total         166       160         326

          Statistics for Table of decision by defrace

    Statistic                    DF       Value       Prob

    Chi-Square                    1       0.2214      0.6379
    Likelihood Ratio Chi-Square   1       0.2215      0.6379
    Continuity Adj. Chi-Square    1       0.0863      0.7689
    Mantel-Haenszel Chi-Square    1       0.2208      0.6385
    Phi Coefficient                      0.0261
    Contingency Coefficient              0.0261
    Cramer's V                           0.0261


                    Fisher's Exact Test

            Cell (1,1) Frequency (F)         149
            Left-sided Pr <= F            0.7412
            Right-sided Pr >= F           0.3843

            Table Probability (P)        0.1255
            Two-sided Pr <= P            0.7246

                    Sample Size = 326
```

The following list gives details about items in the **Chi-Square Tests** table. SAS labels this table as **Statistics for Table of *row-variable* by *column-variable*.** For your data, SAS substitutes the variable names.

Statistic	is the name of the statistical test. **Chi-Square** is the Pearson chi-square test, and **Likelihood Ratio Chi-Square** is the likelihood ratio chi-square test. SAS performs several other tests that are listed in this column, but this book does not discuss them. However, if your data does not meet the conditions for a chi-square test, you should either use Fisher's exact test or consult a statistician because some of these other tests might be appropriate.
DF	are the degrees of freedom for the test. The degrees of freedom are a function of the number of rows and columns in the table. If you have a table with r rows and c columns, the degrees of freedom for the chi-square test are $(r-1)\times(c-1)$. For the **penalty** data, the table has 2 rows and 2 columns, so the degrees of freedom are $(2-1)\times(2-1)=1$.
Value	is the value of the test statistic, which is compared to a critical value to obtain the p-value for the test.
Prob	is the p-value for the test.

Look at the last line in Figure 12.5. This line identifies the sample size used for the tests. This line appears immediately after the **Chi-Square Tests** table when PROC FREQ does not perform Fisher's exact test.

Viewing Expected Cell Frequencies

The upper left corner of the frequency table in Figure 12.5 provides a key to the statistics in each cell. The bottom number in each cell gives the expected cell frequency, assuming the null hypothesis of independence. PROC FREQ prints this number as a result of using the EXPECTED option.

Notice how close the expected cell frequencies are to the actual cell frequencies (the top number in each cell). This leads you to the intuitive conclusion that the defendant's race and the verdict (death penalty or no death penalty) are independent. This intuitive conclusion is supported by the results of the statistical test.

Look again at the values of the expected cell frequencies. The Pearson chi-square test is always valid if there are no empty cells (no cells with a cell frequency of 0), and if the expected cell frequency for all cells is 5 or more. Because these conditions are true for the death **penalty** data, the chi-square test is a valid test. If these two conditions are not met, PROC FREQ prints a warning telling you that the chi-square test might not be valid. There is some disagreement among statisticians about exactly when the test should not be used, and what to do when the test is not valid. One alternative is Fisher's exact test, which is discussed in the next section.

Understanding Fisher's Exact Test Results

Fisher's exact test was developed for the special case of a 2×2 table. This test is very useful when the assumptions for a chi-square test are not reasonable. This test is especially useful for tables with small cell frequencies. SAS automatically performs this test for 2×2 tables. For larger tables, you can request the two-sided test with the **FISHER** option. SAS does not automatically perform this test for larger two-way tables because it can sometimes take a long time to run.

Look under the heading **Fisher's Exact Test** in Figure 12.5. SAS displays results for one-sided tests and the two-sided test. Look at the row labeled **Two-sided Pr <= P**. This *p*-value is **0.7246**, so you fail to reject the null hypothesis that the two variables are independent. This decision is consistent with the decision from the chi-square tests.

In general, to interpret SAS results, look at the *p*-value for **Two-sided Pr <= P**. The one-sided *p*-values might be useful in specific situations. Interpret these *p*-values the same way you do for the two-sided chi-square test. SAS provides only the two-sided test for tables larger than 2×2 tables.

The first row of the **Fisher's Exact Test** table is labeled **Cell (1,1) Frequency (F)**. This is the frequency count for the top left cell of the 2×2 table. For the death penalty data, the value of **149** is the value for the **Black-No** cell. SAS uses this value in computing the one-sided tests. SAS documentation provides details about the computations.

Technical Details: Expected Cell Frequencies

The chi-square test compares observed cell frequencies with expected cell frequencies, assuming the null hypothesis that the variables are independent. Use the EXPECTED option to get these frequencies. To calculate expected cell frequencies, multiply the row total and column total. Then, divide this number by the total number of observations. For the **Black-No** cell, the formula is:

[(row total for **No**)×(column total for **Black**)]/total **N**

=(290×166)/326

=147.67

The chi-square test is always valid if there are no empty cells (no cells with a cell frequency of 0), and if the expected cell frequency for all cells is 5 or more.

Because these conditions are true for the **penalty** data, the chi-square test is a valid test. If these two conditions are not true, SAS prints a warning telling you that the chi-square test might not be valid.

There is some disagreement among statisticians about exactly when the test should not be used, and what to do when the test is not valid. Two practical recommendations are to collect more data, or to combine low-frequency categories.

Combine low-frequency categories only when it makes sense. For example, consider a survey that asked people how often they called a Help Desk in the past month. Suppose the categories are 0 (no calls), 1–2, 3–5, 6–8, 9–11, 12–15, 16–20, and "over 20". Now, suppose the expected cell counts for the last four categories are less than 5. It makes sense to combine these last four categories into a new category of "9 or more".

Creating Measures of Association with Ordinal Variables

When you reject the null hypothesis for either the chi-square test or Fisher's exact test, you conclude that the two variables are not independent. But, you do not know how the two variables are related. When both variables are ordinal, *measures of association* give more insight into your data. (As defined in Chapter 4, ordinal variables have values that provide an implied order.)

SAS provides many measures of association. This chapter discusses only Spearman's rank correlation coefficient and Kendall's *tau-b*. (The Spearman's rank correlation coefficient is similar to correlation coefficients discussed in Chapter 10. This correlation coefficient is essentially the Pearson correlation coefficient applied to ranks instead of to actual values.) Both of these measures can have values between −1 and 1. The closer a measure of association is to 0, the weaker the strength of the relationship. A value of −0.8 or 0.8 indicates a stronger relationship than a value of −0.1 or 0.1.[3]

A positive measure of association indicates an increasing trend between the two variables. As the ordinal levels of one variable increase, the ordinal levels of the other variable also increase.

A negative measure of association indicates a decreasing trend between the two variables. As the ordinal levels of one variable increase, the ordinal levels of the other variable decrease.

SAS also tests that the measure of association is significantly different from 0. For this test, the null hypothesis is that the measure of association is 0. The alternative hypothesis is that the measure of association is different from 0.

An animal epidemiologist tested dairy cows for the presence of a bacterial disease.[4] The disease is detected by analyzing blood samples. Disease severity for each animal was classified as none, low, or high. The size of the herd that each cow belonged to was classified as small, medium, or large. Table 12.4 shows the number of cows in each herd size and in each disease severity category. Both the herd size and disease severity variables are ordinal. The disease is transmitted from cow to cow by bacteria, so the epidemiologist wanted to know whether disease severity is affected by herd size. As the herd size gets larger, is there either an increasing or a decreasing trend in disease severity?

The epidemiologist tested for independence using the chi-square test. However, she still does not know whether there is a trend in disease severity that is related to an increase in herd size. Kendall's *tau-b* or Spearman's rank correlation coefficient can answer this question.

---

[3] See Chapter 10 for a detailed discussion of correlation. Although **PROC CORR** also calculates measures of association, it requires numeric variables.

[4] Data is from Dr. Ramon Littell, University of Florida. Used with permission.

Table 12.4 Cow Disease Data

Disease Severity

Herd Size	None (0)	Low (1)	High (2)
Small	9	5	9
Medium	18	4	19
Large	11	88	136

This data is available in the **cows** data set in the sample data for the book.

To create the data table in SAS, follow the steps in Chapter 3. Your data set has 9 observations, one for each cell in the table. It has 3 columns, one for herd size, one for disease severity, and one for number of cows. The "Example" section at the end of this chapter contains the SAS statements for creating the data set.

To create the measures of association in SAS, use PROC FREQ:

```
proc freq data=cows;
   tables herdsize*disease / measures cl
         nopercent norow nocol expected;
   test kentb scorr;
   weight numcows;
   title 'Measures for Dairy Cow Disease Data';
run;
```

The **MEASURES** option in the TABLES statement requests several measures of association, including Kendall's *tau-b* and Spearman's rank correlation coefficient. The **CL** option requests confidence limits for the measures of association. The other options suppress cell percentages, and request the expected cell frequencies.

The **TEST** statement tests that the measure is significantly different from 0. The options in this statement request the test for the two measures of interest, Kendall's *tau-b* (**KENTB**) and Spearman's rank correlation coefficient (**SCORR**). The WEIGHT statement is needed because the data points represent counts instead of individual values.

Although this example uses the WEIGHT statement, SAS performs exactly the same analyses in situations where you provide the raw data. In those situations, you do not need to use the WEIGHT statement.

Figure 12.6 shows the results.

Figure 12.6 Measures of Association for the Cow Disease Data (first page)

```
               Measures for Dairy Cow Disease Data

                        The FREQ Procedure

                   Table of herdsize by disease

            herdsize         disease

            Frequency│
            Expected │      0│      1│      2│ Total

             large   │    11 │    88 │   136 │   235
                     │29.866 │76.237 │ 128.9 │
                     ├───────┼───────┼───────┤
             medium  │    18 │     4 │    19 │    41
                     │5.2107 │13.301 │22.488 │
                     ├───────┼───────┼───────┤
             small   │     9 │     5 │     9 │    23
                     │2.9231 │7.4615 │12.615 │
                     └───────┴───────┴───────┘
             Total        38      97     164     299


             Statistics for Table of herdsize by disease
```

Statistic	Value	ASE	95% Confidence Limits	
Gamma	-0.4113	0.1009	-0.6091	-0.2135
Kendall's Tau-b	-0.2173	0.0606	-0.3362	-0.0984
Stuart's Tau-c	-0.1482	0.0436	-0.2335	-0.0628
Somers' D C¦R	-0.2762	0.0780	-0.4292	-0.1233
Somers' D R¦C	-0.1710	0.0482	-0.2655	-0.0764
Pearson Correlation	-0.2816	0.0660	-0.4109	-0.1523
Spearman Correlation	-0.2331	0.0656	-0.3617	-0.1045
Lambda Asymmetric C¦R	0.0000	0.0000	0.0000	0.0000
Lambda Asymmetric R¦C	0.1094	0.0794	0.0000	0.2650
Lambda Symmetric	0.0352	0.0264	0.0000	0.0869
Uncertainty Coefficient C¦R	0.0990	0.0256	0.0489	0.1491
Uncertainty Coefficient R¦C	0.1437	0.0375	0.0702	0.2172
Uncertainty Coefficient Symmetric	0.1172	0.0302	0.0581	0.1764

Figure 12.6 shows the first page of output, and Figure 12.7 shows the second page of output.

The TABLES statement creates the first page (Figure 12.6), which shows the frequency table and the **Measures of Association** table that contains several measures of association. SAS labels this table as **Statistics for Table of *row-variable* by *column-variable***. For your data, SAS substitutes the variable names. For each measure of association, SAS lists the name and value of the measure, and a 95% confidence interval for the measure. SAS provides many measures of association. This chapter discusses only Spearman's rank correlation coefficient and Kendall's *tau-b*.

The TEST statement creates the second page (Figure 12.7), which shows details for the two measures listed in the statement.

Figure 12.7 Measures of Association for the Cow Disease Data (second page)

```
           Measures for Dairy Cow Disease Data

                   The FREQ Procedure

      Statistics for Table of herdsize by disease

                   Kendall's Tau-b
        ─────────────────────────────────────────
        Tau-b                            -0.2173
        ASE                               0.0606
        95% Lower Conf Limit             -0.3362
        95% Upper Conf Limit             -0.0984

              Test of H0: Tau-b = 0

        ASE under H0                      0.0639
        Z                                -3.4018
        One-sided Pr <  Z                 0.0003
        Two-sided Pr > |Z|                0.0007


        Spearman Correlation Coefficient
        ─────────────────────────────────────────
        Correlation                      -0.2331
        ASE                               0.0656
        95% Lower Conf Limit             -0.3617
        95% Upper Conf Limit             -0.1045

            Test of H0: Correlation = 0

        ASE under H0                      0.0686
        Z                                -3.3998
        One-sided Pr <  Z                 0.0003
        Two-sided Pr > |Z|                0.0007

              Sample Size = 299
```

Understanding the Results

First, answer the research question, "Is there a trend in disease severity that is related to an increase in herd size?"

Finding the *p*-value

Look in the **Kendall's Tau-b** test table in Figure 12.7 and find the row for **Two-sided Pr > |Z|**. The *p*-value is the result of a test that the measure of association is significantly different from 0. The cows data has a *p*-value of **0.0007**. You conclude that Kendall's *tau-b* is significantly different from 0.

Similarly, look in the **Spearman Correlation Coefficient** test table in Figure 12.7 and find the row for **Two-sided Pr > |Z|**. The *p*-value is the result of a test that the measure of association is significantly different from 0. The cows data has a *p*-value of **0.0007**. You conclude that Spearman's rank correlation coefficient is significantly different from 0.

Based on these results, the epidemiologist can conclude that disease severity increases with increasing herd size. For the cows data, the two *p*-values are the same. This might not be true for your data.

In general, if the *p*-value is less than the significance level, reject the null hypothesis, and conclude that the measure of association is significantly different from 0. If the *p*-value is greater than the significance level, fail to reject the null hypothesis. Conclude that there is not enough evidence to say that the measure of association is significantly different from 0.

Understanding the Measures of Association

Figure 12.6 shows the contingency table for herdsize and disease. Each cell contains the observed cell frequency and the expected cell frequency. The EXPECTED option prints the expected cell frequency, which identifies the number of cows that would be expected in each cell if the herd size and disease severity were independent.

Look at the first row of the contingency table, for the large herd size. The actual cell count (**11**) is less than expected for the case of no disease (**disease=0**). However, the actual cell count (**136**) is more than expected for the case of high disease (**disease=2**). Now, look at the last row, for the small herd size. You see the opposite trend than for large herds. The actual cell count is more than expected for the case of no disease, and less than expected for the case of high disease. This leads you to the intuitive conclusion that disease severity increases with herd size. As discussed in "Finding the *p*-value" above, this intuitive conclusion is correct. The results of the statistical tests show that disease severity increases with herd size for both Kendall's *tau-b* and Spearman's rank correlation coefficient.

In Figure 12.6, look at the row for **Kendall's Tau-b** in the Measures of Association table. The value of **–0.217** means that as one variable decreases, the other variable increases, which seems counter to your intuitive conclusion and the results of the statistical test.

However, PROC FREQ uses the alphabetic ordering of the values of character variables such as herdsize, so that the order is large-medium-small. The negative value for Kendall's *tau-b* means that disease severity increases as herd size changes from small to medium to large. As a result, this matches your intuitive conclusion and the results of the statistical test.

In Figure 12.6, look at the row for **Spearman Correlation** in the Measures of Association table. The value of **–0.233** means that as one variable decreases, the other variable decreases. The same process of alphabetic ordering occurs here. The negative value for Spearman correlation means that disease severity increases as herd size changes from small to medium to large. Again, this matches your intuitive conclusion and the results of the statistical test.

One additional caution about the order of your variables is needed. In the example above, the order of herdsize is large-medium-small. You could use large=1, medium=2, and small=3, and you would obtain the same results. However, you would receive incorrect results if the order of values didn't match an increasing or decreasing trend. For example, if you use large=2, medium=1, and small=3, Kendall's *tau-b* is meaningless. In general, you need to look at the values of your variables (both character and numeric) when interpreting measures of association, and make sure that the order of the values makes sense.

The list below gives details about items in the Measures of Association table (shown in Figure 12.6). SAS labels this table as **Statistics for Table of *row-variable* by *column-variable***. For your data, SAS substitutes the variable names.

Statistic	identifies the measure of association.
Value	gives the value of the measure of association.
ASE	is the asymptotic standard error for the measure of association. This estimate should be used only for large sample sizes.

95% Confidence Limits	are the 95% upper and lower confidence limits for the measure of association. When the confidence interval encloses 0, the measure is not significantly different from 0. A confidence interval that contains 0 says that the measure of association might be 0, so the test is not significant at the 5% alpha level. A confidence interval that doesn't enclose 0 says that the measure of association is not 0, so the test is not significant at the 5% alpha level. Some statisticians prefer confidence intervals to formal statistical tests.

The list below gives details about items in the tables of statistical results on the second page (shown in Figure 12.7). SAS labels these results as **Statistics for Table of *row-variable* by *column-variable***. For your data, SAS substitutes the variable names. For each measure of association, SAS prints the following information:

statistic	identifies the measure of association. SAS prints the name of the measure of association in the title of the table. SAS also prints the name on the line that shows the value of the measure. The information in the tables on the second page matches the information in the **Measures of Association** table.
ASE	is the asymptotic standard error for the measure of association. This value matches the value in the **Measures of Association** table.
95% Lower Conf Limit **95% Upper Conf Limit**	are the 95% upper and lower confidence limits for the measure of association. These values match the values in the **Measures of Association** table.
ASE under H0	is the asymptotic standard error for the measure of association under the null hypothesis that the measure of association is 0.
Z	is the value of the test statistic.

One-sided Pr < Z
or
One-sided Pr > Z

SAS displays only one of these headings, each of which is for a one-sided test.

If the measure of association is less than 0, then SAS displays **One-sided Pr < Z** for a one-sided test that the measure is less than 0.

If the measure of association is greater than 0, then SAS displays **One-sided Pr > Z** for a one-sided test that the measure is greater than 0.

Two-sided Pr > |Z|

is the *p*-value, which is the result of a test that the measure of association is significantly different from 0. See "Finding the *p*-value" for more discussion.

Changing the Confidence Level

SAS automatically creates 95% confidence limits for the measures of association. You can change the confidence level with the **ALPHA=** option in the TABLES statement:

```
proc freq data=cows;
   tables herdsize*disease / measures alpha=0.10 noprint;
   test kentb scorr;
   weight numcows;
   title 'Measures with 90% Confidence Limits';
run;
```

The ALPHA= option in the TABLES statement requests 90% confidence limits. The MEASURES option requests the measures of association. The **NOPRINT** option suppresses the contingency table. Although this example uses the WEIGHT statement, SAS performs exactly the same analyses as it does with raw data. Figure 12.8 shows the results.

Compare the results for measures of association in Figures 12.6 and 12.7 with the results in Figure 12.8. The key difference is that Figure 12.8 shows 90% confidence limits. The output fits on a single page because the NOPRINT option suppresses the contingency table.

Figure 12.8 Measures with 90% Confidence Limits

```
                    Measures with 90% Confidence Limits

                          The FREQ Procedure

                 Statistics for Table of herdsize by disease

                                                          90%
Statistic                        Value        ASE     Confidence Limits

Gamma                          -0.4113      0.1009   -0.5773    -0.2453
Kendall's Tau-b                -0.2173      0.0606   -0.3171    -0.1176
Stuart's Tau-c                 -0.1482      0.0436   -0.2198    -0.0765

Somers' D C|R                  -0.2762      0.0780   -0.4046    -0.1479
Somers' D R|C                  -0.1710      0.0482   -0.2503    -0.0916

Pearson Correlation            -0.2816      0.0660   -0.3901    -0.1731
Spearman Correlation           -0.2331      0.0656   -0.3410    -0.1252

Lambda Asymmetric C|R           0.0000      0.0000    0.0000     0.0000
Lambda Asymmetric R|C           0.1094      0.0794    0.0000     0.2400
Lambda Symmetric                0.0352      0.0264    0.0000     0.0786

Uncertainty Coefficient C|R     0.0990      0.0256    0.0569     0.1411
Uncertainty Coefficient R|C     0.1437      0.0375    0.0821     0.2054
Uncertainty Coefficient Symmetric 0.1172    0.0302    0.0676     0.1669


                            Kendall's Tau-b

            Tau-b                          -0.2173
            ASE                             0.0606
            90% Lower Conf Limit           -0.3171
            90% Upper Conf Limit           -0.1176

                   Test of H0: Tau-b = 0

            ASE under H0                    0.0639
            Z                              -3.4018
            One-sided Pr <  Z               0.0003
            Two-sided Pr > |Z|              0.0007


                   Spearman Correlation Coefficient

            Correlation                    -0.2331
            ASE                             0.0656
            90% Lower Conf Limit           -0.3410
            90% Upper Conf Limit           -0.1252

                  Test of H0: Correlation = 0

            ASE under H0                    0.0686
            Z                              -3.3998
            One-sided Pr <  Z               0.0003
            Two-sided Pr > |Z|              0.0007

                      Sample Size = 299
```

Summarizing Analyses with PROC FREQ

SAS performs many additional tests, and provides many more measures of association than are discussed in this book. If you need a specific test or a specific measure of association, then check SAS documentation because it might be available in SAS.

> **PROC FREQ DATA**=*data-set-name* **PAGE**;
>
>   **TABLES** *row-variable*column-variable I options*;
>
>   **TEST KENTB SCORR**;

Some items were defined earlier. The **PAGE** option prints only one table on each page.

The TABLES statement *options* can be as defined earlier and can be one or more of the following:

EXPECTED	requests expected cell frequencies.
NOPRINT	suppresses the contingency table.
CHISQ	performs chi-square tests.
FISHER	performs Fisher's exact test for tables larger than 2×2.
MEASURES	requests measures of association.
CL	adds confidence limits for the measures of association.
ALPHA=*level*	specifies the alpha level for the confidence limits. Choose a *level* between 0.0001 and 0.9999. SAS automatically uses 0.05.

If any option is used, the slash is required. You can use the options one at a time or together. If you use the CL option, SAS automatically adds the MEASURES option.

The KENTB option in the TEST statement provides statistical tests for Kendall's *tau-b*. The SCORR option in the TEST statement provides statistical tests for Spearman's rank correlation coefficient. You can use these options one at a time or together.

The TABLES statement options and the TEST statement options are also available when you use a WEIGHT statement.

You can use the ODS statement to choose the output tables to print.

Table 12.4 identifies the names of the output tables for PROC FREQ and measures of association discussed in this chapter:

Table 12.4 More ODS Table Names for PROC FREQ

ODS Name	Output Table
ChiSq	Chi-square test results
FishersExact	Fisher's exact test results
Measures	Measures of association
TauB TauBTest	Statistical tests for Kendall's *tau-b*, created by the TEST statement
SpearmanCorr SpearmanCorrTest	Statistical tests for Spearman's rank correlation coefficient, created by the TEST statement

Summary

Key Ideas

- Contingency tables are tables that summarize two or more classification variables. The rows and columns of a contingency table form cells, and the number of observations in each cell is the cell frequency for that cell.

- You can use PROC FREQ to create a table from raw data, create a table when you already know the cell frequencies, print only one table per page, and suppress statistics that are usually printed. PROC FREQ performs statistical tests for independence, and provides measures of association when both variables are ordinal.

- Both the chi-square test and Fisher's exact test are used to test for independence between two classification variables. Generally, the chi-square test should not be used if there are empty cells (cells with a cell frequency of 0), or if any cell has an expected cell frequency of less than 5. One option is to collect more data, which increases the expected cell frequency. Another option is to combine low-frequency categories.

- Regardless of the test, the steps for performing the analysis are:

  1. Create a SAS data set.

  2. Check the data set for errors.

  3. Choose the significance level.

  4. Check the assumptions.

  5. Perform the test.

  6. Make conclusions from the results.

- Regardless of the test, to interpret the results and make conclusions, compare the *p*-value for the test with your significance level.

  - If the *p*-value is less than the significance level, reject the null hypothesis that the two variables are independent.

  - If the *p*-value is greater than the significance level, fail to reject the null hypothesis.

- Kendall's *tau-b* and Spearman's rank correlation coefficient are both measures of association that provide information about the strength of the relationship between the ordinal variables. Use the statistics to decide whether there is an increasing or a decreasing trend in the two variables, or if there is no trend at all. SAS tests whether the measure of association is significantly different from 0, and reports the *p*-value. SAS also provides confidence intervals for these measures of association.

Syntax

To summarize data

- To summarize data for contingency tables, use PROC FREQ. You can summarize raw data or data that already exists in a summary table. See the syntax for "To perform the test (step 5)."

- To check the data for errors, use the tools in Chapter 4.

- If your data already exists in a summary table, use a *weight-variable* when creating the data set. This variable provides the frequency count for each cell in the table. It is used in the WEIGHT statement in PROC FREQ. A sample INPUT statement is shown below:

  INPUT *row-variable column-variable weight-variable*;

To check the assumptions (step 4)

- To check the assumption that data values are counts, review your data. This is not a step where using SAS will answer this question.

- To check the assumption of independent observations, think about your data and whether this assumption seems reasonable. This is not a step where using SAS will answer this question.

- To check the assumption that the observations are a random sample from the population, think about your data and whether this assumption seems reasonable. This is not a step where using SAS will answer this question.

- To check the assumption that the sample size is large enough, use PROC FREQ and the EXPECTED option. See the syntax for "To perform the test (step 5)." Alternatively, review the SAS log. SAS prints warning messages when this assumption is not met.

To perform the test (step 5)

To perform tests for independence, or to create measures of association when both variables are ordinal:

> **PROC FREQ DATA=***data-set-name* **PAGE;**
>   **TABLES** *row-variable*column-variable I options*;
>   **WEIGHT** *weight-variable*;
>   **TEST KENTB SCORR;**

data-set-name	is the name of a SAS data set.
row-variable	is the variable that defines rows.
column-variable	is the variable that defines columns.

If your data already exists in a summary table, use a *weight-variable* in the WEIGHT statement, where *weight-variable* is the variable that provides the frequency count for each cell in the table.

The PAGE option prints only one contingency table on each page.

The TABLES statement *options* can be one or more of the following:

NOFREQ	suppresses frequency counts.
NOCOL	suppresses column percentages.
NOROW	suppresses row percentages.
NOPERCENT	suppresses overall percentages.
EXPECTED	requests expected cell frequencies.
NOPRINT	suppresses the contingency table.
CHISQ	performs chi-square tests.
FISHER	performs Fisher's exact test for tables larger than 2×2.
MEASURES	requests measures of association.
CL	adds confidence limits for the measures of association.
ALPHA=*level*	specifies the alpha level for the confidence limits. Choose a *level* between 0.0001 and 0.9999. SAS automatically uses 0.05.

If any option is used, the slash is required. You can use the options one at a time or together. If you use the CL option, SAS automatically adds the MEASURES option.

The KENTB option in the TEST statement provides statistical tests for Kendall's *tau-b*. The SCORR option in the TEST statement provides statistical tests for Spearman's rank correlation coefficient. You can use these *options* one at a time or together.

The TABLES statement options and the TEST statement options are available whether or not you use a WEIGHT statement.

Example

The program below produces the output shown in this chapter:

```
options ls=80 ps=60 nodate nonumber;
proc format;
    value $gentxt 'M' = 'Male'
                  'F' = 'Female';
    value $majtxt 'S' = 'Statistics'
                  'NS' = 'Other';
run;

data statclas;
    input student gender $ major $ @@;
    format gender $gentxt.;
    format major $majtxt.;
    datalines;
1 M S 2 M NS 3 F S 4 M NS 5 F S 6 F S 7 M NS
8 M NS 9 M S 10 F S 11 M NS 12 F S 13 M S 14 M S
15 M NS 16 F S 17 M S 18 M NS 19 F NS 20 M S
;
run;

proc freq data=statclas;
    tables gender*major;
title 'Major and Gender for Students in Statistics Class';
run;

proc freq data=statclas;
    tables gender*major /  norow nocol nopercent;
title 'Counts for Students in Statistics Class';
run;

data penalty;
    input decision $ defrace $ count @@;
    datalines;
```

```
Yes White 19 Yes Black 17 No White 141 No Black 149
;
run;

proc freq data=penalty;
   tables decision*defrace;
   weight count;
title 'Table for Death Penalty Data';
run;

proc freq data=penalty;
   tables decision*defrace / expected chisq
               norow nocol nopercent;
   weight count;
title 'Death Penalty Data: Statistical Tests';
run;

data cows;
   input herdsize $ disease numcows @@;
   datalines;
large 0 11 large 1 88 large 2 136 medium 0 18
medium 1 4 medium 2 19 small 0 9 small 1 5 small 2 9
;
run;

proc freq data=cows;
   tables herdsize*disease / measures cl
               nopercent norow nocol expected;
   test kentb scorr;
   weight numcows;
title 'Measures for Dairy Cow Disease Data';
run;

proc freq data=cows;
   tables herdsize*disease / measures cl alpha=0.10 noprint;
   test kentb scorr;
   weight numcows;
title 'Measures with 90% Confidence Limits';
run;
```

Special Topic: ODS Statistical Graphics

As described in Chapter 4, you can use **PROC CHART** to create line printer bar charts for nominal or ordinal data. SAS/GRAPH software includes **PROC GCHART**, which provides high-resolution bar charts. For contingency tables, specific bar charts are useful for displaying the data. Although you can create these bar charts using other SAS procedures, PROC FREQ provides a shortcut starting with SAS 9.2. Using ODS Statistical Graphics, you can create these specific bar charts.[5]

ODS Statistical Graphics is a versatile and complex system that provides many features. This topic discusses the basics of using ODS Statistical Graphics for PROC FREQ.

The simplest way to use ODS Statistical Graphics is to use the automatic approach of including the graphs in the SAS output. The key difference between ODS Statistical Graphics and plain ODS is that SAS displays the graphs in a separate window (when using ODS Statistical Graphics) rather than in the Graphics window (when using ODS). Figure ST12.1 shows an example of how ODS Statistical Graphics output is displayed in the Results window when you are in the windowing environment mode.

Figure ST12.1 Results When Using ODS Statistical Graphics

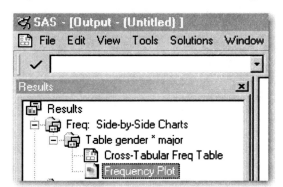

When you double-click on **Frequency Plot** in the Results window, SAS displays the graph in an appropriate viewer. On the PC, SAS automatically uses the Windows Picture and Fax Viewer to display the graph.

---

[5] ODS Statistical Graphics are available in many procedures starting with SAS 9.2. For SAS 9.1, ODS Statistical Graphics were experimental and available in fewer procedures. For releases earlier than SAS 9.1, you need to use alternative approaches like traditional graphics or line printer plots. Also, to use ODS Statistical Graphics, you must have SAS/GRAPH software licensed.

You activate ODS Statistical Graphics before using a procedure, and you deactivate it after the procedure. For PROC FREQ, using ODS Statistical Graphics requires the **PLOTS=** option. The example below creates the graph in Figure ST12.2.

```
ods graphics on;
proc freq data=statclas;
    tables gender*major / plots=freqplot;
title 'Side-by-Side Charts';
run;
ods graphics off;
```

Figure ST12.2 Side-By-Side Bar Charts from ODS Statistical Graphics

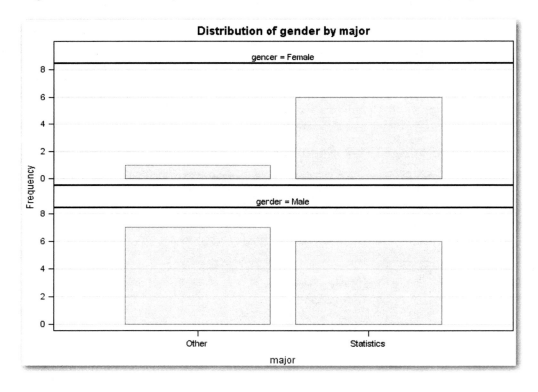

The **PLOTS=FREQPLOT** option creates the side-by-side bar charts of the *column-variable* for each level of the *row-variable*.

Figure ST12.2 shows a side-by-side bar chart for the counts for each gender. You can see that only one of the **Female** students is majoring in **Other**. And, you can view the counts for the cells in the frequency table. While this plot is useful, another PLOTS= option creates a plot that helps you visualize the differences between groups (if they exist).

Using side-by-side stacked bar charts that display percentages helps compare groups. If the pattern for the stacked bars is the same, then you have a visual hint that the groups are the same. If the pattern differs, then you have a visual hint that the groups differ.

```
ods graphics on;
proc freq data=statclas;
   tables gender*major
       / plots=freqplot(twoway=stacked scale=percent);
title 'Stacked Percentage Chart';
run;
ods graphics off;
```

Figure ST12.3 shows the graph.

Compare the stacked bar charts. About half of the students majoring in **Statistics** are **Male**, and about half are **Female**. In contrast, most of the students majoring in **Other** are **Male**. This difference gives a visual hint that the pattern of majoring in **Statistics** or **Other** differs for **Male** and **Female** students. See the section "Performing Tests for Independence" for ways to check whether the results of the statistical tests confirm this visual hint.

This plot provides a visual summary of the two variables. If the pattern for the stacked bars is the same, then the relative percentage of the *row-variable* is the same across the values of the *column-variable*. If the pattern differs, then the relative percentages differ.

Figure ST12.3 Stacked Bar Charts from ODS Statistical Graphics

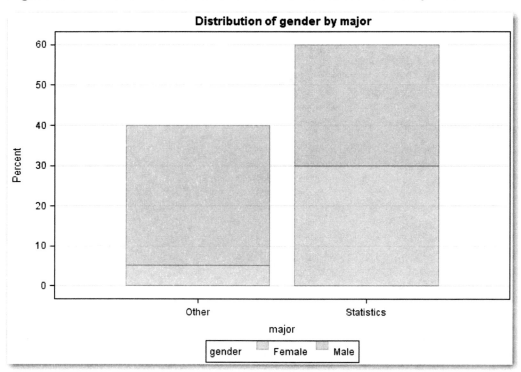

As a contrasting example, when the pattern for the stacked bars is the same, then the relative percentage is the same across the variables. You have a visual hint that the two variables are independent. From Figure 12.5, you concluded that there is not enough evidence to reject the null hypothesis of independence between the defendant's race and the verdict. The side-by-side stacked bar charts can help visualize this result.

```
ods graphics on;
proc freq data=penalty;
   tables decision*defrace;
      / plots=freqplot(twoway=stacked scale=percent);
   weight count;
title 'Stacked Percentage Chart for Penalty Data';
run;
ods graphics off;
```

Figure ST12.4 shows the graph.

Figure ST12.4 Stacked Bar Charts for Penalty Data

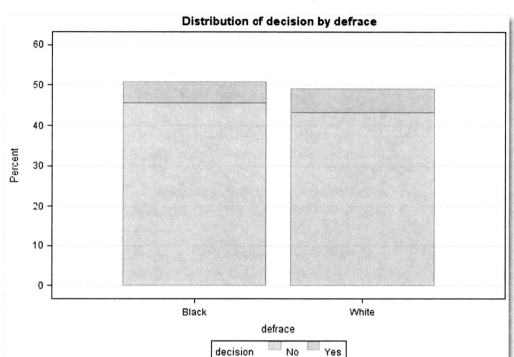

Compare the stacked bars in Figure ST12.4. For each bar, the relative percentage of blue (**No**) and red (**Yes**) is about the same. This similarity gives a visual hint that the pattern of decisions does not differ based on the defendant's race.

The general form of the statements to use ODS Statistical Graphics using PROC FREQ is shown below:

ODS GRAPHICS ON;

PROC FREQ DATA=*data-set-name*;

    TABLES *row-variable*column-variable*

             / options **PLOTS=FREQPLOT(***plot-options***);**

ODS GRAPHICS OFF;

Some items were defined earlier.

Caution: In the TABLES statement, you cannot use the **NOPRINT** option with the PLOTS= option.

plot-options can be one or more of the following:

    TWOWAY=STACKED   creates side-by-side stacked bar charts.

    SCALE=PERCENT   shows percentages instead of frequency counts.

SAS automatically uses **SCALE=FREQ** to display frequency counts.

If you use *plot-options*, the parentheses are required. You can use *plot-options* one at a time or together.

When using ODS Statistical Graphics, you can also use the WEIGHT and TEST statements, and the options in the TEST statement.

As mentioned earlier, ODS Statistical Graphics is a versatile and complex system that provides many features. SAS has many other graphs and options for PROC FREQ. If you need a specific graph, then check SAS documentation because it might be available.

A p p e n d i x **1**

Further Reading

This appendix includes a brief list of books that provide more detail about the statistical topics that are discussed in this book. It is not intended to provide a complete bibliography. Instead, it is intended to provide a few reference books to get you started. Many of the books listed contain exhaustive bibliographies of books and articles for the statistical topics that are discussed in this book.

Statistics References

For more information about using graphs and summary statistics to explore data, see the following:

Cleveland, William S. 1994. *The Elements of Graphing Data, Rev. ed.* AT&T Bell Laboratories.

Hoaglin, David C., Frederick Mosteller, and John W. Tukey. 1985. *Exploring Data Tables, Trends, and Shapes.* New York: John Wiley & Sons, Inc.

Tufte, Edward R. 1983. *The Visual Display of Quantitative Information.* Cheshire, CT: Graphics Press.

Tufte, Edward R. 1990. *Envisioning Information*. Cheshire, CT: Graphics Press.

Tufte, Edward R. 1997. *Visual Explanations: Images and Quantities, Evidence and Narrative*. Cheshire, CT: Graphics Press.

For a non-technical discussion of how statistics can be used incorrectly, see the following:

Gonick, Larry, and Woollcott Smith. 1993. *The Cartoon Guide to Statistics*. New York: Harper Perennial.

Hooke, Robert. 1983. *How to Tell the Liars from the Statisticians*. New York: Marcel Dekker, Inc.

Huff, Darrell. 1954. *How to Lie with Statistics*. New York: W. W. Norton & Company, Inc.

Spirer, Herbert F., Louise Spirer, and A. J. Jaffe. 1998. *Misused Statistics, Second Edition*. New York: Marcel Dekker, Inc.

For a general introduction to statistics and statistical thinking, see the following books. Many of these books also discuss testing for normality.

Brook, Richard J., et al. 1986. *The Fascination of Statistics*. New York: Marcel Dekker, Inc.

Fraenkel, Jack R., Enoch I. Sawin, and Norman E. Wallen. 1999. *Visual Statistics: A Conceptual Primer*. Boston, MA: Allyn & Bacon.

Freedman, David, Robert Pisani, and Roger Purves. 2007. *Statistics, Fourth Edition*. New York: W. W. Norton & Company, Inc.

Gonick, Larry, and Woollcott Smith. 1993. *The Cartoon Guide to Statistics*. New York: Harper Perennial.

Moore, David S., and George P. McCabe. 2006. *Introduction to the Practice of Statistics, Fifth Edition*. W. H. Freeman & Company.

Utts, Jessica M. 1999. *Seeing Through Statistics*. Second Edition. Pacific Grove, CA: Duxbury Press.

For more information about the statistical methods that are discussed in Chapters 5 through 12 of this book, see the following:

De Veaux, Richard D., Paul F. Velleman, and David E. Bock. 2006. *Intro Stats, Second Edition*. Boston, MA: Pearson Addison-Wesley.

D'Agostino, Ralph B., and Michael A. Stephens. 1986. *Goodness-of-Fit Techniques.* New York: Marcel Dekker, Inc.

For more information about the Kruskal-Wallis test and other nonparametric analyses, see the following:

Conover, W. J. 1999. *Practical Nonparametric Statistics, Third Edition.* New York: John Wiley & Sons, Inc.

Hollander, Myles, and Douglas A. Wolfe. 1999. *Nonparametric Statistical Methods, Second Edition.* New York: John Wiley & Sons, Inc.

Siegel, Sidney, and N. John Castellan, Jr. 1988. *Nonparametric Statistics for the Behavioral Sciences, Second Edition.* New York: McGraw-Hill, Inc.

For more information about ANOVA and multiple comparison procedures, see the following:

Box, George E. P., William G. Hunter, and J. Stuart Hunter. 2005. *Statistics for Experimenters: Design, Innovation, and Discovery, Second Edition.* Hoboken, NJ: Wiley-Interscience.

Littell, Ramon C., Walter W. Stroup, and Rudolf J. Freund. 2002. *SAS for Linear Models, Fourth Edition.* Cary, NC: SAS Institute Inc.

Hsu, Jason C. 1996. *Multiple Comparisons: Theory and Methods.* London: Chapman & Hall.

Milliken, George A., and Dallas E. Johnson. 1997. *Analysis of Messy Data Volume I: Designed Experiments.* Boca Raton, FL: Chapman & Hall/CRC.

Muller, Keith E., and Bethel A. Fetterman. 2002. *Regression and ANOVA: An Integrated Approach Using SAS Software.* Cary, NC: SAS Institute Inc.

Westfall, Peter H., et al. 1999. *Multiple Comparisons and Multiple Tests: Using the SAS System.* Cary, NC: SAS Institute Inc.

For more information about correlation, regression, and regression diagnostics, see the following:

Belsley, David A., Edwin Kuh, and Roy E. Welsch. 2004. *Regression Diagnostics: Identifying Influential Data and Sources of Collinearity.* Hoboken, NJ: Wiley-Interscience.

Draper, Norman R., and Harry Smith. 1998. *Applied Regression Analysis, Third Edition.* New York: John Wiley & Sons, Inc.

Freund, Rudolf J., and Ramon C. Littell. 2000. *SAS System for Regression, Third Edition.* Cary, NC: SAS Institute Inc.

Neter, John, et al. 1996. *Applied Linear Statistical Models, Fourth Edition.* Chicago: Irwin.

For more information about tests and measures of association for contingency tables, see the following:

Agresti, Alan. 2002. *Categorical Data Analysis, Second Edition.* Hoboken, NJ: Wiley-Interscience.

Agresti, Alan 1996. *An Introduction to Categorical Data Analysis.* New York: John Wiley & Sons, Inc.

Fleiss, Joseph L., Bruce Levin, and Myunghee Cho Paik. 2003. *Statistical Methods for Rates and Proportions, Third Edition.* Hoboken, NJ: Wiley-Interscience.

Stokes, Maura E., Charles S. Davis, and Gary G. Koch. 2000. *Categorical Data Analysis Using the SAS System, Second Edition.* Cary, NC: SAS Institute Inc.

For more information about practical experimental design, see the following:

Box, George E. P., William G. Hunter, and J. Stuart Hunter. 2005. *Statistics for Experimenters: Design, Innovation, and Discovery, Second Edition.* Hoboken, NJ: Wiley-Interscience.

SAS Press Books and SAS Documentation

The following SAS Press books are especially helpful when learning to use SAS on a PC:

Delwiche, Lora D., and Susan J. Slaughter. 2008. *The Little SAS Book: A Primer, Fourth Edition.* Cary, NC: SAS Institute Inc.

Gilmore, Jodie. 2004. *Painless Windows: A Handbook for SAS Users, Third Edition.* Cary, NC: SAS Institute Inc.

The first book in the previous list discusses the **WHERE** statement and more features of **ODS** statements. (SAS publishes additional manuals on ODS. These manuals are not automatically provided with the software.) The following paper gives an overview of the WHERE statement:

Ma, J. Meimei, and Sandra D. Schlotzhauer. 2000. "How and When to Use WHERE," *Proceedings of the Twenty-Fourth Annual SAS Users Group International Conference.* Cary, NC: SAS Institute Inc.

This paper was available on support.sas.com at the time this book was printed.

SAS provides documentation about using ODS. The following book starts with the basics, much like the very basic discussion of ODS in this book. Then, the book explains how to use the advanced features of ODS. These features include creating RTF, PDF, and HTML output; creating custom reports; modifying fonts and colors; creating and using style templates; working with graphics output; operating system differences; and more.

Haworth, Lauren E., Cynthia L. Zender, and Michele M. Burlew. 2009. *Output Delivery System: The Basics and Beyond*. Cary, NC: SAS Institute Inc.

In addition, SAS publishes companion guides for most operating systems, including Microsoft Windows. These companion guides are not automatically provided with the software, but you can order them online.

For SAS documentation for your release, the best approach is to access the PDF files that are provided with the software. You can search these files much more quickly than the printed books. And, you can print the pages you need—and only those pages. For example, the documentation for **PROC UNIVARIATE** alone is over 180 pages for SAS 9.2, and you might need to print only a few pages.

In addition to hard-copy documentation, you can find technical papers, course notes, books that reference SAS software, and a complete bookstore online at support.sas.com.

A p p e n d i x 2

Summary of SAS Elements and Chapters

The table below identifies the **first** time this book discusses a SAS element, such as a procedure, statement, or option. The SAS element might also be discussed in later chapters.

The exception to the "first-time rule" is that Chapter 1 includes a brief example of the DATA step and PROC MEANS, but it does not discuss the SAS code. Consequently, the table below identifies the chapters where the code for the DATA step and PROC MEANS are discussed.

Chapter	SAS Element
1	TITLE statement
	FOOTNOTE statement
	RUN statement
	OPTIONS statement
	ENDSAS statement
2	Rules for SAS names for data sets and variables
	DATA step
	DATA statement
	INPUT statement
	DATALINES statement
	null statement
	PROC PRINT
	VAR statement
	NOOBS option in PROC PRINT statement
	DOUBLE option in PROC PRINT statement
	PROC SORT
	BY statement
	DESCENDING option
2 Special Topics	LABEL statement
	PROC PRINT with LABEL option
	FORMAT statement (and introduction to selected SAS formats)
	STNAMEL function (and introduction to selected SAS functions)
	PROC FORMAT with VALUE statement

3	LIBNAME statement
	INFILE statement
	PROC IMPORT with DATAFILE=, DBMS=, and OUT= options
	Import Wizard
4	PROC UNIVARIATE
	VAR statement
	ID statement
	HISTOGRAM statement
	NEXTRVAL= option in PROC UNIVARIATE statement
	NEXTROBS= option in PROC UNIVARIATE statement
	NOPRINT option in PROC UNIVARIATE statement
	FREQ option in PROC UNIVARIATE statement
	PLOT option in PROC UNIVARIATE statement
	PROC MEANS
	VAR statement
	Statistical keywords (N, NMISS, MEAN, MEDIAN, STDDEV, RANGE, QRANGE, MIN, MAX) in PROC MEANS statement
	PROC FREQ
	TABLES statement
	MISSPRINT option in TABLES statement
	MISSING option in TABLES statement
	GOPTIONS statement with DEVICE= option
	PATTERN statement with V= and COLOR= options
	PROC GCHART
	VBAR statement
	HBAR statement
	DISCRETE option in VBAR or HBAR statement
	NOSTAT option in HBAR statement
	DESCENDING option in HBAR statement
	SUMVAR= option in HBAR statement
	PROC CHART
	VBAR statement

4 **Special** **Topic**	ODS statement ODS table names for procedures discussed in chapter

5	PROC UNIVARIATE
	  NORMAL option in PROC UNIVARIATE statement
	  NORMAL, COLOR=, and NOPRINT options in HISTOGRAM statement
	  PROBPLOT statement
	  NORMAL, MU=EST, SIGMA=EST, and COLOR= options in PROBPLOT statement
	WHERE statement used with a SAS procedure

6	PROC MEANS
	  CLM statistical keyword in PROC MEANS statement
	  ALPHA= option in PROC MEANS statement
	  MAXDEC= option in PROC MEANS statement

7	Program statements to create new variables in the DATA step PROC TTEST PAIRED statement

8	PROC MEANS
	  CLASS statement
	PROC UNIVARIATE
	  CLASS statement
	PROC BOXPLOT
	  PLOT statement
	PROC TTEST
	  CLASS statement
	  VAR statement
	  ALPHA= option in PROC TTEST statement
	PROC NPAR1WAY
	  WILCOXON option in PROC NPAR1WAY statement
	  CLASS statement
	  VAR statement

9	PROC UNIVARIATE
	NROWS= option in HISTOGRAM statement
	PROC BOXPLOT
	BOXWIDTHSCALE= and BWSLEGEND options in PLOT statement
	PROC ANOVA
	CLASS statement
	MODEL statement
	MEANS statement with HOVTEST, WELCH, T, BON, TUKEY, DUNNETT, ALPHA=, CLDIFF, and LINES options

10	ODS GRAPHICS ON statement
	ODS GRAPHICS OFF statement
	PROC CORR
	VAR statement
	NOMISS option in PROC CORR statement
	PLOTS=SCATTER(NOINSET ELLIPSE=NONE) option in PROC CORR statement for SAS 9.2 and ODS graphics
	PLOTS=MATRIX(HISTOGRAM NVAR=ALL) option in PROC CORR statement for SAS 9.2 and ODS graphics
	PROC REG
	ALPHA= option in PROC REG statement
	ID statement
	MODEL statement with P, CLI, and CLM options
	PRINT statement with P, CLI, and CLM options
	PLOTS(ONLY)=FIT(*fit-options* STATS=NONE) option in PROC REG statement for fitting a straight line for SAS 9.2 and ODS graphics
	NOCLI, NOCLM, or NOLIMITS *fit-options*
	PLOTS(ONLY)=PREDICTIONS(X=*x-variable fit-options* UNPACK) option in PROC REG statement for fitting a curve for SAS 9.2 and ODS graphics
	NOCLI, NOCLM, or NOLIMITS *fit-options*

10 **(cont.)**	PLOT statement for traditional graphics that specifies *y-variables* with one or more statistical keywords (P., LCL., UCL., LCLM., or UCLM.) and uses a single *x-variable*
	NOSTAT, CLINE=, PRED, and CONF options in the PLOT statement when fitting a straight line
	NOSTAT and OVERLAY options in the PLOT statement when fitting a curve
	PLOT statement for line printer graphics that provides multiple *y*x="symbol"* specifications and uses the OVERLAY option; the statement must use the same *x* for all plots specified and the values of *y* can be one or more of the statistical keywords identified for traditional graphics
	SYMBOL*n* statements with VALUE=, COLOR=, and LINE= options
	OPTIONS statement with FORMCHAR= option for line printer graphs
11	PROC REG
	PLOTS(ONLY)=RESIDUALS(UNPACK) option in PROC REG statement for SAS 9.2 and ODS graphics
	PLOTS(ONLY)=RESIDUALBYPREDICTED option in PROC REG statement for SAS 9.2 and ODS graphics
	PLOTS(ONLY)=(RESIDUALHISTOGRAM RESIDUALBOXPLOT QQPLOT) option in PROC REG statement for SAS 9.2 and ODS graphics
	PLOT statement for traditional graphics with statistical keywords R., STUDENT., and OBS. that can be used as *y-variables* in diagnostic plots
	NOMODEL option in the PLOT statement
	VREF= option in the PLOT statement
	CVREF= option in the PLOT statement
	LACKFIT option in MODEL statement
12	PROC FREQ
	PAGE option in PROC FREQ statement
	NOFREQ, NOCOL, NOROW, NOPERCENT, EXPECTED, and NOPRINT options in the TABLES statement
	CHISQ, FISHER, MEASURES, CL, and ALPHA= options in the TABLES statement
	WEIGHT statement
	TEST statement with KENTB and SCORR options
	PLOTS=FREQPLOT with TWOWAY=STACKED and SCALE= options for SAS 9.2 and ODS graphics

A p p e n d i x 3

Introducing the SAS Windowing Environment

This appendix is intended to help you start using SAS in the windowing environment. SAS automatically starts in the windowing environment on a PC. This appendix discusses the following topics:

- viewing initial windows in the environment
- creating a SAS program
- submitting a program
- saving a program
- opening a saved program
- printing a program
- printing and saving output

Note: A PDF version of this appendix, in color, can be found at support.sas.com/publishing/authors/Schlotzhauer.html.

The windowing environment has subtle updates with releases of SAS. The updates might include additional choices in menus or other new features. This appendix describes the windowing environment for SAS 9.2. If you are using a different release of SAS, the appearance of and features for your environment might be different. In this situation, the best source of information is the online Help for your release. Look for a tutorial that introduces you to the SAS windowing environment for your release.

Viewing Initial Windows

When SAS starts in the windowing environment, it automatically opens the Editor, Log, Output, and Explorer windows. After you run a program, the Results window displays a table of contents for your output. Figure A3.1 shows an example of the automatic windows:

Figure A3.1 Automatic Windows in Windowing Environment

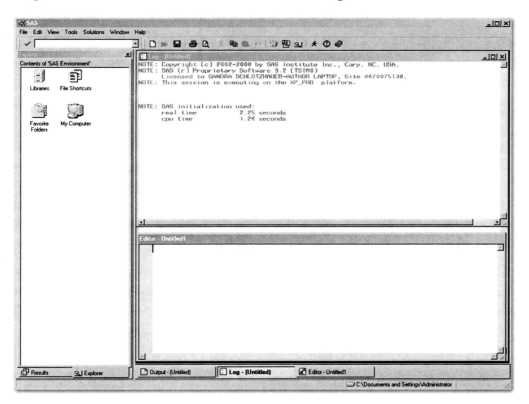

The Log window is active. Only one window is active at a time in the environment. The active window displays the window title in a different color from the other windows.

On a PC, SAS automatically opens the Enhanced Editor window. For other operating systems, SAS displays the Program Editor window, which has fewer features, and the features vary by operating system and release of SAS. The rest of this appendix refers to the Enhanced Editor window as the Editor window.

The Log window shows the SAS log, which includes notes, warnings, and error messages. Figure A3.1 shows the messages that appear when SAS starts.

The Output window is behind the Editor and Log windows when SAS starts. You can activate this window by clicking the **Output** tab at the bottom.

The *command bar*, which appears above the Explorer window, is where you enter commands. This is most useful for experienced users, and is not discussed in this appendix. As you become familiar with SAS, you might want to use this bar. Until then, you can hide it. In the command bar, type the following:

```
command close
```

Then, press ENTER. The command bar disappears.

Several icons appear on the *toolbar* between the top of the windows and the *menu bar*. As you learn SAS, you might want to customize the toolbar. All of the automatic choices in the toolbar are also available from the menu bar. The last toolbar choice (the book with a question mark) displays the SAS Help and Documentation window.

You can use the Explorer window to locate files that contain programs or data sets. Close the window by clicking the **X** in the upper right corner.

Creating a SAS Program

The simplest approach for creating SAS programs is to type the text in the Editor window. To enter text in the window, simply start typing. Press ENTER when you want to start a new line.

This window uses color-coding. Figure A3.2 shows an example of the color-coding, using the program from Chapter 1 of this book.

Figure A3.2 Editor Window

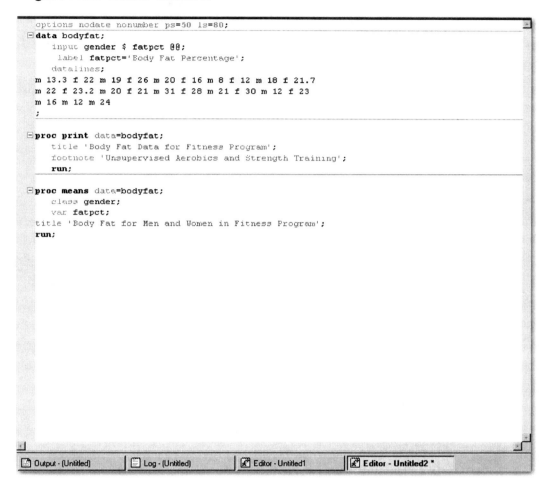

```
 options nodate nonumber ps=50 ls=80;
data bodyfat;
    input gender $ fatpct @@;
    label fatpct='Body Fat Percentage';
    datalines;
 m 13.3 f 22 m 19 f 26 m 20 f 16 m 8 f 12 m 18 f 21.7
 m 22 f 23.2 m 20 f 21 m 31 f 28 m 21 f 30 m 12 f 23
 m 16 m 12 m 24
 ;

proc print data=bodyfat;
    title 'Body Fat Data for Fitness Program';
    footnote 'Unsupervised Aerobics and Strength Training';
    run;

proc means data=bodyfat;
    class gender;
    var fatpct;
 title 'Body Fat for Men and Women in Fitness Program';
 run;
```

| 🗎 Output - (Untitled) | 🗎 Log - (Untitled) | 🗷 Editor - Untitled1 | 🗷 Editor - Untitled2 * |

The list below summarizes some useful features of the Editor window:

- SAS groups statements for DATA and PROC steps. You can collapse or expand the step by clicking on the – or + sign, respectively. The statement that starts each step appears in dark blue. Figure A3.2 shows the words DATA, PROC PRINT, and PROC MEANS in dark blue.

- Data lines are highlighted in yellow.

- Text for titles or labels appears in violet.

- SAS statement names and options appear in medium blue. For example, Figure A3.2 shows the INPUT, LABEL, and DATALINES statements in medium blue.

- Figure A3.3 shows how the Editor window highlights errors in red.

Figure A3.3 Errors in SAS Code

```
data bodyfat;
    inpuat gender $ fatpct @@;
    label fatpct='Body Fat Percentage';
    datalines;
m 13.3 f 22 m 19 f 26 m 20 f 16 m 8 f 12 m 18 f 21.7
m 22 f 23.2 m 20 f 21 m 31 f 28 m 21 f 30 m 12 f 23
m 16 m 12 m 24
```

Viewing the Program Editor Window

For operating systems other than the PC, SAS automatically uses the Program Editor window. Although you can view this window on the PC by selecting **View→Program Editor**, you are likely to find the Editor window more helpful. Figure A3.4 displays the Program Editor window with the program from Chapter 1.

Figure A3.4 Program Editor Window

```
options nodate nonumber ps=50 ls=80;
data bodyfat;
    input gender  fatpct @@;
    label fatpct='Body Fat Percentage';
    datalines;
m 13.3 f 22 m 19 f 26 m 20 f 16 m 8 f 12 m 18 f 21.7
m 22 f 23.2 m 20 f 21 m 31 f 28 m 21 f 30 m 12 f 23
m 16 m 12 m 24
;

proc print data=bodyfat;
    title 'Body Fat Data for Fitness Program';
    footnote 'Unsupervised Aerobics and Strength Training';
    run;

proc means data=bodyfat;
    class gender;
    var fatpct;
title 'Body Fat for Men and Women in Fitness Program';
run;
```

Copying, Cutting, and Pasting

SAS follows the same approach as most PC applications. After selecting text, you can edit, copy, cut, and paste using either menu choices or CTRL keys.

To copy, select the text. Then, either select **Edit→Copy**, or press CTRL+C.

To paste, either select **Edit→Paste**, or press CTRL+V.

To cut, select the text. Then, either select **Edit→Cut**, or press CTRL+X.

To undo your most recent edit, either select **Edit→Undo**, or press CTRL+Z.

Figure A3.5 shows these choices.

Figure A3.5 Edit Menu Choices

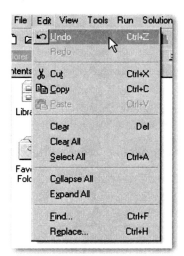

Submitting a Program

When you want to submit the SAS code, select **Run→Submit**, or click the Submit icon, or press the F3 key. (The Submit icon on the toolbar is the "running" icon that shows a person running.)

Figure A3.6 shows the choices from the **Run** menu.

Figure A3.6 Run Menu Choices

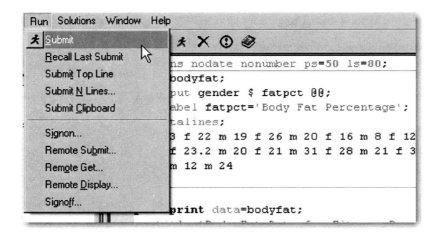

Recalling a Submitted Program

Suppose you submit a program, and then you decide to change it. Perhaps you want a different title, you want to use different options, or you need to fix an error.

On a PC, the Editor window still contains your program after you submit it. You can change the code, and then submit it again.

On other operating systems, the Program Editor window does not contain your program after you submit it. The simplest way to recall your program is to select **Run→Recall Last Submit** as shown in Figure A3.6.

Saving, Including, and Printing Programs

To save a program, select **File→Save**, or click the Save icon on the toolbar (the diskette). This displays the Save As window. Navigate to the location where you want to save the program, enter a filename, and select **Save**. The appearance of the Save As window might differ depending on your operating system.

To include a saved program, select **File→Open Program**, navigate to the location where the program is saved, select the program, and select **Open**. You can also start by clicking the Open icon on the toolbar (the open folder). Alternatively, use the Explorer window to click to the location where the program is saved, and then double-click on the program name. SAS displays the saved program in the Editor window.

To print a program, select **File→Print**, or click the Print icon on the toolbar (the printer). The Print window displays, its appearance might differ depending on your operating system. Select the printer and other options (such as the number of copies to print), and then click **OK**.

Printing and Saving Output

After you run a program, SAS displays the results in the Output window, and displays a table of contents in the Results window. Figure A3.7 shows the windowing environment after running the program for Chapter 1. (The Explorer window is still available. Simply click on the tab, and it displays.)

Figure A3.7 Results and Output Windows

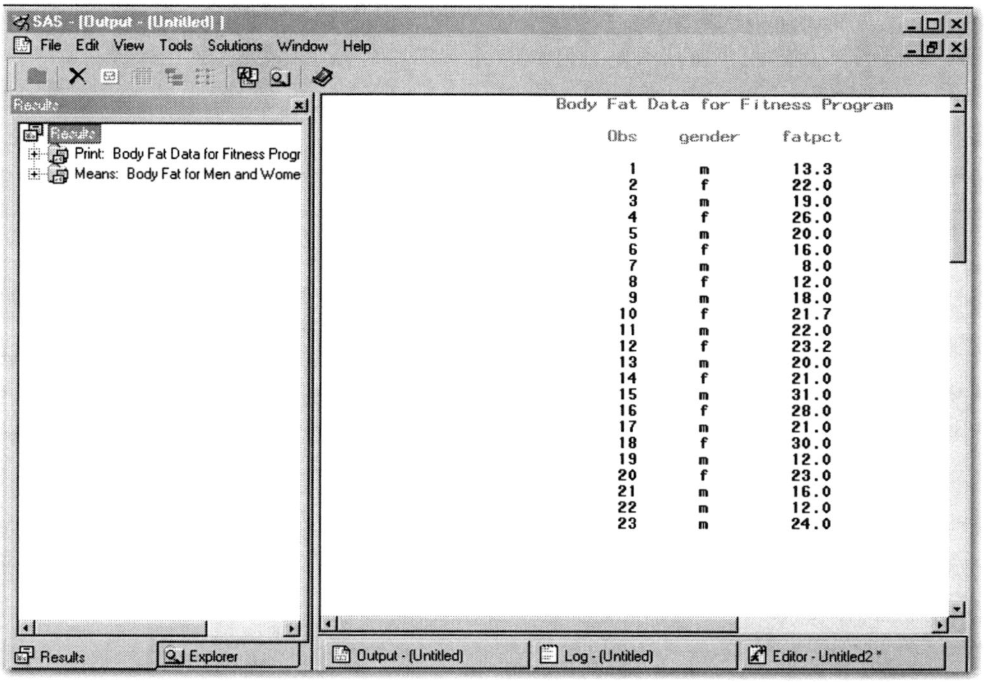

To print all of the output, click in the Output window so that it is the active window. Then, select **File→Print**, and follow the same steps for printing a program.

To print only a portion of the output, click on the output that you want to print in the Results window. Then, select **File→Print**, and follow the same steps for printing a program.

Saving SAS Tables

To save all of the output, click in the Output window so that it is the active window. Then, select **File→Save**, or click the Save icon on the toolbar. The Save As window displays. Navigate to the location where you want to save the output, enter a filename, and select **Save**. The appearance of the Save As window might differ depending on your operating system.

When saving output, SAS provides several choices for the type of file. Selecting a **List** file saves the output in a format specific to SAS. If you plan to include the output in a document or presentation, you might want to save it as an RTF file.

To save only a portion of the output, click on the output that you want to save in the Results window. Then, select **File→Save**, and follow the same steps for saving all of the output.

Saving or Printing Graphs

The simplest way to save graphs is to use the Results window. Expand the table of contents in the Results window to find the graph that you want to save. Click to select the graph, select **File→Save**, and then follow the same steps for saving tables.

Similarly, the simplest way to print graphs is to use the Results window. Expand the table of contents in the Results window to find the graph that you want to print. Click to select the graph, select **File→Print**, and then follow the same steps for printing tables.

A p p e n d i x **4**

Overview of SAS Enterprise Guide

This appendix provides a brief overview of SAS Enterprise Guide.

Purpose and Audience for SAS Enterprise Guide

Some companies have centralized their SAS installation on a server, and they use a client/server environment for individual users. SAS Enterprise Guide is a Microsoft Windows client application designed for this environment. (SAS Enterprise Guide requires a connection to SAS software, either on your local PC or on a remote server.) One of the main purposes of SAS Enterprise Guide is to enable users to use SAS procedures without having to know how to program SAS code.

The audience for SAS Enterprise Guide includes business analysts, programmers, and statisticians.

Summary of Software Structure

SAS Enterprise Guide provides a point-and-click interface to SAS procedures and DATA steps. You can use the Program Editor window in SAS Enterprise Guide if you still want to write your own code. The SAS Enterprise Guide interface includes guides for the procedures in this book. (Availability of specific options might differ depending on release.)

SAS Enterprise Guide uses projects to organize your work. You create a project, which can then contain data sets, programs, output, and graphs. These elements of a project should be familiar to you after working through this book. In addition, you can select an option to turn on a project log file, which is a combined view of all of the log files for all of the activities in the project. SAS automatically updates the project log file any time you perform any activity in the project.

Projects include process flow diagrams that link the elements together. For example, a process flow for a project might show the data set, then the programs for the data set, and then the resulting output, all linked together in an orderly sequence. Some companies use these diagrams as a way to document and validate activity.

Projects can become quite sophisticated. By adding *conditions*, you can control which activities occur based on a Yes or No decision. You can design conditions to run a given program and create output once a week or only when specific issues occur in the data.

Overview of Key Features

The list below provides a brief summary of key features in SAS Enterprise Guide:

- creates and reads SAS data sets.

- accesses all types of data supported by SAS.

- provides the Query Builder, which guides you through the activities of creating subsets of data, joining data, and more.

- creates descriptive statistics and graphs.

- performs analyses discussed in this book, including *t*-tests, ANOVA, correlation, contingency table analyses, and nonparametric alternatives. The software provides a wizard with options for each of these analyses.

- performs more advanced analyses, including multivariate models, survival analysis, and forecasting.

- enables you to deliver results in SAS Report, HTML, RTF, PDF, and text formats. Graphs can be saved as GIFs, JPEGs, and in other formats. Because the results are stored on the server, they can easily be shared with other users.

- enables you to automatically upload results on a periodic basis.

- for advanced users, provides the ability to create custom wizards that guide other users through analysis tasks.

This brief appendix can provide only an overview of the software. If your company uses SAS in a client/server environment, you probably need to understand how to use SAS Enterprise Guide.

References

In addition to the SAS documentation on the software, the following references are especially useful when learning how to use SAS Enterprise Guide. At the time this book was published, the paper was available online at support.sas.com. In addition, this Web site is a good general resource for the latest information about features and releases for SAS Enterprise Guide.

Hemedinger, Chris, and Bill Gibson. 2008. "Find Out What You're Missing: SAS Enterprise Guide for SAS Programmers." *Proceedings of the SAS Global Forum 2008 Conference.* Cary, NC: SAS Institute Inc.

Slaughter, Susan J., and Lora D. Delwiche. 2006. *The Little SAS Book for Enterprise Guide 4.1.* Cary, NC: SAS Institute Inc.

Index

CPSIA information can be obtained at www.ICGtesting.com
Printed in the USA
LVOW111252210812

295285LV00004B/86/P